T0214157

Progress in Mathematics
Vol. 35

Edited by
J. Coates and
S. Helgason

Springer Science+Business Media, LLC

Arithmetic and Geometry

Papers Dedicated to I.R. Shafarevich on the Occasion of His Sixtieth Birthday

Volume I Arithmetic

Michael Artin,
John Tate,
editors

1983 Springer Science+Business Media, LLC

Editors:
Michael Artin
Mathematics Department
Massachusetts Institute of Technology
Cambridge, MA 02139

John Tate
Mathematics Department
Harvard University
Cambridge, MA 02138

This book was typeset at Stanford University using the TEX document preparation system and **computer modern** type fonts by Y. Kitajima. Special thanks go to Donald E. Knuth for the use of this system and for his personal attention in the development of additional fonts required for these volumes. In addition, we extend thanks to the contributors and editors for their patience and gracious help with implementing this system.

Library of Congress Cataloging in Publication Data
Main entry under title:

Arithmetic and geometry.

(Progress in mathematics ; v. 35-36)
Contents: v. 1. Arithmetic — v. 2. Geometry.
1. Algebra – Addresses, essays, lectures. 2. Geometry, Algebraic – Addresses, essays, lectures.
3. Geometry – Addresses, essays, lectures. 4. Shafarevich, I. R. (Igor' Rostislavovich), 1923-
I. Shafarevich, I. R. (Igor' Rostislavovich), 1923-
II. Artin, Michael. III. Tate, John Torrence,
1925- . IV. Series: Progress in mathematics
(Cambridge, Mass.) ; v. 35-36.
QA7.A67 1983 513'.132 83-7124

ISBN 978-0-8176-3132-1 ISBN 978-1-4757-9284-3 (eBook)
DOI 10.1007/978-1-4757-9284-3

CIP-Kurztitelaufnahme der Deutschen Bibliothek

Arithmetic and geometry : papers dedicated to
I. R. Shafarevich on the occasion of his 60.
birthday / Michael Artin ; John Tate, ed.
(Progress in mathematics ; ...)

NE: Artin, Michael (Hrsg.); Safarevic, Igor' R.:
Festschrift

Vol. 2. Geometry

Originally published by Birkhäuser Boston, Inc., in 1983

ISBN 978-0-8176-3132-1

Igor Rostislavovich Shafarevich has made outstanding contributions in number theory, algebra, and algebraic geometry. The flourishing of these fields in Moscow since World War II owes much to his influence. We hope these papers, collected for his sixtieth birthday, will indicate to him the great respect and admiration which mathematicians throughout the world have for him.

Michael Artin
Igor Dolgachev
John Tate
A.N. Todorov

Spencer Bloch
Barry Mazur
Arthur Ogus
William E. Lang
[illegible]
Steve Lichtenbaum
A. Weil
[illegible]
Lang
[illegible]
V. Arnold
Tetsuji Shioda
Noboru Aoki
Michael Artin
M. F. Atiyah
[illegible]
[illegible]
[illegible] Dolgachev
Mark Spivakovsky
F. Hirzebruch
[illegible]
[illegible] Zariski
C. Musili

David Eisenbud
Heisuke Hironaka
Hyman Bass
Oscar Zariski
Dale Peterson

Поздравляем с днём рождения!

[illegible]
Jan Denef
A. P. Ogg
[illegible]
Nicholas M. Katz
Miles Reid
Victor Kac
J S Milne
Andrew Pressley
John Tate
Robert Bryant
J. W. Masser
Seshadri
John Coates

photo by M. Artin

I.R. Shafarevich, Moscow, 1966

Volume I Arithmetic

N. Aoki and **T. Shioda**, Generators of the Néron-Severi Group of a Fermat Surface 1

S. Bloch, p-adic Etale Cohomology 13

J.W.S. Cassels, The Mordell-Weil Group of Curves of Genus 2 27

G.V. Chudnovsky, Number Theoretic Applications of Polynomials with Rational Coefficients Defined by Extremality Conditions 61

J. Coates, Infinite Descent on Elliptic Curves with Complex Multiplication 107

N.M. Katz, On the Ubiquity of Pathology in Products 139

S. Lang, Conjectured Diophantine Estimates on Elliptic Curves 155

S. Lichtenbaum, Zeta-Functions of Varieties over Finite Fields at $s = 1$ 173

B. Mazur and **J. Tate**, Canonical Height Pairings via Biextensions 195

J.S. Milne, The Action of an Automorphism of **C** on a Shimura Variety and its Special Points 239

N.O. Nygaard, The Torelli Theorem for Ordinary K3 Surfaces over Finite Fields 267

A.P. Ogg, Real Points on Shimura Curves 277

I.I. Piatetski-Shapiro, Special Automorphic Forms on $PGSp_4$ 309

M. Raynaud, Courbes sur une variété abélienne et points de torsion 327

A. Weil, Euler and the Jacobians of Elliptic Curves 353

Volume II Geometry

V.I. Arnold, Some Algebro-Geometrical Aspects of the Newton Attraction Theory 1

M. Artin and **J. Denef**, Smoothing of a Ring Homomorphism Along a Section 5

M.F. Atiyah and **A.N. Pressley**, Convexity and Loop Groups 33

H. Bass, The Jacobian Conjecture and Inverse Degrees 65

R. Bryant and **P. Griffiths**, Some Observations on the Infinitesimal Period Relations for Regular Threefolds with Trivial Canonical Bundle 77

H. Hironaka, On Nash Blowing-Up 103

F. Hirzebruch, Arrangements of Lines and Algebraic Surfaces 113

V.G. Kac and **D.H. Peterson**, Regular Functions on Certain Infinite-dimensional Groups 141

W.E. Lang, Examples of Surfaces of General Type with Vector Fields 167

Yu.I. Manin, Flag Superspaces and Supersymmetric Yang-Mills Equations 175

B. Moishezon, Algebraic Surfaces and the Arithmetic of Braids, I 199

D. Mumford, Towards an Enumerative Geometry of the Moduli Space of Curves 271

C. Musili and **C.S. Seshadri**, Schubert Varieties and the Variety of Complexes 327

A. Ogus, A Crystalline Torelli Theorem for Supersingular K3 Surfaces 361

M. Reid, Decomposition of Toric Morphisms 395

M. Spivakovsky, A Solution to Hironaka's Polyhedra Game 419

A.N. Tjurin, On the Superpositions of Mathematical Instantons 433

A.N. Todorov, How Many Kähler Metrics Has a K3 Surface? 451

O. Zariski, On the Problem of Irreducibility of the Algebraic System of Irreducible Plane Curves of a Given Order and Having a Given Number of Nodes 465

Generators of the Néron-Severi Group of a Fermat Surface

Noboru Aoki and Tetsuji Shioda

To I.R. Shafarevich

1. Introduction

The Néron-Severi group of a (nonsingular projective) variety is, by definition, the group of divisors modulo algebraic equivalence, which is known to be a finitely generated abelian group (cf. [2]). Its rank is called the Picard number of the variety. Thus the Néron-Severi group is defined in purely algebro-geometric terms, but it is a rather delicate invariant of arithmetic nature. Perhaps, because of this reason, it usually requires some nontrivial work before one can determine the Picard number of a given variety, let alone the full structure of its Néron-Severi group. This is the case even for algebraic surfaces over the field of complex numbers, where it can be regarded as the subgroup of the cohomology group $H^2(X, \mathbb{Z})$ characterized by the Lefschetz criterion.

Now the purpose of the present paper is to find certain explicitly defined curves on the complex Fermat surface

$$(1.1) \qquad X_m^2 : \quad x^m + y^m + z^m + w^m = 0$$

whose cohomology classes (equivalently, algebraic equivalence classes) form (a part of) generators of the Néron-Severi group $NS(X_m^2) \otimes \mathbb{Q}$. The Picard number $\rho(X_m^2)$ has recently been determined and is given by the following formula:

$$(1.2) \qquad \rho(X_m^2) = 3(m-1)(m-2) + 1 + \delta_m + 24(m/3)^* + 48(m/2)^* + 24\epsilon(m)$$

where

$$\delta_m = \begin{cases} 1 & (m: \quad \text{even}) \\ 0 & (m: \quad \text{odd}) \end{cases}$$
$$(m/3)^* = \begin{cases} m/3 & (\text{if} \quad 3 \mid m) \\ 0 & (\text{if} \quad 3 \nmid m) \end{cases}$$
$$(m/2)^* = \begin{cases} m/2 & (m: \quad \text{even}) \\ 0 & (m: \quad \text{odd}) \end{cases}$$

and where $\epsilon(m)$ is a bounded function of m which is expressed as a certain sum over divisors d of m such that $(d, 6) \neq 1$ and $d \leq 180$ (see Shioda [4], Aoki[1]). It is known that $3(m-1)(m-2)+1+\delta_m$ is the rank of the subspace of $NS\ (X^2_m)$ which is spanned by the *lines* of the ambient space $\mathbb{P}^3$ lying on X^2_m; in particular, if $(m, 6) = 1$, then $NS\ (X^2_m) \otimes \mathbb{Q}$ is spanned by the classes of these lines (cf. [4,Thm.7]).

The main results of this paper can be stated roughly as follows:

(i) One half of the term $48(m/2)^*$ in the above formula of $\rho(X^2_m)$ corresponds to the subspace of $NS\ (X^2_m) \otimes \mathbb{Q}$ spanned by the classes of curves lying on (the intersection of X^2_m with) certain *quadric surfaces* of the form

$$z^2 - c \cdot xy = 0$$

(up to permutation of coordinates x, y, z, w).

(ii) The term $24(m/3)^*$ corresponds, in a similar sense, to curves lying on *cubic surfaces* of the form

$$w^3 - c \cdot xyz = 0.$$

(iii) Another half of the term $48(m/2)^*$ corresponds similarly to curves lying on *quartic surfaces* of the form

$$w^4 - c \cdot xyz^2 = 0.$$

More precise statements will be given in §3.

Concerning the structure of the Néron-Severi group of the (complex) Fermat surface X^2_m, the following problems remain to be studied:

i) to find curves corresponding to the "exceptional" term $\epsilon(m)$;

ii) to find generators of $NS\ (X^2_m)$ over $\mathbb{Z}$.

Futhermore, it is expected that a similar approach should be applicable to Fermat varieties of higher dimension for the explicit construction of algebraic cycles corresponding to given Hodge classes. Combined with the method based on the inductive structure of Fermat varieties [3], this might lead to the verification of the Hodge Conjecture for all Fermat varieties.

2. Preliminaries

First recall that the Hodge classes on the Fermat variety $X^n_m : \sum_{i=0}^{n+1} x_i^m = 0$ are described in terms of the characters of the group $G^n_m = (\mu_m)^{n+2}/(\text{diagonal})$ acting on $H^n(X^n_m, \mathbb{C})$ (see [3,Thm. I]). In the special case $n = 2$, this gives

$$NS(X^2_m) \otimes_{\mathbb{Z}} \mathbb{C} = \bigoplus_{\alpha \in \mathcal{B}^2_m} V(\alpha) \oplus V(0) \tag{2.1}$$

where the index set $\mathcal{B}^2_m =$

$$\left\{ (a_0, a_1, a_2, a_3) \mid 1 \leq a_i \leq m-1, \ \sum_{i=0}^{3} \langle t a_i \rangle = 2m \quad (\forall t \in (\mathbb{Z}/m)^\times) \right\} \tag{2.2}$$

is naturally regarded as a subset of the character group of G^2_m and $V(\alpha)$ (resp. $V(0)$) is the eigenspace of $H^2(X^2_m, \mathbb{C})$ with character α (resp. trivial character) which is known to be 1-dimensional. In particular, the Picard number of the (complex) Fermat surface X^2_m is

$$\rho(X^2_m) = |\mathcal{B}^2_m| + 1. \tag{2.3}$$

Next we recall the following structure theorem of $\mathcal{B}^2_m$, which has been formulated and partially proven in Shioda [4] and recently fully proven by Aoki [1]. The formula (1.2) is a consequence of this result in view of (2.3). We call an element $\alpha = (a_0, a_1, a_2, a_3) \in \mathcal{B}^2_m$ decomposable if $a_i + a_j \equiv 0 \pmod{m}$ for some $i \neq j$, and indecomposable otherwise.

Theorem ($\mathcal{B}^2_m$). (i) *If* $(m, 6) = 1$, *then* $\mathcal{B}^2_m$ *consists of decomposable elements.*

(ii) *If* $(m,6) > 1$, *then every indecomposable element* $\alpha = (a_0, a_1, a_2, a_3)$ *of* $\mathcal{B}^2_m$ *with* $\mathrm{GCD}(a_i) = 1$ *is equal (up to permutation) to one of the "standard" elements* α_i, β_i *or* γ_j *below, except for finitely many "exceptional" elements which exist only for* $m \leq 180$:

a) $m = 2d$, $\alpha_i = (i, d+i, m-2i, d)$, $1 \leq i < d$, $(i,d) = 1$, $i \neq m/4$.
b) $m = 2d$, $\beta_i = (i, d+i, d+2i, m-4i)$, $1 \leq i < d$, $(i,d) = 1$, $i \neq m/3, m/4, m/6$.
c) $m = 3d$, $\gamma_j = (j, d+j, 2d+j, m-3j)$, $1 \leq j < d$, $(j,d) = 1$, $j \neq m/6$.

If $\alpha = (a_0, a_1, a_2, a_3)$ is an element of $\mathcal{B}^2_m$ with $\mathrm{GCD}(a_i) = d > 1$, set $a'_i = a_i/d$ and $m' = m/d$. Then $\alpha' = (a'_0, a'_1, a'_2, a'_3)$ is an element of $\mathcal{B}^2_{m'}$ with $\mathrm{GCD}(a'_i) = 1$, and one checks easily that the morphism $f\colon X^2_m \longrightarrow X^2_{m'}$ given by $(x_i) \mapsto (x_i^d)$ induces the map

$$f^*\colon H^2(X^2_{m'}) \longrightarrow H^2(X^2_m)$$

such that $f^*V(\alpha') = V(\alpha)$. Thus, in order to construct explicit curves on X^2_m corresponding to $V(\alpha)$, it suffices to consider the case $\mathrm{GCD}(a_i) = 1$.

Now define, for any character α of G^n_m,

$$\text{(2.4)} \quad p_\alpha(\xi) = \frac{1}{m^{n+1}} \sum_{g \in G^n_m} \alpha(g)(g^{-1})^*(\xi) \qquad (\xi \in H^n_{\mathrm{prim}}(X^n_m)).$$

Then p_α is the projector of $H^n_{\mathrm{prim}}(X^n_m)$ to $V(\alpha)$. In particular, if $\alpha \in \mathcal{B}^n_m$ and if $\xi = cl.(C)$ is the class of an algebraic cycle C on X^n_m, then $p_\alpha(\xi)$ *is an algebraic cohomology class generating the subspace* $V(\alpha)$ *provided that* $p_\alpha(\xi) \neq 0$. Note that $p_\alpha(\xi) \neq 0$ if and only if the "intersection number" $p_\alpha(\xi) \cdot \overline{p_\alpha(\xi)} \neq 0$.

Definition. Given $\alpha \in \mathcal{B}^2_m$ and a curve C on X^2_m, we say (for short) that C *represents the Hodge class* α if the following two conditions are satisfied. Let $G = G^2_m$.

$$\text{(2.5)} \qquad G_C \overset{\mathrm{def}}{=} \{g \in G \mid g(C) = C\} \subset \mathrm{Ker}(\alpha).$$

$$\text{(2.6)} \qquad \omega_C \overset{\mathrm{def}}{=} \sum_{g \in G/G_C} \alpha(g) cl.(g(C)) \neq 0, \quad \text{i.e.,} \quad \omega_C \cdot \overline{\omega}_C \neq 0.$$

Observe that ω_C is a multiple of $p_\alpha(cl.(C))$ such that

$$\omega_C \cdot \overline{\omega}_C = [G : G_C] \sum_{g \in G/G_C} \alpha(g)(C \cdot g(C)), \tag{2.7}$$

as is clear from the invariance of the intersection number under an automorphism.

Theorem 0. *For any $m > 1$, let L denote a line of $\mathbb{P}^3$ lying on X_m^2. Then L represents a decomposable element of $\mathcal{B}_m^2$, and conversely every decomposable element is represented by some lines. Further*

$$\omega_L \cdot \overline{\omega}_L = -m^3. \tag{2.8}$$

Proof. See the proof of [4, Thm. 7].

In the next section, we shall exhibit certain curves on X_m^2 which represent the "standard" indecomposable elements of $\mathcal{B}_m^2$ stated in Theorem $(\mathcal{B}_m^2)$. Note that, for this purpose, it suffices to consider the case $\alpha = \alpha_1$, β_1, and γ_1, because in general we have $V(\alpha)^\sigma = V(t\alpha)$ for any automorphism σ of $\mathbb{C}$ inducing $\sigma_t : \varsigma \mapsto \varsigma^t$ on the subfield of m-th roots of unity ($t \in (\mathbb{Z}/m)^\times$), while $cl.(C)^\sigma = cl.(C^\sigma)$, C^σ being the conjugate of C under σ (this makes sense since our surface X_m^2 is defined over $\mathbb{Q}$).

3. Main Results

Theorem 1. *Let $m = 2d (d > 1)$ and $\alpha_1 = (1, d+1, m-2, d)$. Let C denote the curve of degree m in $\mathbb{P}^3$ defined by*

$$z^2 - \sqrt[4]{2}xy = 0, \ x^d + y^d + \sqrt{-1}w^d = 0. \tag{3.1}$$

Then C is a nonsingular irreducible curve lying on X_m^2 which represents the Hodge class α_1 and which satisfies

$$\omega_C \cdot \overline{\omega}_C = -2m^3. \tag{3.2}$$

Theorem 2. *Let $m = 3d (d \geq 1)$ and $\gamma_1 = (1, d+1, 2d+1, m-3)$. Let C denote the curve of degree m in $\mathbb{P}^3$ defined by*

$$w^3 - \sqrt[4]{-3}xyz = 0, \quad x^d + y^d + z^d = 0. \tag{3.3}$$

Then C is a nonsingular irreducible curve lying on X_m^2 which represents the Hodge class γ_1 and which satisfies

$$\omega_C \cdot \overline{\omega}_C = -3m^3. \tag{3.4}$$

Theorem 3. *Let $m = 2d(d > 2)$ and $\beta_1 = (1, d+1, d+2, m-4)$.*

(i) *In case d is odd, let C denote the curve of degree $2m$ on X_m^2 defined (in addition to the Fermat equation) by*

$$\begin{aligned} &w^4 - \sqrt[d]{-8}xyz^2 = 0, \\ &(x^d + y^d + \sqrt{-1}z^d)z - aw^2(xy)^{(d-1)/2} = 0 \\ &(a = (-1)^{-1/2d} \cdot 2^{(d-3)/2d}) \end{aligned} \tag{3.5}$$

Then C is a nonsingular irreducible curve which represents the Hodge class β_1 and which satisfies

$$\omega_C \cdot \overline{\omega}_C = -4m^3. \tag{3.6}$$

(ii) *In case d is even, let C' denote the curve of degree $2m$ in $\mathbb{P}^3$ defined by*

$$w^4 - \sqrt[d]{-8}xyz^2 = 0,\ x^d + y^d - \sqrt{2}(xy)^{d/2} + \sqrt{-1}z^d = 0. \tag{3.7}$$

Then C' is a singular irreducible curve lying on X_m^2 which represents the Hodge class β_1 and which satisfies

$$\omega_{C'} \cdot \overline{\omega}_{C'} = -4m^3 - 2m^2. \tag{3.8}$$

Remark. (a) The value of $\sqrt[d]{2}$, $\sqrt[d]{-3}$, etc. in the defining equations of C are fixed once and for all. (b) Theorems 1,2,3 (and Theorem 0 in §2) remain valid for the Fermat surface $X_m^2(p)$ in characteristic p, provided p does not divide m. This will be clear from the proof.

4. Proof of Theorems 1 and 2

First of all look at the polynomial identities:

(4.1) $x^2 + y^2 + z^2 + w^2 = (x + y + \sqrt{-1}w)(x + y - \sqrt{-1}w) + (z^2 - 2xy)$.

(4.2) $$x^3 + y^3 + z^3 + w^3 = (x + y + z)(x + \rho y + \rho^2 z)(x + \rho^2 y + \rho z)$$
$$+(w^3 + 3xyz) \qquad (\rho^3 = 1, \rho \neq 1).$$

Replacing $x, y, \ldots$ by $x^d, y^d, \ldots$, it is clear that the curve C defined by (3.1) or (3.3) in $\mathbb{P}^3$ actually lies on the Fermat surface X_m^2. It is easy to check that C is nonsingular. Since a nonsingular complete intersection curve in $\mathbb{P}^3$ of multidegree (d,e) has the genus $1 + de(d + e - 4)/2$, we have

(4.3) $$\text{genus}(C) = (d-1)^2 \quad \text{or} \quad (3d^2 - 3d + 2)/2.$$

By the adjunction formula, the self-intersection number of C is given respectively by

(4.4) $$(C^2) = -2d^2 + 4d \quad \text{or} \quad -6d^2 + 9d.$$

In what follows, we shall prove only Theorem 1, since Theorem 2 can be proven in the same way.

For $g = [1 : \varsigma_1 : \varsigma_2 : \varsigma_3] \in G = G_m^2$, let

(4.5) $$\sigma(g) = \varsigma_1 \varsigma_2^{-2}, \quad \tau(g) = (\varsigma_1 \varsigma_3)^d, \quad \rho(g) = \varsigma_3^d.$$

Then $\alpha_1 = (1, d + 1, m - 2, d) \in \mathcal{B}_m^2$ is expressed as $\alpha_1(g) = \sigma(g) \cdot \tau(g)$. For simplicity, we set $\varsigma_i^d = \epsilon_i (i = 1, 2, 3)$ so that $\sigma^d(g) = \epsilon_1$, $\rho(g) = \epsilon_3$ and $\tau(g) = \epsilon_1 \epsilon_3 (\epsilon_i = \pm 1)$. Now the curve $g(C)$ is defined by the following equations:

(4.6) $$z^2 - \overline{\sigma}(g)\sqrt[d]{2}xy = 0, \quad x^d + \epsilon_1 y^d + \epsilon_3 \sqrt{-1} w^d = 0.$$

Hence g stabilizes C if and only if $\sigma(g) = \epsilon_1 = \epsilon_3 = 1$. Thus

(4.7) $$G_C = \text{Ker}(\sigma) \cap \text{Ker}(\tau).$$

On the other hand, we see easily:

(4.8) $$C \cap g(C) \neq \phi \quad \Leftrightarrow \quad g \in \text{Ker}(\sigma) \cup \text{Ker}(\tau) \cup \text{Ker}(\rho).$$

Since G_C is contained in $\mathrm{Ker}(\alpha_1)$, we can define the element ω_C of $NS(X_m^2)\otimes\mathbb{C}$ as in (2.6):

$$\omega_C = \sum_{g\in G/G_C} \alpha_1(g)cl.(g(C)). \tag{4.9}$$

In order to compute $\omega_C\cdot\overline{\omega}_C$ by (2.7), we need to know the intersection numbers $(C\cdot g(C))$ for $g\in\mathrm{Ker}(\sigma)\cup\mathrm{Ker}(\tau)\cup\mathrm{Ker}(\rho)$. We distinguish the three cases: (a) $g\in\mathrm{Ker}(\sigma)$, $g\notin\mathrm{Ker}(\tau)$, (b) $g\notin\mathrm{Ker}(\sigma)$, $g\in\mathrm{Ker}(\tau)$, and (c) $g\notin\mathrm{Ker}(\sigma)\cup\mathrm{Ker}(\tau)$, $g\in\mathrm{Ker}(\rho)$.

Case (a). In this case, we have $\epsilon_1=1$, $\epsilon_3=-1$. By (4.6), $C\cap g(C)$ consists of the 2d points $P_\varsigma=(1,\varsigma^2,\sqrt[2d]{2}\varsigma,0)$ $(\varsigma^{2d}=-1)$. For each $P=P_\varsigma$, let O_P denote the local ring of X_m^2 at P, regarded as a localization of the affine ring $k[y,z,w]$ of $X_m^2-\{x=0\}$ $(x=1)$. The intersection multiplicity of $C\cap g(C)$ at P is equal to the length of the quotient ring

$$\begin{aligned}&O_P/(1+y^d+\sqrt{-1}w^d,1+y^d-\sqrt{-1}w^d,z^2-\sqrt[d]{2}y)O_P\\ &\simeq k[y,z,w]_P/(1+y^d,w^d,z^2-\sqrt[d]{2}y)\\ &\simeq k[w]/(w^d).\end{aligned}$$

Therefore we have

$$(C\cdot g(C))=d\cdot 2d=2d^2. \tag{4.10}$$

Case (b). In this case, we have $\sigma(g)\neq 1$, $\epsilon_1=\epsilon_3=\pm1$. The intersection $C\cap g(C)$ consists of the $2d$ points $P_\eta=(0,\eta,0,1)$ and $Q_\eta=(\eta,0,0,1)$ $(\eta^d=-\sqrt{-1})$ if $\epsilon_1=1$, and of the d points P_η only if $\epsilon_1=-1$. The intersection multiplicity of $C\cap g(C)$ at each P_η or Q_η is computed as before and is equal to 2. Therefore $(C\cdot g(C))=4d$ or $2d$, according as $\epsilon_1=1$ or -1.

Case (c). In this case, we have $\sigma(g)\neq 1$, $\epsilon_1=-1$ and $\epsilon_3=1$. There are d intersection points $(1,0,0,\varsigma)$ $(\varsigma^d=-1)$ in $C\cap g(C)$, and the multiplicity at each of them is equal to 2. Thus we have $(C\cdot g(C))=2d$.

Since $\epsilon_1=\sigma^d(g)$, the case (b) and (c) can be summarized as follows:

$$(C\cdot g(C))=(3+\sigma^d(g))d. \tag{4.11}$$

Now we are ready to compute $\omega_C \cdot \overline{\omega}_C$ by (2.6). Let $K(\sigma)$ denote the image of $\mathrm{Ker}(\sigma)$ in G/G_C, and let $K(\sigma)^* = K(\sigma) - \{id.\}$. Similarly we define $K(\tau)$, $K(\rho)$, etc. Then

$$\sum_{g \in G/G_C} \alpha(g)(C \cdot g(C)) = (C^2) + \sum_{K(\sigma)^*} \alpha(g)(2d^2) + \sum_{\substack{K(\tau)^* \cup K(\rho)^* \\ -K(\tau)^* \cap K(\rho)^*}} \alpha(g)(3 + \sigma^d(g))d.$$

Note that $\alpha (= \alpha_1)$ and $\alpha \cdot \sigma^d$ induce nontrivial characters on each of the subgroups $K(\sigma)$, $K(\tau)$, $K(\rho)$, and $K(\tau) \cap K(\rho)$ of G/G_C, and so we have $\sum_{K(\sigma)^*} \alpha(g) = -1$, etc. Hence the above is equal to

$$(-2d^2 + 4d) + 2d^2(-1) + 3d(-1 - 1 + 1) + d(-1 - 1 + 1) = -4d^2 = -m^2.$$

Since $[G : G_C] = 2m$, this implies

$$\omega_C \cdot \overline{\omega}_C = 2m \cdot (-m^2) = -2m^3,$$

completing the proof of Theorem 1. Q.E.D.

5. Proof of Theorem 3

First we note that the curve C' in $\mathbb{P}^3$ defined by (3.7) actually lies on X_m^2 in case $m = 2d$ with d even. This follows from the elementary identity

$$x^4 + y^4 + z^4 + w^4 =$$
$$(x^2 + y^2 - \sqrt{2}xy + \sqrt{-1}z^2)(x^2 + y^2 + \sqrt{2}xy - \sqrt{-1}z^2) + (w^4 - \sqrt{-8}xyz^2),$$

by substituting $x, \ldots$ by $x^{d/2}, \ldots$. The curve C' has the d singular points $P_\eta = (1, \eta, 0, 0)$ where $\eta^{d/2} = (-1 \pm \sqrt{-1})/\sqrt{2}$. To find the normalization of C', observe that xy is a square on C' as the first equation of (3.7) shows. Suggested by this, we consider the following curve $\tilde{C}$ in $\mathbb{P}^4$:

$$(5.1)\; x^d + y^d + \sqrt{-1}(z^d + t^d) = 0, \quad t^2 - \sqrt[d]{-2}xy = 0, \quad w^2 - \sqrt[d]{2}tz = 0.$$

It is easy to check that C is nonsingular and that the projection of $\mathbb{P}^4$ to $\mathbb{P}^3$ $(x, y, z, w, t) \mapsto (x, y, z, w)$ defines a birational morphism of $\tilde{C}$ onto C' in case d is even, and an isomorphism of $\tilde{C}$ to C in case d is odd. As a nonsingular complete intersection curve in $\mathbb{P}^4$, the genus of $\tilde{C}$ is given by $d \cdot 2 \cdot 2 \cdot (d+2+2-5)/2+1 = 2d^2 - 2d + 1$. Therefore we have by the adjunction formula

$$(5.2) \qquad \begin{cases} genus(C) = 2d^2 - 2d + 1, \\ (C^2) = -m^2 + 6m \end{cases}$$

and

$$(5.3) \qquad \begin{cases} p_a(C') = 2d^2 - d + 1 \quad (p_a : \text{arithmetic genus}) \\ (C'^2) = -m^2 + 7m. \end{cases}$$

The rest of the proof is simillar to that of Theorem 1 given in §4. We shall sketch it only for the case d odd, omitting details.

For $g = [1 : \varsigma_1 : \varsigma_2 : \varsigma_3] \in G$, we set

$$(5.4) \qquad \sigma(g) = \varsigma_1 \varsigma_2^2 \varsigma_3^{-4}, \quad \tau(g) = \varsigma_1^d \varsigma_2^d, \quad \rho(g) = \varsigma_1^{(d-1)/2} \varsigma_2^{-1} \varsigma_3^2.$$

Then we have

$$(5.5) \qquad G_C = \mathrm{Ker}(\sigma) \cap \mathrm{Ker}(\tau), \quad [G : G_C] = 2m.$$

For each $g \in G - G_C$, the intersection of the curve C and its transform $g(C)$ is described as follows:

(I) Assume $g \notin \mathrm{Ker}(\sigma)$. (1) If $g \in \mathrm{Ker}(\tau)$, there are d points $(0, y, z, 0)$, $y^d + \sqrt{-1} z^d = 0$, with intersection multiplicity 4. (2) If $g \in \mathrm{Ker}(\sigma^d \tau)$, there are d points $(x, 0, z, 0), x^d + \sqrt{-1} z^d = 0$, with multiplicity 4. (3) If $g \in \mathrm{Ker}(\sigma^d)$ or $g \in \mathrm{Ker}(\sigma^2)$, then there are m points $(x, y, 0, 0)$, $x^m + y^m = 0$, with multiplicity 2. (4) If $g \in G - \mathrm{Ker}(\sigma^d) \cup \mathrm{Ker}(\sigma^2)$, there are m points $(x, y, 0, 0)$, $x^m + y^m = 0$, with multiplicity 4.

(II) Assume $g \in \mathrm{Ker}(\sigma)$, $g \notin \mathrm{Ker}(\tau)$. Then there are m points $(x, y, 0, 0)$, $x^m + y^m = 0$, with multiplicity 2, and m^2 points of the form (x, y, z, w), $x^d + y^d = 0, \quad zw \neq 0$, with multiplicity 1.

Therefore, with the same notation as in the proof of Theorem 1 (§4), we have (write $\beta = \beta_1$):

$$\sum_{g \in G/G_C} \beta(g)(C \cdot g(C))$$

$$= (C^2) + 4d \sum_{K(\tau)^*} \beta(g) + 4d \sum_{K(\sigma^d \tau)^*} \beta(g) + 2m \sum_{K(\sigma^d) - K(\sigma)} \beta(g)$$

$$+ 2m \sum_{K(\sigma^2) - K(\sigma)} \beta(g) + 4m \sum_{\substack{G/G_C \\ -K(\sigma^d) \cup K(\sigma^2)}} \beta(g)$$

$$+ (2m + m^2) \sum_{K(\sigma)^*} \beta(g)$$

$$= (-m^2 + 6m) + 2m(-1) + 2m(-1) + (2m + m^2)(-1)$$

$$= -2m^2.$$

By (2.7) and (5.5), we have

$$\omega_C \cdot \overline{\omega}_C = 2m(-2m^2) = -4m^3.$$

This proves part (i) of Theorem 3. Part (ii) is similar (and even simpler), so it will be omitted. Q.E.D.

References

[1] Aoki, N.: Properties of Dirichlet characters and its application to Fermat varieties. Master Thesis, University of Tokyo, 1982 (in Japanese).

[2] Lang, S.: Diophantine geometry, Intersc. Publishers, New York-London, 1962.

[3] Shioda, T.: The Hodge Conjecture for Fermat varieties, Math. Ann. 245 (1979), 175–184.

[4] Shioda, T.: On the Picard number of a Fermat surface, J. Fac. Sci. Univ. Tokyo, Sec. IA, 28 (1982), 725–734.

Received June 18, 1982

N. Aoki, Department of Mathematics
Faculty of Science
University of Tokyo
Hongo, Tokyo 113, Japan

Professor Tetsuji Shioda
Department of Mathematics
Faculty of Science
University of Tokyo
Hongo, Tokyo 113, Japan

p-adic Etale Cohomology

S. Bloch

To I.R. Shafarevich

What follows is a report on joint work with O. Gabber and K. Kato.* A manuscript with complete proofs exists and is currently being revised. For compelling physical reasons (viz. time, space, and distance) however, I will give here only statements of results; and my coauthors have not had the opportunity to correct any stupidities which may have slipped in. The conjectures in §3 are my own. I like to think that this research has been strongly influenced by the work of Shafarevich, both by his work on algebraic geometry in characteristic p and by his work on arithmetical algebraic geometry. In fact, recently Ogus has used these results to apply the basic Rudakov-Shafarevich result on existence and smoothness of moduli for K3 surfaces in characteristic p to the study of the moduli space when $p = 2$.

§0 Global Results

For purposes of this report let the phrase "p-adic field" mean a field K of characteristic 0 which is the quotient field of a henselian discrete valuation ring Λ with perfect residue field k of characteristic $p > 0$. For example, $\mathbf{Q}_p$ is a p-adic field. Let V be a smooth, complete variety over a p-adic field K. Let $\overline{K}$ be the algebraic closure of K, and write $G = Gal(\overline{K}/K)$. The étale cohomology groups

$$H^*(V_{\overline{K}}, \mathbf{Q}_p) \underset{\text{def}}{=} [\varprojlim_n H^*(V_{\overline{K}}, \mathbf{Z}/p^n\mathbf{Z})] \otimes_{\mathbf{Z}_p} \mathbf{Q}_p$$

are representation spaces for G which are known to give the correct betti numbers for V.

Concerning the structure of these cohomology groups one has a basic conjecture formulated by Tate and proved by him for H^1 when V has good reduction. Let $\mathbf{C}$ be the completion of $\overline{K}$. By continuity G acts on $\mathbf{C}$ and hence G acts diagonally on

$$H^*(V_{\overline{K}}, \mathbf{C}) \underset{\text{def}}{=} H^*(V_{\overline{K}}, \mathbf{Q}_p) \otimes_{\mathbf{Q}_p} \mathbf{C}$$

Conjecture (Tate [25]). *There is a canonical isomorphism of G-modules*

$$H^n(V_{\overline{K}}, \mathbf{C}) \cong \bigoplus_{l+m=n} H^l(V_{\mathbf{C}}, \Omega^m)(-m). \tag{0.1}$$

For a G-module M, we write $M(m)$ for the G-module obtained by twisting M by the m-th power of the dual of the cyclotomic character. Also Ω^m denotes the sheaf of Kähler m-forms for $V_{\mathbf{C}}$ over $\mathbf{C}$. Recent results on Tate's conjecture have been obtained by Fontaine [9].

We now fix a *model* X of V over $S = Sp\,\Lambda$, i.e. we fix a diagram

$$\begin{array}{ccccc} V = X_\eta & \xrightarrow{j} & X & \xleftarrow{\quad i \quad} & X_s = Y \\ \downarrow & & \downarrow & & \downarrow \\ Sp\,K = \eta & \rightarrow & S = Sp\,\Lambda & \leftarrow & s = Sp\,k \end{array} \tag{0.2}$$

with all vertical arrows flat and proper. A bar will either indicate algebraic or integral closure (viz. $\overline{K}$, $\overline{\Lambda}$, $\overline{K}$) or base extension ($\overline{X} = X_{\overline{S}}$, $\overline{V} = V_{\overline{K}}$, $\overline{i}$, $\overline{j}$, etc.). There is a spectral sequence

$$E_2^{s,t} = H^s_{\acute{e}t}(\overline{Y}, \overline{i}^* R^t \overline{j}_* \mathbf{Z}_p) \Rightarrow H^{s+t}_{\acute{e}t}(\overline{V}, \mathbf{Z}_p) \tag{0.3}$$

inducing a G-stable filtration

$$F^\bullet H^*_{\acute{e}t}(\overline{V}, \mathbf{Z}_p), \quad gr^n H^q = F^n H^q / F^{n+1} H^q. \tag{0.4}$$

Since the p-cohomological dimension of $\overline{Y}$ is equal to $\dim \overline{Y}$ [11] one has

$$F^0 H^*_{\acute{e}t} = H^*_{\acute{e}t}$$

$$F^n H^q_{\acute{e}t} = 0, \quad n > \min(q, \dim \overline{Y}).$$

This construction is analogous to one used in the study of vanishing cycles [12] and degenerating families of varieties [23]. We assume now and henceforth that X is smooth over S, so V has good reduction and Y is smooth over k. Associated to $\overline{Y}$ one has crystalline cohomology groups $H^*_{\text{crys}}(\overline{Y}/W)$ [1] and deRham-Witt cohomology groups $H^s(\overline{Y}, W\Omega^t)$ [14]. Both are $W(\overline{K})$-modules with given σ-linear frobenius endomorphisms f, where $W(\overline{K})$ is the ring of p-Witt vectors over $\overline{K}$ and $\sigma : W(\overline{K}) \to W(\overline{K})$ is induced by the p-th power map on $\overline{K}$, There is a spectral sequence ("slope spectral sequence")

$$(0.5) \qquad E_1^{s,t} = H^t(\overline{Y}, W\Omega^s) \Rightarrow H^{s+t}_{crys}(\overline{Y}/W)$$

which degenerates at E_1 up to torsion [2], [14].

It turns out that the "p-geometry" of the closed fibre $\overline{Y}$ as expressed in the deRham-Witt cohomology profoundly influences the structure of the G- modules $gr^n H^q_{\acute{e}t}(\overline{V}, \mathbf{Q}_p)$.

Case 1. Some $H^t(\overline{Y}, W\Omega^s)$ is *not* finitely generated over $W(\overline{K})$. In this worst (or "most interesting" in the sense of the old Chinese proverb) case, we don't even formulate a conjecture about $gr^n H^q_{\acute{e}t}(\overline{V}, \mathbf{Q}_p)$.

Case 2. All $H^t(\overline{Y}, W\Omega^s)$ are finitely generated. $\overline{Y}$ is said to be *Hodge-Witt* (op. cit.). In this case the slope spectral sequence (0.5) is known to degenerate at E_1 [15]. The conjecture (discussed in more detail in §3 below) is that writing $d = \dim Y$,

$$\mathrm{Hom}_{\mathbf{Q}_p}(gr^n_F H^q_{\acute{e}t}(\overline{V}, \mathbf{Q}_p), \mathbf{Q}_p(n-q+1))$$

is the Tate module associated to a lifting over S of the p-divisible formal group with covariant Dieudonné module $H^{d-n}(\overline{Y}, W\Omega^{d+n-q})/(\text{tors})$.

Case 3. Let $B^n \subset \Omega^n_Y$ be the sheaf of locally exact Kähler n-forms on Y. Y is said to be *ordinary* if $H^q(Y, B^n) = (0)$ for all q and n. Roughly speaking, "ordinary" is equivalent to "Hodge-Witt" plus equality of Hodge and Newton polygons. For a more careful discussion see §2 below. Let $H^q_{crys}(\overline{Y}/W)^{(i)} = \{x \in H^q_{crys}(\overline{Y}/W) \mid fx = p^i x\}$. Note G acts on this group via the action of $Gal(\overline{k}/k)$ on $H^q_{crys}(\overline{Y}/W)$.

Theorem (0.6). *Let notation be as above and assume Y is ordinary. Then for all integers q, i there exist functorial G-isomorphisms*

(i) $gr^{q-i}_F H^q_{\acute{e}t}(\overline{V}, \mathbf{Q}_p) \cong H^q_{crys}(\overline{Y}/W)^{(i)}_{\mathbf{Q}}(-i)$.

(ii) $gr_F^{q-i} H^q_{\acute{e}t}(\overline{V}, \mathbf{Q}_p) \otimes W(\overline{k}) \cong H^{q-i}(\overline{Y}, W\Omega^i)_{\mathbf{Q}}(-i)$.
(iii) $gr_F^{q-i} H^q_{\acute{e}t}(\overline{V}, \mathbf{Q}_p) \otimes \mathbf{C} \cong H^{q-i}(V_{\mathbf{C}}, \Omega^i)(-i) \cong H^{q-i}(V, \Omega^i) \otimes_K \mathbf{C}(-i)$.

The spectral sequence (0.3) degenerates up to torsion at E_2.

Corollary (0.7). *When Y is ordinary, Tate's conjecture (0.1) holds, i.e.*

$$H_{\acute{e}t}(\overline{V}, \mathbf{C}) \cong \bigoplus_{l+m=n} H^l(V_{\mathbf{C}}, \Omega^m)(-m).$$

The corollary follows from (0.6) and the fact proved in [25] that

$$H^1(G, \mathbf{C}(i)) = (0)$$

for $i \neq 0$, so there are no extensions between $\mathbf{C}(r)$ and $\mathbf{C}(s)$ for $r \neq s$.

§ 1 Local Results

Let notation be as in (0.2), and define étale sheaves on Y

$$(1.1) \qquad M^r_n = i^* R^r j_* (\mathbf{Z}/p^n\mathbf{Z}(r)).$$

To analyse the spectral sequence (0.3) it is necessary to understand the local structure of the sheaves M^r_n. Our results in this area rest on Kato's calculation of the Galois cohomology of quotient fields of henselian discrete valuation rings [17] and Gabber's theorem [10]. We need Y to be smooth but make no hypothesis about Y ordinary. Note that the geometric stalks of M^r_n are the étale cohomology groups

$$H^r_{\acute{e}t}\left(Sp(A[\tfrac{1}{p}]), \mathbf{Z}/p^n\mathbf{Z}(r)\right),$$

where A is the strict henselization of the local ring on X at a point $y \in Y$. For example, taking y to be a geometric generic point, A is a henselian discrete valuation ring with quotient field $A[\frac{1}{p}]$ and the result can be read off from [17].

In fact it is possible and occasionally technically convenient to work with a finer object. we view i^* as a functor from étale sheaves on X to *Zariski sheaves* on Y, and we define

(1.2) L_n^r = Zariski sheaf associated to the presheaf

$$U \to H^r(U, i^*\mathrm{R}j_*(\mathbb{Z}/p^n\mathbb{Z}(r))).$$

For example, the generic stalk of L_n^r is the galois cohomology of the quotient field of the (*non-strict*) henselization of X at the generic point of Y. M_n^r is the sheafification of L_n^r for the étale topology.

The structure of L_n^r is closely related to the K-thoery of O_V in the sense of Milnor (i.e. using symbols [20]). the Kummer sequence

$$0 \to \mathbb{Z}/p^n\mathbb{Z}(1) \to O_V^* \xrightarrow{p^n} O_V^* \to 0$$

induces an exact sequence of Zariski sheaves on Y

$$i^*j_*O_V^* \xrightarrow{p^n} i^*j_*O_V^* \to L_n^1 \to 0.$$

For local sections $x_1, \dots, x_r$ of $i^*j_*O_V^*$, let

$$\{x_1, \dots, x_r\} \tag{1.3}$$

be the local section of L_n^r defined by the cup product of the images of the x_i in L_n^1. These "symbols" satisfy the K-theory type relations

$$0 = \{x, -x\} = \{x, y\} + \{y, x\} = \{z, 1-z\}$$

for local sections x, y, z such that $1 - z$ is invertible.

Define $U^m L_n^1$ for $m \geq 1$ to be the image

$$1 + \pi^m i^* O_X \subset i^*j_*O_V^* \to L_n^1,$$

where $\pi \in \Lambda$ is now (and ever shall be) a uniformizing element. Define

$$U^m L_n^r = Image(U^m L_n^1 \otimes L_n^{r-1} \xrightarrow{\text{cup}} L_n^r).$$

Note $L_n^0 = (\mathbb{Z}/p^n\mathbb{Z})_Y$ and $U^m L_n^0 = (0)$ for $m \geq 1$.

One wants to understand the $gr_U^m L_n^r$. At the moment this is done only in the two cases $n = 1$ (which together with the case $m = 0$ suffices for the proof of (0.6), but probably will not suffice to understand the "non-ordinary" case) or $m \leq \frac{ep}{p-1}$ where $e = ord_\pi(p)$ is the index of ramification.

For simplicity I will restrict attention to the case $n = 1$. Define

$$B^r = \mathrm{Im}(d : \Omega_Y^{r-1} \to \Omega_Y^r)$$

$$Z^r = \mathrm{Ker}(d : \Omega_Y^r \to \Omega_Y^{r+1})$$

$$C\colon Z^r \longrightarrow \Omega^r \quad \text{Cartier operator} \quad [5]$$

$$\Omega^r_{Y,\log} = \mathrm{Ker}(C - 1 : Z^r \to \Omega^r).$$

Theorem (1.4). (i) *The sheaves L_n^r are generated Zariski locally by symbols (1.3).*

(ii) $gr_U^0(L_1^r) \cong \Omega^r_{Y,\log} \oplus \Omega^{r-1}_{Y,\log}$

(iii) *If* $0 < m < \frac{ep}{p-1}$ *and* $(m,p) = 1$, *then* $gr_U^m(L_1^r) \cong \Omega_Y^{r-1}$

(iv) *If* $0 < m < \frac{ep}{p-1}$ *and* $p|m$, *then* $gr_U^m(L_1^r) \cong B_Y^r \oplus B_Y^{r-1}$

(v) *Assume* $e' = \frac{ep}{p-1}$ *is an integer, and let* $a \in k$ *be the residue class of* $p\pi^{-e}$. *Then*

$$gr_U^{e'}(L_1^r) \cong (\Omega^{r-1}/(1 + aC)Z^{r-1}) \oplus (\Omega^{r-2}/(1 + aC)Z^{r-2}).$$

(iv) $U^m L_1^r = (0)$ *for* $m > e'$.

Remark (1.5). The deRham-Witt complex [14] is a complex of pro-sheaves $W^\bullet\Omega^q = \{W_n\Omega^q\}_{n\geq 1}$. The pro-object $W^\bullet\Omega^q_{Y\log}$ is defined to be the kernel of 1-frobenius on $W^\bullet\Omega^q$. Multiplication by p^n induces exact sequences (op.cit.)

$$0 \to W_n\Omega^q_{Y,\log} \to W_{m+n}\Omega^q_{Y,\log} \xrightarrow{\text{“}p^n\text{”}} W_m\Omega^q_{Y,\log} \to 0.$$

The isomorphism in (1.4) (ii) above extends to

$$gr_U^0(L_n^r) \cong W_n\Omega^r_{Y,\log} \oplus W_n\Omega^{r-1}_{Y,\log}. \tag{1.6}$$

The maps in (1.4) can be explicated in terms of symbols as follows (here $x_i \in O_X^*$, $y \in O_X$, $\overline{z} = z \,(\mathrm{mod}\,\pi)$):

(1.4)(ii) $\{x_1, \ldots, x_r\} \mapsto \mathrm{dlog}(\overline{x}_1) \wedge \cdots \wedge \mathrm{dlog}(\overline{x}_r) \in \Omega^r_{Y,\log}$

$$\{\pi, x_1, \dots, x_{r-1}\} \mapsto \mathrm{dlog}(\bar{x}_1) \wedge \cdots \wedge \mathrm{dlog}(\bar{x}_{r-1}) \in \Omega^{r-1}_{Y,\log}$$

(1.4)(iii) $\{1 + \pi^m y, x_1, \dots, x_{r-1}\} \mapsto \bar{y}\, \mathrm{dlog}(\bar{x}_1) \wedge \cdots \wedge \mathrm{dlog}(\bar{x}_{r-1})$

(1.4)(iv) $\{1 + \pi^m y, x_1, \dots, x_{r-1}\} \mapsto d\bar{y} \wedge \mathrm{dlog}(\bar{x}_1) \wedge \cdots \wedge \mathrm{dlog}(\bar{x}_{r-1}) \in B^r$

$\{\pi, 1 + \pi^m y, x_1, \dots, x_{r-1}\} \mapsto$

$$d\bar{y} \wedge \mathrm{dlog}(\bar{x}_1) \wedge \cdots \wedge \mathrm{dlog}(\bar{x}_{r-2}) \in B^{r-1}$$

(1.4)(v) $\{1 + \pi^m y, x_1, \dots, x_{r-1}\} \mapsto \bar{y}\, \mathrm{dlog}(\bar{x}_1) \wedge \cdots \wedge \mathrm{dlog}(\bar{x}_{r-1})$

$\{\pi, 1 + \pi^m y, x_1, \dots, x_{r-2}\} \mapsto \bar{y}\, \mathrm{dlog}(\bar{x}_1) \wedge \cdots \wedge \mathrm{dlog}(\bar{x}_{r-2})$

Note that the étale sheaves M^r_n have similar filtrations $U^\bullet M^r_n$ and the structure of $gr^m_U M^r_n$ is deduced from that of $gr^m_U L^r_n$ by passage to the étale topology. For example

$$gr^{e'}_U M^r_1 = (0), \quad e' = \frac{ep}{p-1}.$$

The following result (due to Kato) should be important in studying the non-ordinary case.

Theorem (1.7). *There exists a unique homomorphism of sheaves* $M^r_n \to (\Omega^r_{X/S})/p^n(\Omega^r_{X/S})$ given by $\{x_1, \dots, x_r\} \mapsto \mathrm{dlog}(x_1) \wedge \cdots \wedge \mathrm{dlog}(x_r)$ *for x_i local sections of $\mathcal{O}^*_X$ and trivial on symbols with one element constant (i.e. in K).*

Applied to the spectral sequence (0.3), this result gives a map

$$E_2^{s,t} = H^s_{\acute{e}t}(\bar{Y}, \bar{i}^* R^t \bar{j}_* \mathbf{Q}_p) \to H^s(V_{\mathbf{C}}, \Omega^t_{V_{\mathbf{C}}}(-t)).$$

When Y is ordinary, tensoring on the left with $\mathbf{C}$ yields the isomorphism (0.6)(iii).

In addition to the Galois cohomological results of Kato, the main idea in the proof of (1.4) is a general theorem of Gabber [10].

Theorem (1.8). *Let $M \xrightarrow{\pi} S$ be a smooth morphism of schemes, $s \in S$, Φ a bounded below complex of torsion étale sheaves on a retrocompact open set $V \subset S$. Write*

$$K = (Rj_* \pi^* \Phi)\,|_{M_s}$$

where $j\colon \pi^{-1}(V) \hookrightarrow M$. *Let* $\underline{\underline{H}}^d_{zar}(K)$ *be the Zariski sheaf on* M_s *associated to the presheaf*

$$U \mapsto H^d(U_{\acute{e}t}, K\,|_U).$$

For $x \in M_s$, *let* $i_x : Sp(k(x)) \to M_s$ *and write* L_i *for the Zariski sheaf*

$$L_i = \bigoplus_{\substack{x \in M_s \\ \mathrm{cod}(x) = i}} i_{x_*}\big(H_x^{d+i}(M_s, K)\big).$$

Then the Cousin complex gives a flasque resolution

$$(1.8.1) \qquad 0 \to \underline{\underline{H}}^d_{zar}(K) \xrightarrow{\epsilon} L_0 \to L_1 \to L_2 \to \cdots$$

In our situation, if A is the henselization of the local ring of X at a point y of Y and E is the quotient field of the henselization of X at the generic point of Y, then the injectivity of ϵ in (1.8.1) implies

$$M^r_{n,y} = H^r\Big(Sp(A[\tfrac{1}{p}]), \mu_{p^n}^{\otimes r}\Big) \to H^r(SpE, \mu_{p^n}^{\otimes r}).$$

In this way, the stalks of M^r_n at points other than the generic point can be controlled.

Concerning the proof of (0.6), we have by (1.6)

$$0 \to U^1M^r_n \to M^r_n \to W_n\Omega^r_{Y,\log} \oplus W_n\Omega^{r-1}_{Y,\log} \to 0.$$

adjoining $\pi^{p^{-n}}$ to A "kills" symbols $\{\pi, x_1, \ldots, x_{r-1}\}$, so writing $\overline{M}^r_\bullet = \{\overline{M}^r_n\}_{n\geq 1}$ for the pro-system associated to the limiting situation $\overline{X}/\overline{S}$, we get an exact sequence (defining $U\overline{M}^r_\bullet$) of pro-systems of étale sheaves on $\overline{Y}$,

$$0 \to U\overline{M}^r_\bullet \to \overline{M}^r_\bullet \to W_\bullet\Omega^r_{\overline{Y},\log} \to 0.$$

The key result is then

Proposition (1.9). *Assume* $\overline{Y}$ *is ordinary. Then*

$$H^m(\overline{Y}, U\overline{M}^r_\bullet) = (0), \quad \textit{all} \quad m, r.$$

One reduces to showing the map

$$H^m(Y, U^1M^r_1) \to H^m(\overline{Y}, U\overline{M}^r_1)$$

is zero and then one argues using the structure theory (1.4). An analysis of the structure of $H^m(\overline{Y}, U\overline{M}^r_\bullet)$ in the non-ordinary case is the key to further progress in this area.

§2 Ordinary Varieties

Recall the variety Y is *ordinary* if $H^m(Y, B^n) = (0)$ for all m and n. The following proposition overlaps with results of Illusie, and we would like to acknowledge considerable inspiration from conversations with him. The formulation here is due to Kato.

Proposition (2.1). *The following conditions (1)-(5) are equivalent.*

(1) *Y is ordinary.*

(2) $H^q(\overline{Y}, \Omega^r_{\overline{Y},\log}) \otimes_{\mathbf{F}_p} \overline{k} \to H^q(\overline{Y}, \Omega^r_{\overline{Y}})$ *is an isomorphism for any q, r.*

(3) $H^q(\overline{Y}, W_n\Omega^r_{\overline{Y},\log}) \otimes_{\mathbf{Z}/p^n\mathbf{Z}} W_n(\overline{k}) \to H^q(\overline{Y}, W_n\Omega^r)$ *is an isomorphism for any q, r, n.*

(4) $H^q(\overline{Y}, W\Omega^r_{\overline{Y},\log}) \otimes_{\mathbf{Z}_p} W(\overline{k}) \to H^q(\overline{Y}, W\Omega^r)$ *is an isomorphism for any q, r.*

(5) *The frobenius f: $H^q(Y, W\Omega^r_Y) \longrightarrow H^q(Y, W\Omega^r_Y)$ is an isomorphism for any q, r.*

Moreover, Y is ordinary and $H^q_{\mathrm{crys}}(Y/W)$ is torsion-free for all q if and only if the following holds:

(6) *For any q, the Newton polygon defined by the slopes of frobenius on $H^q_{\mathrm{crys}}(Y/W)$ coincides with the Hodge polygon defined by the $h^{q-i}(Y, \Omega^i_{Y/k})$.*

If $H^q(Y, W\Omega^r)$ is torsion-free for any q, r, the conditions (1)-(6) are also equivalent to

(7) *For any q, the slopes of frobenius on $H^q_{\mathrm{crys}}(Y/W)$ are all integers.*

By way of example, it is possible to show that ordinary hypersurfaces of degree d fill out a non-empty open set in the moduli space. This result is due to Deligne (unpublished). As a second example, a general polarized abelian variety is ordinary. (Gabber points out that no one has published a proof of connectedness for the moduli space of polarized abelian varieties. Failing this, the above assertion is to some extent conjectural.)

§3 Open Questions and Conjectures

Here are some directions one might try to explore.

p-adic functions on moduli: It follows from (0.6) (i) that when Y is ordinary, the G-module $gr_F^n H^q(\overline{V}, \mathbf{Q}_p)$ is independent of the lifting X of Y. The extensions

$$0 \to gr_F^{n+1} H^q \to F^n H^q / F^{n+2} H^q \to gr_F^n H^q \to 0$$

do, however, depend on the lifting. Assume for simplicity $k = \overline{k}$. We obtain a map

$$\{\text{liftings of } Y\} \to \mathrm{Ext}^1(gr_F^n H^q, gr_F^{n+1} H^q) \cong$$

$$H^1(G, \mathbf{Q}_p(1)) \otimes \mathrm{Hom}\left(H^q_{\mathrm{crys}}(Y/W)^{\langle q-n\rangle}, H^q_{\mathrm{crys}}(Y/W)^{\langle q-n-1\rangle}\right)$$
$$\cong \mathbf{Q} \otimes \widehat{K^*} \otimes \mathrm{Hom},$$

where $(\widehat{K^*} = \varprojlim K^*/K^{*p^n})$. Note that the group Hom is independent of lifting. Choosing a basis for the dual Hom^*, we get functions

$$\{\text{liftings of } Y\} \to \widehat{K^*} \otimes \mathbf{Q}.$$

The subgroup $1 + \pi\Lambda \subset K^*$ is p-adically complete. I would conjecture that these functions take values in $1 + \pi\Lambda \subset \widehat{K^*} \otimes \mathbf{Q}$. This, in any event, is what happens for abelian varieties and K3 surfaces [16], [6]. It would be very interesting to have a general theory of these "functions on p-adic moduli".

Galois representations associated to modular forms [7]: I am endebted to K. Ribet for suggesting the following possible application of the theory. Let M be the compactified modular curve associated to the principal congruence subgroup $\Gamma \subset SL_2(\mathbf{Z})$ of some level $N > 3$. We view M as a scheme over $\mathbf{Q}_p$ for some p prime to N. Let $E \xrightarrow{f} M$ be the universal curve with level structure. For some integer $k > 0$, consider

$$\tilde{H}^1_{\acute{e}t}(M_{\overline{\mathbf{Q}}_p}, Sym^k(R^1 f_* \mathbf{Q}_p)) = W^k$$

where $\tilde{H} = Image\,(H^1_{compact}(M - cusps) \to H^1(M))$. Let T_l be the Hecke operator corresponding to l. Let g be a modular form of weight $k+2$ for Γ which is an eigenfunction for Hecke and suppose $T_l(g) = a_l g$ for $l \not| N$ with $p \not| a_p$. Let $W^k(g) = \bigcap_l \mathrm{Ker}(T_l - a_l) \subset W^k$. The idea is that $W^k(g)$ should correspond to a submotive of

$$H^{k+1}(\overbrace{E \times_M \cdots \times_M E}^{\text{k times}}) = H^{k+1}(E^k)$$

which has Hodge type $(k+1, 0) + (0, k+1)$ and for which the characteristic polynomial of the conjugacy class of inverse frobenius $[\phi_p^{-1}] \subset Gal(\overline{\mathbf{Q}}/\mathbf{Q})$ is

$$T^2 - a_p T + p^{k+1}.$$

The hypothesis $p \not| a_p$ implies that the Hodge and Newton polygons associated to this motive coincide. The challenge now is to establish some version of (0.6) involving projections, together with an appropriate modification of (2.1) (6) or (7) sufficient to prove

Conjecture. *Let* $W^k(g)_{\mathrm{crys}} = \bigcap \mathrm{Ker}(T_l - a_l)$ *on* $H^{k+1}_{crys}(E^k/W(\overline{k}))$. *Then* $F^\bullet H^{k+1}(E^k, \mathbf{Q}_p)$ *induces a filtration* $F^\bullet W^k(g)$ *with*

$$gr^0_F W^k(g) \cong W^k(g)^{(k+1)}_{\mathrm{crys}}(-k-1); \quad gr^{k+1}_F W^k(g) \cong W^k(g)^{(0)}_{\mathrm{crys}}$$

$$gr^n_F W^k(g) = 0, \quad n \neq 0, k+1.$$

In particular, the p-adic galois representation $W^k(g)$ should have a Hodge-Tate decomposition as in (0.1).

Structure of $H^*(V, \mathbf{Q}_p)$ when Y is Hodge-Witt: To begin with, we conjecture in this case that the spectral sequence (0.3) degenerates up to torsion at E_2. If so, we obtain from (1.6) a map

$$\phi^{r,s}\colon F^r H^s(\overline{V}, \mathbf{Q}_p) \longrightarrow H^r(\overline{Y}, W\Omega^{s-r}_{Y,\log})(r-s) \otimes \mathbf{Q}$$

Define

$$U^{r+1} H^s(\overline{V}, \mathbf{Q}_p) = \mathrm{Ker}\,\phi^{r,s} \subset F^r H^s.$$

We obtain

$$H^s = U^0H^s = F^0H^s \supset U^1H^s \supset F^1H^s \supset \cdots \supset U^s \supset F^s \supset U^{s+1} = 0.$$

Let $d = \dim \overline{V}$, so

$$H^{2d}(\overline{V}, \mathbf{Q}_p) \cong gr_F^d H^{2d} \cong \mathbf{Q}_p(-d)$$

The maps $\phi^{r,s}$ above are compatible with cup product, from which one deduces that under cup product

$$F^rH^s \times U^{d+1-r}H^{2d-s} \to U^{d+1}H^{2d} = (0),$$

so there is a pairing

$$(*) \qquad gr_F^r H^s(\overline{V}, \mathbf{Q}_p) \times gr_U^{d-r} H^{2d-s}(\overline{V}, \mathbf{Q}_p) \to \mathbf{Q}_p(-d),$$

which we conjecture to be perfect. Let C be the category of pointed artinian local Λ-algebras, and let $\underline{\underline{K}}_n$ denote the Zariski sheaf associated to the functor K_n of Quillen. (Maybe it would be better to consider the Milnor K_n.) Define for $A \in ob\, C$

$$H^m(K_n)(A) = \mathrm{Ker}\big(H^m(X_{Sp\,\Lambda} \times Sp\,A, \underline{\underline{K}}_n) \to H^m(Y, \underline{\underline{K}}_n)\big).$$

Conjecture. *The functor $H^m(K_n) : C \to$ (abelian groups) has (under the hypothesis Y Hodge-Witt) a pro-representable "piece" (for more details on this point, see [24]) which is a p-divisible formal group with Dieudonné module $H^m(Y, W\Omega^{n-1})/tors$ and Tate module $gr_U^m H^{m+n-1}(\overline{V}, \mathbf{Q}_p(n))$.*

* Added in proof: O. Gabber has decided not to collaborate, so these results will be given in a joint paper with K. Kato. Also, Ogus asks me to stress that his result mentioned below, which is given in his paper in these volumes, came in response to a question of Shafarevich, and that the results on the moduli of K3 surfaces are due to Rudakov-Shafarevich.

References

[1] Berthelot, P., Cohomologie cristalline des schémas de caractéristique $p > 0$. Lecture Notes in Math. 407, Springer-Verlag (1974).

[2] Bloch, S., Algebraic K-theory and crystalline cohomology, Pub. Math. I.H.E.S. 47, 187–268 (1974).

[3] Bloch. S., Some formulas pertaining to the K-theory of commutative group schemes, J. Algebra, 53, 304–326 (1978).

[4] Bloch, S., Gabber, O., and Kato, K., p-adic étale cohomology (manuscript).

[5] Cartier, P., Une nouvelle opération sur les formes différentielles, C. R. Acad. Sci. Paris 244, 426–428 (1957).

[6] Deligne, P., (with Illusie, L.) Cristaux ordinaires et coordonnés canoniques, in *Surfaces algébriques*, Lecture Notes in Math. 868, Springer-Verlag (1981).

[7] Deligne, P., Formes modulaires et représentations l-adiques, *Sém. Bourbaki*, 1968/69, exp. 355, Lecture Notes in Math. 179, Springer-Verlag (1971).

[8] Dwork, B., Normalized period matrices, Ann. Math., 94, 337–388 (1971).

[9] Fontaine, J. M., Formes différentielles et modules de Tate ..., Inv. Math. 65, 379–409 (1982).

[10] Gabber, O., Gersten's conjecture for some complexes of vanishing cycles (manuscript).

[11] Grothendieck, A., et.al., *SGA4*, Lecture Notes in Math. 269, 270, 305, Springer-Verlag (1972–73).

[12] Grothendieck, A., Deligne, P., Katz, N., *SGA7*, Lecture Notes in Math. 288, 340, Springer-Verlag (1972–73).

[13] Hartshorne, R., *Residues and Duality*, Lecture Notes in Math. 20, Springer-Verlag (1966).

[14] Illusie, L., Complexe de deRham-Witt et cohomologie cristalline, Ann. Sci. Ec. Norm. Sup. 4e sér. 12, 501–661 (1979).

[15] Illusie, L., Raynaud, M., Les suites spectrales associées au complexe de deRham-Witt (manuscript).

[16] Katz, N., Serre-Tate local moduli, in *Surfaces algébriques*, Lecture Notes Math. 868, Springer-Verlag (1981).

[17] Kato, K., Galois cohomology of complete discrete valuation fields (preprint).

[18] Kato, K., A generalization of local class field theory by using K-groups. I, II, J. Fac. Sci. Univ. Tokyo, 26, 303–376 (1979), and 27, 603–683 (1980).

[19] Koblitz, N., p-adic variation of the zeta function over families of varieties defined over finite fields, Comp. Math. 31, 119–218 (1975).

[20] Milnor, J., Algebraic K-theory and quadratic forms, Inv. Math. 9, 318–344 (1970).

[21] Oort,F., Lenstra, H., Simple abelian varieties having prescribed formal isogeny type, J. Pure Appl. Algebra 4, 47–53 (1974).

[22] Serre, J-P., and Tate, J., Good reduction of abelian varieties, Ann. Math. 88, 492–517 (1968).

[23] Steenbrink, J., Limits of Hodge structures, Inv. Math. 31, 229–257 (1975).

[24] Stienstra, J., The formal completion of the second chow group; a K-theoretic approach, in *Proceedings of the Rennes Conference in Algebraic Geometry*.

[25] Tate, J., p-divisible groups, in *Proceedings of a Conference on local fields*, Springer-Verlag (1967).

Received July 19, 1982
Supported in part by N.S.F.

Professor Spencer Bloch
Department of Mathematics
University of Chicago
Chicago, Illinois 60637

The Mordell-Weil Group of Curves of Genus 2

J. W. S. Cassels

To I.R. Shafarevich

0.0 Introduction.

In 1922 Mordell [2] proved Poincaré's conjecture that the group of rational points on an abelian variety of dimension 1 (= elliptic curve with rational point) is finitely generated. His proof was somewhat indirect. In 1928 Weil [5] in his thesis generalized Mordell's result to abelian varieties of any dimension and to any algebraic number field as ground field. At the same time, Weil [6] gave a very simple and elegant proof of Mordell's original result. I observed some time ago that Weil's simple proof admits a further simplification. He uses the explicit form of the duplication and addition theorems on the abelian variety, but these can be avoided by, roughly speaking, the observation that every element of a cubic field-extension $k(\theta)$ of a field k can be put in the shape $(a\theta + b)/(c\theta + d)\,(a, b, c, d \in k)$. This additional simplification has little interest in itself, but suggested that similar ideas might be usefully exploited in investigating the Mordell-Weil group of (the Jacobians of) curves of genus greater than 1. This is done here for curves of genus 2.

The literature dealing with individual curves of genus 1 is vast. Very little, if anything, has been done with curves of higher genus. In fact the machinery which has been developed, while admirable for deep and general investigations, appears (*pace* Mumford [3,4]) little suited to the explicit treatment of explicitly given curves. I cannot even find in the literature an explicit set of equations for the Jacobian of a curve of genus 2 together with explicit expressions for the group operation in a form amenable to calculation: nor do I give one, since it is not needed for my present purposes. Although I do not give any numerical examples, I hope and believe that the theorems proved here will make it practicable to compute the Mordell-Weil group of explicitly given curves of genus 2.

0.1 It is useful to recall the main steps of Weil's simple proof of Mordell's theorem in a shape suited to later generalization. The ground field is k, and an abelian variety of dimension 1 is given by

$$(0.1.1) \qquad C: \quad Y^2 = F(X),$$

where

$$(0.1.2) \qquad F(X) = X^3 + a_2X^2 + a_1X + a_0 \in k[X]$$

does not have multiple factors. There is a single point $\mathfrak{o}$ at infinity. Every divisor class of degree 0 defined over k contains a uniquely defined divisor $\mathfrak{a} - \mathfrak{o}$, where $\mathfrak{a}$ is a point on C defined over k. The group $\mathcal{G}$ of divisor classes of degree 0 defined over k (the Mordell-Weil group) is thus also the group of points on C defined over k with the induced law of composition [C is its own Jacobian].

Weil's proof that $\mathcal{G}$ is finitely generated when k is an algebraic number field falls into three parts:

(I) He constructs an abelian group $\mathcal{M}$ (depending on C and k) and a homomorphism

$$(0.1.3) \qquad \mu \colon \mathcal{G} \longrightarrow \mathcal{M}$$

whose kernel is precisely $2\mathcal{G}$. This step is valid for any field k of characteristic 0.

(II) If k is an algebraic number field, the image $\mu(G)$ is a finite group. Hence $\mathcal{G}/2\mathcal{G}$ is finite. ("Weak Mordell-Weil Theorem.")

(III) Finally, the proof that $\mathcal{G}$ is finitely generated uses an "infinite descent."

In this paper we are concerned with the generalization of (I), and so our ground field is any field k of characteristic 0. Given the definition of μ, the generalization of (II) is immediate: we do not prove it because it can be confidently left to the reader. Finally, the generalization of (III) in a shape amenable to practical calculation might well be difficult: I have not yet attempted it.

0.2 We shall be concerned with curves

$$(0.2.1) \qquad C: \quad Y^2 = F(X)$$

where $F(X) \in k[X]$ and k is the ground field. When F has degree 5 or 6, the genus of C is 2. Lines $X = \text{const.}$ meet C in effective divisors of degree 2. They are all linearly equivalent, and their equivalence class will be denoted by $\mathfrak{O}$. The Mordell-Weil group $\mathcal{G}$ is defined to be the group of divisor classes of degree 0 defined over k. By the Riemann-Roch theorem, every element of $\mathcal{G}$ contains a divisor $\mathfrak{A} - \mathfrak{O}$, where $\mathfrak{A}$ is effective, defined over k, of degree 2. The divisor $\mathfrak{A}$ is uniquely determined by the element of $\mathcal{G}$, except that the whole divisor class $\mathfrak{O}$ corresponds to the zero element. We shall show that there is a map μ from $\mathcal{G}$ to an abelian group $\mathcal{M}$ giving a situation very similar to that used by Weil [6].

In a sense, the case when $F(X)$ is of degree 5 is very special. A bilinear transformation of X and the appropriate change of Y takes a curve (0.2.1) with F of degree 5 into one with F of degree 6. However, the case of degree 5 is somewhat simpler, so we treat it separately first.

The case when $F(X)$ has degree 4 also fits into the picture, at least when the coefficient of X^4 is 1. Then C is of genus 1 and is birationally equivalent to a C with F of degree 3. We shall, however, give this case separately, as it motivates some features of the degree 6 case and because of the unified picture presented by degrees 3, 4, 5, 6 together.

0.3 It is convenient to introduce here notation which will be common to all four cases. There follows a synopsis which will bring out the common pattern, although detailed enunciations will have to await the relevant later sections.

The ground field is any field k of characteristic 0. Its algebraic closure is $\bar{k}$. We are concerned with curves

$$C: \quad Y^2 = F(X), \tag{0.3.1}$$

where

$$F(X) = a_N X^N + a_{N-1} X^{N-1} + \cdots + a_o \in k[X], \quad a_N \neq 0 \tag{0.3.2}$$

has no multiple factors. We write

$$k[\Theta] = k[T]/F(T), \tag{0.3.3}$$

where Θ is the image of the indeterminate T in the quotient ring, so

$$k[\Theta] = \bigoplus_{j=1}^{J} K_j \tag{0.3.4}$$

is the direct sum of fields K_j corresponding to the irreducible factors of $F(X)$.

Let $k[\Theta]^*$ be the group of invertible elements of $k[\Theta]$. Our homomorphisms μ will be into the group

$$\mathcal{L} = \begin{cases} k[\Theta]^*/\{k[\Theta]^*\}^2 & (N \text{ odd}) \\ k[\Theta]^*/k^*\{k[\Theta]^*\}^2 & (N \text{ even}). \end{cases} \tag{0.3.5}$$

More precisely, the image will be in the kernel $\mathcal{M}$ of the map

$$\text{Norm: } \mathcal{L} \longrightarrow k^*/(k^*)^2 \tag{0.3.6}$$

induced in the obvious way by the Norm map of $k[\Theta]$ into k.

When $N = 3$ or 4, so C is of genus 1, the general element of $\mathcal{G}$ is given by $\mathfrak{a} = (a, d)$ on C, where $a, d \in k$. If $d \neq 0$, then $a - \Theta$ is in $k[\Theta]^*$ and we define $\mu(\mathfrak{a})$ to be its image in $\mathcal{L}$. For elements of $\mathcal{G}$ not given by such $\mathfrak{a}$, the definition of μ is given later.

When $N = 5$ or 6, so C is of genus 2, the general element of $\mathcal{G}$ is given by an effective divisor $\mathfrak{A} = \{(a_1, d_1), (a_2, d_2)\}$ $(a_1, a_2, d_1, d_2 \in \bar{k})$ defined over k. If $d_1 d_2 \neq 0$, then $(a_1 - \Theta)(a_2 - \Theta) \in k[\Theta]^*$ and $\mu(\mathfrak{A})$ is defined to be its image in $\mathcal{L}$. For elements of $\mathcal{G}$ not of this type, the definition of μ is given later.

Then in all the four cases listed below, μ is a homomorphism of $\mathcal{G}$ into $\mathcal{M}$:

(A) $N = 3$, $a_3 = 1$. The zero element of $\mathcal{G}$ is the point at infinity. The kernel is $2\mathcal{G}$.

(B) $N = 4$, $a_4 = 1$. The double point at infinity has two branches $\mathfrak{o}_+$, $\mathfrak{o}_-$. Either can be taken as zero element of $\mathcal{G}$. The kernel is generated by $2\mathcal{G}$ and $\mathfrak{o}_+$, $\mathfrak{o}_-$.

(C) $N = 5$, $a_5 = 1$. Lines $X =$ const. meet C in linearly equivalent divisors. Their class $\mathfrak{O}$ is taken as the zero of $\mathcal{G}$. The kernel is $2\mathcal{G}$.

(D) $N = 6$, $a_6 \in k^*$. The zero of $\mathcal{G}$ is as in (C). The kernel is generated by $2\mathcal{G}$ and by divisors linearly equivalent to $2\mathfrak{C} - 2\mathfrak{O}$, where $\mathfrak{C}$ is an effective divisor of degree 3 defined over k.

1.0 Genus 1, Degree = 3.

After multiplying X, Y by appropriate elements of k^*, we may suppose that

(1.0.1) $$F(X) = X^3 + a_2X^2 + a_1X + a_o \in k[X].$$

This is (effectively) the case treated by Weil [6]. For a point $\mathfrak{a}$ on $C : Y^2 = F(X)$ defined over k we define

(1.0.2) $$\mu(\mathfrak{a}) \in \mathcal{L} = k[\Theta]^*/\{k[\Theta]^*\}^2$$

(in the notation of §0.3) as follows:

(i) $\mu(\mathfrak{o}) = 1$, where $\mathfrak{o}$ is the point at infinity.

(ii) $\mathfrak{a} = (a, d)$, where $a, d \in k$ and $d \neq 0$. Then $a - \Theta \in k[\Theta]^*$ and $\mu(\mathfrak{a})$ is its image in $\mathcal{L}$. We note that

(1.0.3) $$\mathrm{Norm}_{k[\Theta]/k}(a - \Theta) = d^2,$$

and so $\mu(\mathfrak{a}) \in \mathcal{M}$, where $\mathcal{M}$ is (as always) the kernel of the norm map (0.3.6).

(iii) $\mathfrak{a} = (a, 0)$, where $a \in k$. Then $k[\Theta]$ is a sum (0.3.4) of fields. One of them, K_1 say, is isomorphic to k, the isomorphism being given by $\Theta \to a$. The image of $a - \Theta$ is non-zero in $K_j (j \neq 1)$. We replace the K_1-component 0 of $a - \Theta$ by an element of k^*, leaving the other components unchanged, in such a way that the norm for $k[\Theta]/k$ is in $(k^*)^2$. The image in $\mathcal{L}$ is then unique and is necessarily in $\mathcal{M}$: it is defined to be $\mu(\mathfrak{a})$. We can make this construction explicit. Let

(1.0.4) $$F(X) = (X - a)F_0(X).$$

Then the norm of $a - \Theta$ from $\bigoplus_{j \neq 1} K_j$ is $F_0(a)$ $(= F'(a))$. Hence the K_1-component of $\mu(a, 0)$ is defined to be

(1.0.5) $$F_0(a)(k^*)^2.$$

For an alternative discussion see §1.3.

1.1 Theorem 1.1. *The map*

(1.1.1) $$\mu\colon \mathcal{G} \longrightarrow \mathcal{M}$$

given by

$$\mathfrak{a} \to \mu(\mathfrak{a}) \tag{1.1.2}$$

is a homomorphism.

Proof. If $\mathfrak{a} = (a, d)$, then $-\mathfrak{a} = (a, -d)$, so clearly

$$\mu(-\mathfrak{a}) = \mu(\mathfrak{a}). \tag{1.1.3}$$

Hence it is enough to show that

$$\mathfrak{a} + \mathfrak{b} + \mathfrak{c} = 0 \tag{1.1.4}$$

implies

$$\mu(\mathfrak{a})\mu(\mathfrak{b})\mu(\mathfrak{c}) = 1. \tag{1.1.5}$$

If any of $\mathfrak{a}, \mathfrak{b}, \mathfrak{c}$ is $\mathfrak{o}$, the implication is trivial, so without loss of generality

$$\mathfrak{a} = (a, d); \quad \mathfrak{b} = (b, e); \quad \mathfrak{c} = (c, f), \tag{1.1.6}$$

where $a, b, c, d, e, f \in k$.

The condition (1.1.4) is equivalent to the statement that $\mathfrak{a}, \mathfrak{b}, \mathfrak{c}$ are the intersections of C with a line:

$$Y = l_1 X + l_0 \in k[X]. \tag{1.1.7}$$

Hence

$$F(X) - (l_1 X + l_0)^2 = (X - a)(X - b)(X - c) \tag{1.1.8}$$

and

$$(a - \Theta)(b - \Theta)(c - \Theta) = (l_1 \Theta + l_0)^2. \tag{1.1.9}$$

If $def \neq 0$, this gives (1.1.5) at once by the definition of μ. The proof remains valid when two or all three of $\mathfrak{a}, \mathfrak{b}, \mathfrak{c}$ coincide.

If the images of $a - \Theta$, $b - \Theta$, $c - \Theta$ vanish in only one of the fields K_j of (0.3.4), say they do not vanish in $K_j (j \neq 1)$, then (1.1.5) holds for all the components except, possibly, the K_1-components. Since the values of μ are in $\mathcal{M}$, they are determined by the K_j components ($j \neq 1$), and so (1.1.5) holds also for the K_1 components. There remains the case when, say, $d = e = 0$ and $a \neq b$. Then $\mathfrak{a}, \mathfrak{b}, \mathfrak{c}$ are the intersection of $\mathcal{C}$ with $Y = 0$, and (1.1.5) follows by direct calculation. [Alternative arguments for these special cases are in §1.3.]

1.2 **Theorem 1.2.** *The kernel of* μ: $\mathcal{G} \longrightarrow \mathcal{M}$ *is precisely* $2\mathcal{G}$.

Proof. Since $\mathcal{M} \subset \mathcal{L}$ is of exponent 2, the kernel certainly contains $2\mathcal{G}$. We have therefore to show that $\mu(\mathfrak{a}) = 1$ implies $\mathfrak{a} = 2\mathfrak{b}$ for some $\mathfrak{b} \in \mathcal{G}$. Since $\mathfrak{o} = 2\mathfrak{o}$, we need consider only

(1.2.1) $$\mathfrak{a} = (a, d), \quad \mu(\mathfrak{a}) = 1.$$

Hence

(1.2.2) $$a - \Theta = (p_2\Theta^2 + p_1\Theta + p_0)^2$$

for some p_0, p_1, $p_2 \in k$. For s_1, $s_2 \in k$ we have

(1.2.3) $$(s_1\Theta + s_0)(p_2\Theta^2 + p_1\Theta + p_0) = r_2\Theta^2 + r_1\Theta + r_0$$

(using $F(\Theta) = 0$), where r_2, r_1, r_0 are linear forms in s_1, s_0. We may therefore find s_1, s_0 not both 0 such that

(1.2.4) $$r_2 = 0.$$

Suppose, first, that $s_1 = 0$, so $s_0 \neq 0$. Then $p_2 = 0$ and

(1.2.5) $$(a - \Theta) = (p_1\Theta + p_0)^2.$$

This is impossible, since the defining relation for Θ is of degree 3.

Hence, by homogeneity, we may suppose that $s_1 = -1$, and so by (1.2.2), (1.2.3) we have

(1.2.6) $$(s_0 - \Theta)^2(a - \Theta) = (r_1\Theta + r_0)^2.$$

On replacing Θ by an indeterminate T we have

(1.2.7) $$(r_1T + r_0)^2 - (s_0 - T)^2(a - T) = wF(T)$$

for $w \in k$. In fact

$$w = 1 \tag{1.2.8}$$

on comparing coefficients of T^3. Hence, on changing the signs of r_1, r_0 if necessary, the line

$$Y = r_1 X + r_0 \tag{1.2.9}$$

passes through $\mathfrak{a} = (a, d)$. The residual intersection is $2\mathfrak{b}$, where $\mathfrak{b} = (s_0, t)$ and t is given by (1.2.9). Hence $\mathfrak{a} \in 2\mathcal{G}$, as required.

1.3 The previous treatment of the case when $a - \Theta$ is not invertible was somewhat *ad hoc*. We now sketch a rather more highbrow treatment which has the advantage that it carries over automatically to the cases we shall consider later.

Let $\mathfrak{a} = (a, 0)$ and choose the notation in (0.3.4) so that the projection of $k[\Theta]$ onto the field K_1 is given by $\Theta \to a$. Let $\mathfrak{x}$ be a generic point of C over k and choose t to be a local uniformizer for $\mathfrak{x}$ at $\mathfrak{a}$. Then $X(\mathfrak{x}) - a$ has a double zero at $\mathfrak{x} \to \mathfrak{a}$; and so

$$X(\mathfrak{x}) - a = t^2 M(\mathfrak{x}), \tag{1.3.1}$$

where $M(\mathfrak{x})$ specializes to a finite non-zero $M(\mathfrak{a}) \in k^*$. Replacement of t by another local uniformizer multiplies $M(\mathfrak{a})$ by an element of $(k^*)^2$. We show that $\mu(\mathfrak{a})$ can be defined by replacing the K_1- component 0 of $a - \Theta$ by $M(\mathfrak{a})$ before taking the image in $\mathcal{L}$.

In fact, $Y(\mathfrak{x})$ has a simple zero at $\mathfrak{a}$, and so

$$Y(\mathfrak{x}) = tN(\mathfrak{x}), \tag{1.3.2}$$

where $N(\mathfrak{a}) \in k^*$. We have

$$F(X) = (X - a)F_0(X), \tag{1.3.3}$$

where

$$F_0(X) \in k[X], \quad F_0(a) \neq 0. \tag{1.3.4}$$

On taking the equation $Y^2 = F(X)$ of C, dividing by t^2 and specializing $\mathfrak{x}$ to $\mathfrak{a}$, we get

$$\left(N(\mathfrak{a})\right)^2 = M(\mathfrak{a})F_0(a). \tag{1.3.5}$$

Hence

$$F_0(a)(k^*)^2 = M(\mathfrak{a})(k^*)^2. \tag{1.3.6}$$

Comparison with (1.0.5) shows that the new definition of $\mu(\mathfrak{a})$ coincides with the old.

We now show how to prove Theorem 1.1 using the new definition of μ. This time we show directly that the K_1-components satisfy (1.1.5). Let $\mathfrak{a} = (a, 0)$ be as above, let $\mathfrak{b} \neq \mathfrak{a}$ and let $\mathfrak{c}$ satisfy $\mathfrak{a} + \mathfrak{b} + \mathfrak{c} = 0$. Let $\mathfrak{x}$ be a generic point, and work with the ground field $k(\mathfrak{x})$. As in the proof of Theorem 1.1, there are $l_1(\mathfrak{x})$, $l_0(\mathfrak{x}) \in k(\mathfrak{x})$ such that the line

$$Y = l_1(\mathfrak{x})X + l_0(\mathfrak{x}) \tag{1.3.7}$$

meets C in $\mathfrak{x}$ and $\mathfrak{b}$. Let the third intersection be $\mathfrak{p}$. Then, as before,

$$\big(X(\mathfrak{x}) - a\big)(b - a)\big(X(\mathfrak{p}) - a\big) = \big(l_1(\mathfrak{x})a + l_0(\mathfrak{x})\big)^2.$$

Let $\mathfrak{x}$ specialize to $\mathfrak{a}$, so $\mathfrak{p}$ specializes to $\mathfrak{c}$. The left-hand side has a double zero and so, if we divide by t^2 before specializing, where t is a local uniformizer, we have

$$M(\mathfrak{a})(b - a)(c - a) \in (k^*)^2.$$

This is just the assertion of the K_1-component of (1.1.5), as required.

2.0 Genus 1, Degree = 4.

Here we consider

$$F(X) = X^4 + a_3X^3 + a_2X^2 + a_1X + a_0 \in K[X]. \tag{2.0.1}$$

There is a double point at infinity. We denote the two branches by $\mathfrak{o}_+$, $\mathfrak{o}_-$, where

$$\left.\begin{aligned} Y/X^2 &= +1 \text{ at } \mathfrak{o}_+ \\ &= -1 \text{ at } \mathfrak{o}_- \end{aligned}\right\}. \tag{2.0.2}$$

We take $\mathfrak{o}_+$ as the zero of the Mordell-Weil group $\mathcal{G}$. Let $\mathcal{E}$ be the subgroup generated by $\mathfrak{o}_-$. We shall generally be concerned with $\mathcal{G}/\mathcal{E}$, and so the distinction between $\mathfrak{o}_+$ and $\mathfrak{o}_-$ disappears.

Here the definition of $\mathcal{L}$ is different from that in §1. For $\mathfrak{a} \in \mathcal{G}$ we define

$$\mu(\mathfrak{a}) \in \mathcal{L} = k[\Theta]^*/k^*\{k[\Theta]^*\}^2 \tag{2.0.3}$$

as follows:

(i) $\mu(\mathfrak{o}_+) = \mu(\mathfrak{o}_-) = 1$.

(ii) If $\mathfrak{a} = (a, d)$ with $d \neq 0$, so $a - \Theta \in k[\Theta]^*$, then $\mu(\mathfrak{a})$ is the image of $a - \Theta$ in $\mathcal{L}$. As before,

$$\mu(\mathfrak{a}) \in \mathcal{M}, \tag{2.0.4}$$

where $\mathcal{M} \subset \mathcal{L}$ is the kernel of the norm map (0.3.6).

(iii) If $\mathfrak{a} = (a, 0)$, we define $\mu(\mathfrak{a})$ either as in (iii) of §1.0 or, equivalently, as in §1.3.

2.1. **Theorem 2.1.** *The map*

$$\mu\colon \mathcal{G} \longrightarrow \mathcal{M} \tag{2.1.1}$$

given by

$$\mathfrak{a} \to \mu(\mathfrak{a}) \tag{2.1.2}$$

is a group homomorphism. The kernel contains the subgroup $\mathcal{E}$ generated by $\mathfrak{o}_-$.

Proof. Let $\mathfrak{a} = (a, d)$. The function $X - a$ has zeros $\mathfrak{a}$ and $\mathfrak{a}' = (a, -d)$ and poles $\mathfrak{o}_+$, $\mathfrak{o}_-$. Hence $\mathfrak{a}' + \mathfrak{a} = \mathfrak{o}_-$ in $\mathcal{G}$. Clearly $\mu(\mathfrak{a}') = \mu(\mathfrak{a})$, so

$$\mu(\mathfrak{o}_- - \mathfrak{a}) = \mu(\mathfrak{a}). \tag{2.1.3}$$

Now consider the intersections of

$$Y = X^2 + b_1 X + b_0 \in k[X] \tag{2.1.4}$$

with C. They are given by

$$(X^2 + b_1 X + b_0)^2 = F(X) = X^4 + a_3 X^3 + \cdots + a_0. \tag{2.1.5}$$

Here one of the four intersections is always at $\mathfrak{o}_+$. We suppose, first, that

$$2b_1 = a_3 \tag{2.1.6}$$

so $\mathfrak{o}_+$ is a double intersection. Let $\mathfrak{a} = (a, d)$ be arbitrary. We can choose b_0 so that (2.1.4) passes through $\mathfrak{a}$. Let $\mathfrak{b} = (b, e)$ be the fourth intersection. Then

$$(X^2 + b_1X + b_0)^2 - F(X) = l(X - a)(X - b) \tag{2.1.7}$$

for some $l \in k^*$. On putting $X \to \Theta$ and recalling the definition (2.0.3) of $\mathcal{L}$, we obtain

$$\mu(\mathfrak{a}) = \mu(\mathfrak{b}), \tag{2.1.8}$$

at least when $de \neq 0$, so $a - \Theta$, $b - \Theta$ are invertible. The truth of (2.1.8) when $de = 0$ can be verified either by an *ad hoc* argument using (2.0.4) as in §1.1 or by the technique of §1.3. We have ignored the possibility that $\mathfrak{b} = \mathfrak{o}_+$: but then the right-hand side of (2.1.7) becomes $l(X - a)$, so $\mu(\mathfrak{a}) = 1 = \mu(\mathfrak{o}_+)$. The divisor of poles of

$$Y - (X^2 + b_1X + b_0) \tag{2.1.9}$$

is $2\mathfrak{o}_-$, so $\mathfrak{a} + \mathfrak{b} = 2\mathfrak{o}_-$ in $\mathcal{G}$, and hence

$$\mu(2\mathfrak{o}_- - \mathfrak{a}) = \mu(\mathfrak{a}). \tag{2.1.10}$$

From (2.1.3) and (2.1.10) we have

$$\mu(\mathfrak{a} + \mathfrak{o}_-) = \mu(\mathfrak{a}), \tag{2.1.11}$$

and so $\mu(\mathfrak{a})$ depends only on $\mathfrak{a}$ modulo $\mathcal{E}$.

We now abandon the assumption (2.1.6). Let $\mathfrak{a}=(a, d)$ and $\mathfrak{b} = (b, e) \neq (a, -d)$. Then we can choose b_1, b_0 so that (2.1.4) meets C in $\mathfrak{a}$ and $\mathfrak{b}$ [a double intersection at $\mathfrak{a}$ if $\mathfrak{b} = \mathfrak{a}$]. Let $\mathfrak{c} = (c, f)$ be the residual intersection. Then

$$(X^2 + b_1X + b_0)^2 - F(X) = l(X - a)(X - b)(X - c).$$

On putting $X \to \Theta$ we have

$$\mu(\mathfrak{a})\mu(\mathfrak{b})\mu(\mathfrak{c}) = 1,$$

at least if $def \neq 0$. When $def = 0$, one can either argue *ad hoc* or as in §1.3. Since

$$\mathfrak{a} + \mathfrak{b} + \mathfrak{c} = 2\mathfrak{o}_- \equiv 0 \pmod{\mathcal{E}},$$

we have completed the proof that μ is a homomorphism.

2.2. **Theorem 2.2.** *The kernel of $\mu : \mathcal{G} \to \mathcal{M}$ is the subgroup generated by $2\mathcal{G}$ and $\mathcal{E}$.*

Proof. The kernel contains $\mathcal{E}$ by Theorem 2.1. It contains $2\mathcal{G}$ since $\mathcal{M}$ is of exponent 2. We have therefore only to prove that the kernel is not larger than stated.

Let $\mu(\mathfrak{a}) = 1$. If $\mathfrak{a} = \mathfrak{o}_+$ or $\mathfrak{o}_-$, the conclusion holds trivially, so we may suppose that

$$\mathfrak{a} = (a, d) \qquad a, d \in k. \tag{2.2.1}$$

By hypothesis,

$$l(a - \Theta) = (c_3\Theta^3 + c_2\Theta^2 + c_1\Theta + c_0)^2 \tag{2.2.2}$$

for some $l \in k^*$ and some $c_3, c_2, c_1, c_0 \in k$. We cannot have $c_3 = c_2 = 0$, since the defining relation $F(\Theta) = 0$ of Θ is of degree 4.

Suppose, first, that $c_3 = 0$. Then without loss of generality $c_2 = 1$ and

$$(T^2 + c_1T + c_0)^2 + l(T - a) = F(T). \tag{2.2.3}$$

Hence $\mathfrak{a} = (a, d)$ lies on

$$Y \pm (X^2 - c_1X + c_0) = 0 \tag{2.2.4}$$

for one choice of sign. The only points other than $\mathfrak{a}$ occurring as zeros or poles of the left-hand side of (2.2.4) are $\mathfrak{o}_+$ and $\mathfrak{o}_-$. Hence the conclusion of the theorem holds.

We are left with

$$c_3 \neq 0. \tag{2.2.5}$$

For $s_1, s_0 \in k$ we have

$$(s_1\Theta + s_0)(c_3\Theta^3 + c_2\Theta^2 + c_1\Theta + c_0) = t_3\Theta^3 + t_2\Theta^2 + t_1\Theta + t_0, \tag{2.2.6}$$

where $t_3, \ldots, t_0$ are linear in s_0, s_1. We may choose s_1, s_0 so that

$$t_3 = 0. \tag{2.2.7}$$

Then we get $s_1 \neq 0$ by (2.2.5) and, without loss of generality,

$$s_1 = -1. \tag{2.2.8}$$

Substituting (2.2.6) into (2.2.2) we have

$$l(\Theta - s_0)^2(a - \Theta) = (t_2\Theta^2 + t_1\Theta + t_0)^2. \tag{2.2.9}$$

Here $t_2 \neq 0$ because the defining relation for Θ is of degree 4, and so, without loss of generality, $t_2 = 1$.

Then

$$(T^2 + t_1T + t_0)^2 + l(T - s_0)^2(T - a) = F(T). \tag{2.2.10}$$

Hence $\mathfrak{a} = (a, d)$ is a zero of

$$Y \pm (X^2 + t_1X + t_0) \tag{2.2.11}$$

for one choice of sign. There is a double zero at (s_0, r) for appropriate $r \in k$. The remaining zeros and poles are at $\mathfrak{o}_+$, $\mathfrak{o}_-$. Hence the conclusion of the theorem holds.

3.0 Genus 2. Generalities.

A curve

$$\mathcal{C}: \quad Y^2 = F(X) \in k[X], \tag{3.0.1}$$

where F has degree 5 or 6, is of genus 2. Any line $X =$ constant cuts $\mathcal{C}$ in a divisor which is linearly equivalent to the divisor cut out by the line at infinity. This class of divisors of degree 2 defined over k will be denoted by $\mathfrak{O}$.

The differentials of the first kind are

$$\omega = Y^{-1}(c_1X + c_0)\,dX \qquad (c_1, c_0 \in k). \tag{3.0.2}$$

Hence the divisors in $\mathfrak{O}$ are the only divisors which can be the zeros of such a differential. It follows from the Riemann-Roch theorem that every divisor class of degree 2 defined over k, other than $\mathfrak{O}$, contains precisely one effective divisor, and it is defined over k. An effective divisor $\mathfrak{A}$ of degree 2 defined over k is an unordered pair of points $\{\mathfrak{a}_1, \mathfrak{a}_2\}$ such that either (i) $\mathfrak{a}_1$ and $\mathfrak{a}_2$ are each defined over k or (ii) $\mathfrak{a}_1, \mathfrak{a}_2$ are defined over a quadratic extension of k and conjugate over k. Either $\mathfrak{a}_1$ or $\mathfrak{a}_2$ may be at infinity, and $\mathfrak{a}_2 = \mathfrak{a}_1$ is permitted.

The Mordell-Weil group $\mathcal{G}$ is, by definition, the group of divisor classes of degree 0 defined over k. If $\alpha \in \mathcal{G}$, $\alpha \neq 0$, we represent it by the unique effective divisor $\mathfrak{A}$ in $\alpha + \mathfrak{O}$. The set of such $\mathfrak{A}$ together with $\mathfrak{O}$ thus give an isomorphic image of $\mathcal{G}$. Here

$$\mathfrak{A} + \mathfrak{B} = \mathfrak{C}, \tag{3.0.3}$$

where $\mathfrak{A}$, $\mathfrak{B}$, $\mathfrak{C}$ are effective of degree 2, means that $\mathfrak{A} + \mathfrak{B} - \mathfrak{C}$ is a (not necessarily effective) divisor in the class $\mathfrak{O}$. If

$$\mathfrak{A} = \{(a, d), (b, f)\}, \tag{3.0.4}$$

then, clearly,

$$-\mathfrak{A} = \{(a, -d), (b, -f)\}. \tag{3.0.5}$$

We shall often by abuse of language write $\mathfrak{O}$ for any effective divisor in $\mathfrak{O}$. They all correspond to the zero element of $\mathcal{G}$.

We note that the divisor class $\mathfrak{O}$ is canonically distinguished. This is contrary to the genus 1 case.

3.1 The curve (3.0.1) may be transformed birationally into other curves of the same shape but with in general different F. Let

$$r_{11}, r_{12}, r_{21}, r_{22} \in k, \quad r_{11}r_{22} \neq r_{21}r_{12}. \tag{3.1.1}$$

Put

$$X = (r_{11}X_1 + r_{12})/(r_{21}X_1 + r_{22})$$

and

$$Y_1 = (r_{21}X_1 + r_{22})^3 Y.$$

Then

$$\begin{aligned} Y_1^2 &= (r_{21}X_1 + r_{22})^6 F((r_{11}X_1 + r_{12})/(r_{21}X_1 + r_{22})) \\ &= F_1(X_1) \quad \text{(say)} \\ &\in k[X_1]. \end{aligned}$$

If F is of degree 5, then in general F_1 is of degree 6. Conversely, If F is of degree 6 and has a zero in k, then we can use such a transformation to obtain F of degree 5.

3.2 If $a, b, \in k$ with $a \neq b$ are zeros of $F(X)$, then the divisor $\{(a,0),(b,0)\}$ is of order 2 in the Mordell-Weil group since

$$2\{(a,0),(b,0)\} = \{(a,0),(a,0)\} + \{(b,0),(b,0)\}, \tag{3.2.1}$$

and both the divisors on the right-hand side are in $\mathfrak{O}$. If F is of degree 5, we may take the point $\mathfrak{o}$ at infinity instead of one of $(a,0),(b,0)$.

If F splits completely over k (or if we work over the algebraic closure $\overline{k}$), we thus get

$$\binom{6}{2} = 15 = 2^4 - 1 \tag{3.2.2}$$

divisors of order 2. We have therefore obtained them all.

4.0 Genus 2. Quintic Case.

Here we consider

$$\mathcal{C}: \quad Y^2 = F(X) \in k[X] \tag{4.0.1}$$

where, after multiplying X, Y by elements of k^* if necessary, we have

$$F(X) = X^5 + a_4X^4 + a_3X^3 + a_2X^2 + a_1X + a_0. \tag{4.0.2}$$

There is a single point $\mathfrak{o}$ at infinity. The poles of X, Y are $2\mathfrak{o}$ and $5\mathfrak{o}$, respectively. We retain the notation of §0.3. In particular,

$$\mathcal{L} = k[\Theta]^*/\{k[\Theta]^*\}^2. \tag{4.0.3}$$

For an effective divisor $\mathfrak{A}$ of degree 2 defined over k, we define $\mu(\mathfrak{A}) \in \mathcal{L}$ as follows:

(i) $\mu\{\mathfrak{o}, \mathfrak{o}\} = 1$.

(ii) $\mathfrak{A} = \{\mathfrak{o}, \mathfrak{a}\}$, where $\mathfrak{a} = (a, d)$ and $d \neq 0$. Then $\mu(\mathfrak{A})$ is the image of $a - \Theta$ in $\mathcal{L}$.

(iii) $\mathfrak{A} = \{\mathfrak{a}_1, \mathfrak{a}_2\}$, where $\mathfrak{a}_j = (a_j, d_j)$ and $d_1 d_2 \neq 0$. Then

$$(a_1 - \Theta)(a_2 - \Theta) \in k[\Theta]^*$$

and $\mu(\mathfrak{A})$ is its image in $\mathcal{L}$.

There remain the cases like (ii),(iii) but where $d = 0$ or $d_1 d_2 = 0$, respectively. Here one must use the expression (0.3.4) of $k[\Theta]$ as a direct sum of fields K_j. When $K_j = k$ we can proceed as in §1.3, and the details are left to the reader. When the degree of K_j over k is ≥ 3, the projection of $a - \Theta$ or $(a_1 - \Theta)(a_2 - \Theta)$ into K_j is non-zero, and so no difficulty arises.

The only case left is when K_1 (say) is of degree 2 and $(a_1 - \Theta)(a_2 - \Theta)$ projects into 0 in K_1. Then $k(a_1) = k(a_2) = k_1$ (say), $d_1 = d_2 = 0$, and there are two isomorphisms λ_j of k_1 onto K_1 given by

$$\lambda_j \colon a_j \longrightarrow \Theta. \tag{4.0.4}$$

Adjoin two independent generic points $\mathfrak{x}_1, \mathfrak{x}_2$ over k_1. We extend the non-trivial automorphism σ (say) of k_1/k to $k_1(\mathfrak{x}_1, \mathfrak{x}_2)$ by $\sigma\mathfrak{x}_1 = \mathfrak{x}_2$. Let t_1 be a local uniformizer for $\mathfrak{x}_1$ at $\mathfrak{a}_1 = (a_1, 0)$, so $\sigma t_1 = t_2$ (say) is a local uniformizer for $\mathfrak{x}_2$ at $\sigma\mathfrak{a}_1 = \mathfrak{a}_2$.

Now

$$(X(\mathfrak{x}_1) - a_j)(X(\mathfrak{x}_2) - a_j) = t_j^2 N_j(\mathfrak{x}_1, \mathfrak{x}_2), \tag{4.0.5}$$

where $N_j(\mathfrak{a}_1, \mathfrak{a}_2) \in k_1^*$. Clearly $\sigma N_1(\mathfrak{a}_1, \mathfrak{a}_2) = N_2(\mathfrak{a}_2, \mathfrak{a}_1)$, and so

$$\lambda_1 N_1(\mathfrak{a}_1, \mathfrak{a}_2) = \lambda_2 N_2(\mathfrak{a}_2, \mathfrak{a}_1) \in K_1^*. \tag{4.0.6}$$

This provides us with the missing K_1-component. Clearly choice of another local uniformizer instead of t_1 multiplies (4.0.6) by an element of $(K_1^*)^2$, and so $\mu(\mathfrak{A})$ is well defined.

We can give an explicit form of (4.0.6). Let

$$F(X) = (X - a_1)(X - a_2)F_0(X), \tag{4.0.7}$$

where

$$F_0(X) \in k[X]. \tag{4.0.8}$$

We may take $Y(\mathfrak{x}_1) = t_1$, when $\big(X(\mathfrak{x}_1) - a_1\big)/t_1^2$ takes the value

$$\{(a_1 - a_2)F_0(a_1)\}^{-1}$$

at $\mathfrak{x}_1 = \mathfrak{a}_1$. Hence (4.0.6) may be replaced by

$$-\lambda_1 F_0(a_1) = -\lambda_2 F_0(a_2) = -F_0(\Theta) \tag{4.0.9}$$

as the missing component.

4.1 **Theorem 4.1.** *The map*

$$\mu\colon \mathcal{G} \longrightarrow \mathcal{M} \tag{4.1.1}$$

defined by

$$\mathfrak{A} \to \mu(\mathfrak{A}) \tag{4.1.2}$$

is a group homomorphism.

Proof. We note, first, that the definition of μ gives $\mu(\mathfrak{A}) = 1$ for all $\mathfrak{A} \in \mathfrak{G}$.

The automomorphism $Y \to -Y$ of $k(X,Y)$ over $k(X)$ clearly maps $\mathfrak{A}$ into $-\mathfrak{A}$. Hence

$$\mu(-\mathfrak{A}) = \mu(\mathfrak{A}). \tag{4.1.3}$$

Now let

$$\mathfrak{A} = \{(a_1, d_1), (a_2, d_2)\}, \quad \mathfrak{B} = \{(b_1, e_1), (b_2, e_2)\} \tag{4.1.4}$$

be defined over k. If, say,

$$a_1 = b_1, \qquad d_1 = -e_1, \tag{4.1.5}$$

then

$$\mathfrak{A} + \mathfrak{B} = \{(a_2, d_2), (b_2, e_2)\}$$

and

$$\mu(\mathfrak{A} + \mathfrak{B}) = \mu(\mathfrak{A})\mu(\mathfrak{B}) \tag{4.1.6}$$

directly from the definition of μ. Hence we may suppose that the situation exemplified by (4.1.5) does not occur. We may then find $h_3, h_2, h_1, h_0 \in k$ such that

$$Y = h_3X^3 + h_2X^2 + h_1X + h_0 \tag{4.1.7}$$

meets C in $\mathfrak{A}$ and $\mathfrak{B}$. The total intersection is of degree 6, and so the residual intersection $\mathfrak{C}$ (say) is of degree 2. Clearly $\mathfrak{C}$ is defined over k, and

$$\mathfrak{A} + \mathfrak{B} + \mathfrak{C} = 0. \tag{4.1.8}$$

Suppose, first , that

$$h_3 \neq 0. \tag{4.1.9}$$

Then

$$\begin{aligned} &(h_3X^3 + h_2X^2 + h_1X + h_0)^2 - F(X) \\ &\quad = h_3^2(X - a_1)(X - a_2)(X - b_1)(X - b_2)(X - c_1)(X - c_2) \end{aligned} \tag{4.1.10}$$

for some $c_1, c_2 \in \bar{k}$, and

$$\mathfrak{C} = \{(c_1, f_1), (c_2, f_2)\}, \tag{4.1.11}$$

where f_1, f_2 are determined by (4.1.7). If

$$d_1d_2e_1e_2f_1f_2 \neq 0, \tag{4.1.12}$$

we get

$$\mu(\mathfrak{A})\mu(\mathfrak{B})\mu(\mathfrak{C}) = 1 \tag{4.1.13}$$

directly by substituting Θ for X in (4.1.10). If (4.1.12) fails, we still get (4.1.13) by arguing in the separate fields K_j as in §1.3.

Now suppose that

$$h_3 = 0. \tag{4.1.14}$$

Then

$$(h_2X^2 + h_1X + h_0)^2 - F(X)$$
$$= -(X - a_1)(X - a_2)(X - b_1)(X - b_2)(X - c),$$

where $c \in k$. Then $\mathfrak{C} = \{\mathfrak{o}, (c, f)\}$, where f is given by (4.1.7), and (4.1.13) follows agan in the obvious way.

There remains the case when $\mathfrak{A}$ or $\mathfrak{B}$ contains $\mathfrak{o}$. If $\mathfrak{o}$ occurs twice, we argue as for (4.1.5). If $\mathfrak{o}$ occurs only once, we make the obvious modifications to the above argument with $h_3 = 0$. This concludes the proof.

4.2 Theorem 4.2. *The kernel of*

$$\mu\colon \mathcal{G} \longrightarrow \mathcal{M} \tag{4.2.0}$$

is $2\mathcal{G}$.

Proof. Since $\mathcal{M}$ is of exponent 2, the kernel contains $2\mathcal{G}$. We have to show that it is not any bigger.

We shall say that the divisor $\mathfrak{A}$ is in "general position" if

$$\mathfrak{A} = \{\mathfrak{a}_1, \mathfrak{a}_2\}, \quad \mathfrak{a}_1 = (a_1, d_1), \quad \mathfrak{a}_2 = (a_2, d_2) \tag{4.2.1}$$

and

$$d_1 d_2 \neq 0. \tag{4.2.2}$$

We have to show that

$$\mu(\mathfrak{A}) = 1 \tag{4.2.3}$$

implies $\mathfrak{A} \in 2\mathcal{G}$, and will do this first for $\mathfrak{A}$ in general position.

Put

$$G(T) = (a_1 - T)(a_2 - T) \in k[T], \tag{4.2.4}$$

where T is an indeterminate. then (4.2.3) implies that

$$G(\Theta) = \rho^2, \tag{4.2.5}$$

where

$$\rho = r_4\Theta^4 + r_3\Theta^3 + r_2\Theta^2 + r_1\Theta + r_0 \in k[\Theta]. \tag{4.2.6}$$

Suppose, first, that

$$r_4 = r_3 = 0. \tag{4.2.7}$$

Then

$$G(T) = (r_2T^2 + r_1T + r_0)^2, \tag{4.2.8}$$

since the defining relation for Θ is of degree 5. Hence $r_2 = 0$ and $a_1 = a_2 \in k$. There are now two cases: (i) $d_1 + d_2 = 0$. Then $\mathfrak{A} \in \mathfrak{O} \in 2\mathcal{G}$. (ii) $d_1 = d_2$. Then $\mathfrak{a}_1 = \mathfrak{a}_2 = \mathfrak{a}$ (say), and

$$\mathfrak{A} = \{\mathfrak{a}, \mathfrak{a}\} = 2\{\mathfrak{o}, \mathfrak{a}\} \in 2\mathcal{G}. \tag{4.2.9}$$

We shall therefore suppose that (4.2.7) does not hold.

We now consider polynomials

$$L(T) = l_2T^2 + l_1T + l_0 \in k[T] \tag{4.2.10}$$

$$M(T) = m_3T^3 + m_2T^2 + m_1T + m_0 \in k[T] \tag{4.2.11}$$

such that

$$L(\Theta)\rho = M(\Theta). \tag{4.2.12}$$

The condition that the coefficient of Θ^4 in (4.2.12) vanishes is a single homogeneous linear condition on l_2, l_1, l_0, which is non-vacuous since (4.2.7) does not hold. Hence the set of pairs $\{L, M\}$ is a 2-dimensional k-linear space. On eliminating ρ between (4.2.5) and (4.2.12), they all satisfy

$$L(\Theta)^2G(\Theta) = M(\Theta)^2. \tag{4.2.13}$$

Suppose, first, that we can find $L \neq 0$, M such that

$$l_2 = m_3 = 0. \tag{4.2.14}$$

Then (4.2.13) implies

$$L(T)^2G(T) = M(T)^2 \tag{4.2.15}$$

since both sides have degree ≤ 4. Hence $G(T)$ is a perfect square, a case we have already eliminated.

There is, however, a pair $(L, M) = (L_1, M_1)$ (say) for which $l_2 = 0$, since this is a single condition. Then $m_3 \neq 0$ and, by homogeneity, we may take $m_3 = 1$. Similarly, there is a pair $(L, M) = (L_2, M_2)$ for which $m_3 = 0$ and $l_2 = 1$. Hence in general

$$L = uL_1 + vL_2, \quad M = uM_1 + vM_2, \quad (u, v \in k), \tag{4.2.16}$$

where

$$\begin{aligned} L_1 &= \text{degree } 1 \\ M_1 &= T^3 + \text{degree } 2 \\ L_2 &= T^2 + \text{degree } 1 \\ M_2 &= \text{degree } 2. \end{aligned} \tag{4.2.17}$$

Now (4.2.13) gives

$$M(T)^2 - G(T)L(T)^2 = \{(u^2 - v^2)T + w\}F(T) \tag{4.2.18}$$

for some $w = w(u, v) \in k$.

Put

$$u = 1, \quad v = \pm 1, \tag{4.2.19}$$

where we consider both choices of sign together for the moment. Then

$$M(T)^2 - G(T)L(T)^2 = wF(T). \tag{4.2.20}$$

If $w = 0$, then $G(T)$ is a square, which we have already excluded. Hence

$$w \in k^*. \tag{4.2.21}$$

Now specialize $T \to a_1$. Then $G(a_1) = 0$, $F(a_1) = d_1^2$; and so

$$wd_1^2 = M(a_1)^2. \tag{4.2.22}$$

Here $d_1 \neq 0$ by general position, and so

$$0 \neq \sqrt{w} \in k(a_1, d_1) = k(a_2, d_2). \tag{4.2.23}$$

Suppose, first, that

$$w \in (k^*)^2. \tag{4.2.24}$$

Then we can choose $s \in k$ with $s^2 = w$ so that

$$sY = M(X) \tag{4.2.25}$$

passes through $\mathfrak{a}_1$. We subdivide cases.

(i) If (4.2.25) also passes through $\mathfrak{a}_2$, the intersection of (4.2.24) with C is $\mathfrak{A} + 2\mathfrak{B}$, where $\mathfrak{B}$ is given by $L(X) = 0$ and (4.2.25). Hence the conclusion of the theorem holds.

(ii) If (4.2.25) does not pass through $\mathfrak{a}_2$, then it passes through $\mathfrak{a}_2' = (a_2, -d_2)$. This can happen only if $\mathfrak{a}_1, \mathfrak{a}_2$ are not conjugate over k, and so are each defined over k. Arguing as in (i), we have

$$\{\mathfrak{a}_1, \mathfrak{a}_2'\} + 2\mathfrak{B} \doteq 0. \tag{4.2.26}$$

However,

$$\{\mathfrak{o}, \mathfrak{o}\} + \mathfrak{A} + \{\mathfrak{a}_1, \mathfrak{a}_2'\} = \{\mathfrak{a}_2, \mathfrak{a}_2'\} + 2\{\mathfrak{o}, \mathfrak{a}_1\}. \tag{4.2.27}$$

Here $\{\mathfrak{o}, \mathfrak{o}\}$ and $\{\mathfrak{a}_2, \mathfrak{a}_2'\}$ are both in $\mathfrak{O}$, and again the conclusion of the theorem holds.

If (4.2.24) does not hold, then by (4.2.23)

$$k(a_1, d_1) = k(a_2, d_2) = k(\sqrt{g}) \text{ say, } g \notin (k^*)^2 \tag{4.2.28}$$

and

$$w = gs^2, \quad s \in k^*. \tag{4.2.29}$$

Then by (4.2.20) and (4.2.23) we have

$$M(a_1) = qd_1\sqrt{g}, \quad q \in k. \tag{4.2.30}$$

Hitherto we have allowed both choices of sign in (4.2.19). We should have reached the conclusion of the theorem if (4.2.24) holds for either choice of sign. Hence we are left with the situation when(4.2.30) holds for $u = +1$ and both $v = +1$ and $v = -1$. It follows from (4.2.16) and (4.2.30) that

$$M_1(a_1) = pd_1\sqrt{g}, \quad p \in k. \tag{4.2.31}$$

On substituting $u = 1, v = 0, T = a_1$ in (4.2.18), we get

$$p^2 g d_1^2 = (a_1 + w)d_1^2, \tag{4.2.32}$$

where $w = w(1,0) \in k$. Here $d_1 \neq 0$ by general position, so

$$a_1 = p^2 g - w \in k, \tag{4.2.33}$$

a contradiction. This concludes the proof when $\mathfrak{A}$ is in general position. [We note in passing that we should also have reached a contradiction from supposing that (4.2.24) holds for both choices of sign in (4.2.19).]

Now suppose that $\mu(\mathfrak{A}) = 1$ but $\mathfrak{A}$ is not in general position. Suppose that there is a $\mathfrak{B}$ such that

$$\mathfrak{A} + 2\mathfrak{B} = \mathfrak{C} \tag{4.2.34}$$

is in general position. Then $\mu(\mathfrak{C}) = 1$, and the general position argument shows that $\mathfrak{C} \in 2\mathcal{G}$. Hence $\mathfrak{A} \in 2\mathcal{G}$, and we are done. Unfortunately there may not exist such a $\mathfrak{B}$, and we must extend the ground field.

Let $\mathfrak{x}_1, \mathfrak{x}_2$ be independent generic points of C over k and let $\underset{\sim}{k}$ be the subfield of $k(\mathfrak{x}_1, \mathfrak{x}_2)$ fixed under the interchange of $\mathfrak{x}_1$ and $\mathfrak{x}_2$. Then

$$\underset{\sim}{k}[\Theta] = k[\Theta] \otimes_k \underset{\sim}{k}. \tag{4.2.35}$$

Put

$$\underset{\sim}{\mathcal{L}} = \underset{\sim}{k}[\Theta]^*/\{\underset{\sim}{k}[\Theta]^*\}^2. \tag{4.2.36}$$

There is a natural injection of $\mathcal{L}$ into $\underset{\sim}{\mathcal{L}}$, and we regard $\mathcal{L}$ as a subset of $\underset{\sim}{\mathcal{L}}$. We denote by $\underset{\sim}{\mu}$ the analogue of μ, but for the ground field $\underset{\sim}{k}$. If $\mathfrak{A}$ is defined over k, then clearly

$$\underset{\sim}{\mu}(\mathfrak{A}) = \mu(\mathfrak{A}). \tag{4.2.37}$$

Let $\underset{\sim}{\mathcal{G}}$ be the Mordell-Weil group for $\underset{\sim}{k}$. On arguing as for (4.2.34) but with $\tilde{\mathfrak{B}} = \{\mathfrak{x}_1, \mathfrak{x}_2\}$, we obtain $\mathfrak{A} \in 2\underset{\sim}{\mathcal{G}}$. But $2\underset{\sim}{\mathcal{G}} \cap \mathcal{G} = 2\mathcal{G}$, and we are done.

5.0 Genus 2. The Sextic Case.

Here

$$F(X) = a_6X^6 + \cdots + a_0 \in k[X], \quad a_6 \neq 0. \tag{5.0.1}$$

There is a double point at infinity, with two branches $\mathfrak{o}_+$, $\mathfrak{o}_-$ (say) defined over $\bar{k}$. If $a_6 \in (k^*)^2$, then $\mathfrak{o}_+$, $\mathfrak{o}_-$ are individually defined over k, but in any case $\{\mathfrak{o}_+, \mathfrak{o}_-\}$ is defined over k.

We retain the notation of §0.3. In particular, now

$$\mathcal{L} = k[\Theta]^*/k^*\{k[\Theta]^*\}^2. \tag{5.0.2}$$

For any effective divisor $\mathfrak{A}$ of degree 2 defined over k, we define

$$\mu(\mathfrak{A}) \in \mathcal{M} \subset \mathcal{L}$$

as follows:

(i) $\mu(\mathfrak{o}_+, \mathfrak{o}_-) = 1$. Further, $\mu(\mathfrak{o}_+, \mathfrak{o}_+) = \mu(\mathfrak{o}_-, \mathfrak{o}_-) = 1$ if $\mathfrak{o}_+$, $\mathfrak{o}_-$ are defined over k.

(ii) If $\mathfrak{o} = \mathfrak{o}_+$ or $\mathfrak{o}_-$ and $\mathfrak{a} = (a, d)$ with $d \neq 0$ are defined over k, then $\mu(\mathfrak{o}, \mathfrak{a})$ is the image of $a - \Theta$ in $\mathcal{L}$.

(iii) If $\mathfrak{A} = \{(a_1, d_1), (a_2, d_2)\}$ with $d_1 d_2 \neq 0$, then $\mu(\mathfrak{A})$ is the image of

$$(a_1 - \Theta)(a_2 - \Theta) \tag{5.0.3}$$

in $\mathcal{L}$.

When $d = 0$ in (ii) or $d_1 d_2 = 0$ in (iii), the definition is extended precisely as in the quintic case [§4.0]. As always, $\mu(\mathfrak{A}) \in \mathcal{M}$.

5.1 **Theorem 5.1.** *The map*

$$\mu\colon \mathcal{G} \longrightarrow \mathcal{M} \tag{5.1.1}$$

given by

$$\mathfrak{A} \to \mu(\mathfrak{A}) \tag{5.1.2}$$

is a group homomorphism.

Proof. The proof is very similar to that of Theorem 4.1 for the quintic case, so we only briefly indicate how the different definition of $\mathcal{L}$ comes in.

Let

$$\mathfrak{A} = \{(a_1, d_1), (a_2, d_2)\}, \quad \mathfrak{B} = \{(b_1, e_1), (b_2, e_2)\} \tag{5.1.3}$$

where, for simplicity, a_1, a_2, b_1, b_2 are distinct. There are $h_3, h_2, h_1, h_0 \in k$ such that

$$Y = h_3X^3 + h_2X^2 + h_1X + h_0 \tag{5.1.4}$$

passes through $\mathfrak{A}$ and $\mathfrak{B}$. Let $\mathfrak{C}$ be the residual intersection, so

$$\mathfrak{A} + \mathfrak{B} + \mathfrak{C} = 0. \tag{5.1.5}$$

If $h_3^2 \neq a_6$, we have

$$\begin{aligned}(h_3X^3 + \cdots + h_0)^2 - F(X) \\ = n(X - a_1)(X - a_2)(X - b_1)(X - b_2)(X - c_1)(X - c_2)\end{aligned} \tag{5.1.6}$$

for some $n \in k^*$ and $c_1, c_2 \in \overline{k}$. Here

$$\mathfrak{C} = \{(c_1, f_1), (c_2, f_2)\} \tag{5.1.7}$$

for appropriate f_1, f_2. On putting $X = \Theta$ in (5.1.6), we get

$$\mu(\mathfrak{A})\mu(\mathfrak{B})\mu(\mathfrak{C}) = 1, \tag{5.1.8}$$

as required.

The remaining cases are left to the reader.

5.2 The kernel of μ contains $2\mathcal{G}$ as usual. In general it contains other elements. Let

$$\mathfrak{A} \doteq \{\mathfrak{a}, \mathfrak{a}\}, \tag{5.2.1}$$

where the point $\mathfrak{a}$ is defined over k. Then $\mu(\mathfrak{A}) = 1$. Two divisors $\mathfrak{A}$ and $\mathfrak{B} = \{\mathfrak{b}, \mathfrak{b}\}$ of this type are in the same class of $\mathcal{G}$ modulo $2\mathcal{G}$ since

$$\{\mathfrak{a}, \mathfrak{a}\} + \{\mathfrak{b}, \mathfrak{b}\} = 2\{\mathfrak{a}, \mathfrak{b}\}. \tag{5.2.2}$$

One has curves C for which (5.2.1) is in $2\mathcal{G}$ and others for which it is not. Indeed, let $a_6 = 1$ and let $\mathfrak{o}$ be one of the two points at infinity defined over k. Then it is easy to see that $\{\mathfrak{o}, \mathfrak{o}\} \in 2\mathcal{G}$ precisely when

$$F(X) = H(X)^2 - lG(X)^2 \tag{5.2.3}$$

for some

$$H(X), \quad G(X) \in k[X], \quad l \in k^* \tag{5.2.4}$$

and $H(X), G(X)$ are of degree 3 and ≤ 2, respectively.

It is possible for there to be no points on C defined over k, and so no divisors of type (5.2.1). We now describe further elements of the kernel. Let $\mathfrak{C}$ be an effective divisor of degree 3 defined over k. Then by Riemann-Roch there is an effective divisor $\mathfrak{D}$ of degree 2 defined over k such that

$$2\mathfrak{C} + \mathfrak{D} \in 4\mathfrak{O}. \tag{5.2.5}$$

We show below that $\mu(\mathfrak{D}) = 1$. An argument similar to (5.2.2) shows that different $\mathfrak{D}$ lie in the same class of $\mathcal{G}$ modulo $2\mathcal{G}$ and also in the same class as the $\mathfrak{A}$ of type (5.2.1), if they exist.

It follows that the $\mathfrak{D}$ of type (5.2.5) are in the kernel whenever there are points $\mathfrak{a}$ on C defined over k. We intend to show that the $\mathfrak{D}$ are in the kernel even if no such $\mathfrak{a}$ exist. One can construct a proof by adjoining a generic point much as at the end of §4, but we give a different argument, since we need the details later.

Lemma 5.2. *The $\mathfrak{A}$ of type (5.2.1) and the $\mathfrak{D}$ given by (5.2.5) are in the kernel. They all lie in the same class of $\mathcal{G}$ modulo $2\mathcal{G}$, which may or may not be $2\mathcal{G}$ itself.*

Proof. After what has been said above, all that is needed is to show that $\mathfrak{D}$ is in the kernel. We may suppose that there are no points on C defined over k, and so that $\mathfrak{C} = \{\mathfrak{t}_1, \mathfrak{t}_2, \mathfrak{t}_3\}$ where the $\mathfrak{t}_j$ are conjugate over

k. In particular, none of the $\mathfrak{t}_j$ are at infinity, and the $\mathfrak{t}_j$ are distinct, where $\mathfrak{t}_j = (c_j, f_j)$.

We may choose $i_1, i_0, h_4, \ldots, h_0 \in k$, not all 0, such that

$$I(X)Y = H(X) \tag{5.2.6}$$

meets C in $2\mathfrak{C}$; where

$$\begin{aligned} I(X) &= i_1 X + i_0 \\ H(X) &= h_4 X^4 + \cdots + h_0. \end{aligned} \tag{5.2.7}$$

Clearly not both i_1, i_0 vanish. The intersection of (5.2.6) with C has degree 8, and so the residual intersection is $\mathfrak{D}$.

Suppose, first, that both points of $\mathfrak{D} = \{(d_1, g_1), (d_2, g_2)\}$ are finite. Then

$$H(X)^2 - I(X)^2 F(X) = l\{\textstyle\prod_j (X - c_j)\}^2 (X - d_1)(X - d_2)$$

for some $l \in k^*$. Hence $\mu(\mathfrak{D}) = 1$. If a point of $\mathfrak{D}$ is at infinity, the modification to the proof is obvious.

5.3 **Theorem 5.3.** *The kernel of $\mu : \mathcal{G} \to \mathcal{M}$ is generated by $2\mathcal{G}$ and the divisors described in Lemma 5.2.*

Proof. Let $\mathfrak{A}$ be a divisor with

$$\mu(\mathfrak{A}) = 1. \tag{5.3.1}$$

We suppose, first, that $\mathfrak{A}$ is in "general position," by which we mean

$$\mathfrak{A} = \{\mathfrak{a}_1, \mathfrak{a}_2\}, \quad \mathfrak{a}_j = (a_j, d_j) \tag{5.3.2}$$

with

$$a_1 \neq a_2 \qquad d_1 d_2 \neq 0. \tag{5.3.3}$$

Put

$$G(T) = (a_1 - T)(a_2 - T). \tag{5.3.4}$$

Then (5.3.1) implies that

$$G(\Theta) = n\rho^2 \tag{5.3.5}$$

for some $n \in k^*$ and some

$$\rho = r_5\Theta^5 + \cdots + r_0 \quad (r_j \in k). \tag{5.3.6}$$

There are

$$L(T) = l_2T^2 + l_1T + l_0 \neq 0, \quad \in k[T] \tag{5.3.7}$$

$$M(T) = m_3T^3 + m_2T^2 + m_1T + m_0 \in k[T] \tag{5.3.8}$$

such that

$$L(\Theta)\rho = M(\Theta), \tag{5.3.9}$$

since the absence of terms in Θ^5, Θ^4 imposes two linear conditions on l_2, l_1, l_0. Hence

$$L(\Theta)^2G(\Theta) = nM(\Theta)^2, \tag{5.3.10}$$

and so

$$nM(T)^2 - L(T)^2G(T) = wF(T) \tag{5.3.11}$$

for some $w \in k$. If $w = 0$, then $G(T)$ is a square contrary to the general position hypothesis, so

$$w \in k^*. \tag{5.3.12}$$

On comparing terms in T^6 in (5.3.11), this gives

$$nm_3^2 \neq l_2^2. \tag{5.3.12 bis}$$

On substituting a_1 for T, we have

$$nM(a_1)^2 = wd_1^2, \tag{5.3.13}$$

so

$$w/n \in \{k(a_1, d_1)^*\}^2. \tag{5.3.14}$$

We now distinguish two cases:

(I) Suppose that

$$w/n \in (k^*)^2. \tag{5.3.15}$$

Then for one choice of u with $w = nu^2$ the curve

$$Y = uM(X) \tag{5.3.16}$$

passes through $\mathfrak{a}_1$. Suppose, first, that it passes through $\mathfrak{a}_2$. Then, by a now familiar argument, the total intersection of (5.3.16) is $\mathfrak{A} + 2\mathfrak{B}$, where $\mathfrak{B}$ is given by $L(X) = 0$ and (5.3.16). Hence $\mathfrak{A} \in 2\mathcal{G}$.

If (5.3.16) does not pass through $\mathfrak{a}_2$, then it passes through $\mathfrak{a}_2' = (a_2, -d_2)$. This can happen only if $k(a_1, d_1) = k(a_2, d_2) = k$, since otherwise, $\mathfrak{a}_1, \mathfrak{a}_2$ are conjugate over k. The preceding argument shows that

$$\{\mathfrak{a}_1, \mathfrak{a}_2'\} \in 2\mathcal{G}. \tag{5.3.17}$$

Now

$$\begin{aligned} \mathfrak{A} + \{\mathfrak{a}_1, \mathfrak{a}_2'\} &= \{\mathfrak{a}_1, \mathfrak{a}_1\} + \{\mathfrak{a}_2, \mathfrak{a}_2'\} \\ &= \{\mathfrak{a}_1, \mathfrak{a}_1\}, \end{aligned} \tag{5.3.18}$$

since $\{\mathfrak{a}_2, \mathfrak{a}_2'\} \in \mathfrak{O}$: and again $\mathfrak{A}$ is in the specified kernel.

(II) If (5.3.15) does not hold, we must have

$$k(a_1, d_1) = k(a_2, d_2) = k(a_1) = k(a_2) = k(\sqrt{g}), \quad g \notin (k^*)^2 \tag{5.3.19}$$

(using $a_1 \neq a_2$ by "general position"), and

$$w/n = gu^2, \quad u \in k^*. \tag{5.3.20}$$

We note for later use that the substitution of a_1 for T in (5.3.11) gives

$$M(a_1) = ud_1\sqrt{g} \tag{5.3.21}$$

for appropriate choice of u.

We shall show that we now have the situation described in the proof of Lemma 5.2. Our first objective is to define a bilinear transformation of X which, as it will turn out, will ensure that $I(X)$ (in the notation of the proof of the lemma) can be taken to be a constant.

On comparing the coefficients of T^6 in (5.3.11) and using (5.3.20), we have

$$nm_3^2 - l_2^2 = w = gnu^2. \tag{5.3.22}$$

Hence n is a norm for $k(\sqrt{g})/k$ and so by (5.3.18) is representable by the homogenized version of $G(T)$, say

$$n = (\alpha - \gamma a_1)(\alpha - \gamma a_2) = \alpha^2 + g_1\alpha\gamma + g_2\gamma^2 \quad (\alpha, \gamma \in k). \tag{5.3.23}$$

Choose $\beta, \delta \in k$ so that $\alpha\delta \neq \beta\gamma$, and put

$$X = (\alpha X_1 + \beta)/(\gamma X_1 + \delta). \tag{5.3.24}$$

Then

$$(\gamma X_1 + \delta)^2 G(X) = nG_1(X_1) \tag{5.3.25}$$

for some

$$G_1(X_1) = X_1^2 + g_{11}X_1 + g_{21} \in k[X_1]. \tag{5.3.26}$$

On putting

$$Y_1 = Y/(\gamma X_1 + \delta)^3, \tag{5.3.27}$$

the equation of C becomes

$$Y_1^2 = F_1(X_1) \in k[X_1], \tag{5.3.28}$$

where

$$(\gamma X_1 + \delta)^6 F((\alpha X_1 + \beta)/\gamma X_1 + \delta)) = F_1(X_1). \tag{5.3.29}$$

Here F_1 is of degree 6 in general, but is of degree 5 if $F(\alpha/\gamma) = 0$. We can avoid this exceptional case, since (5.3.23) has a 1-parameter family of solutions (α, γ).

We now go over to the (X_1, Y_1) coordinates and, for simplicity of notation, drop the suffix "1." It is clear that the change of coordinates does not alter the definition of the μ function. Further, by (5.3.5) and (5.3.25) we may now suppose that

$$n = 1. \tag{5.3.30}$$

After a substitution $X \to X +$ constant, we may eliminate the linear term in $G(X)$, and so

$$G(X) = X^2 + g_2 = X^2 - g. \tag{5.3.31}$$

We now consider polynomials

$$L'(T) = l'_3T^3 + l'_2T^2 + l'_1T + l'_0 \neq 0, \quad \in k[T], \tag{5.3.32}$$

$$M'(T) = m'_4T^4 + \cdots + m'_0 \in k[T] \tag{5.3.33}$$

such that

$$L'(\Theta)\rho = M'(\Theta) \tag{5.3.34}$$

and, in addition,

$$m'_4 = l'_3, \quad m'_3 = l'_2. \tag{5.3.35}$$

There are three linear conditions on the four l'_j, so an $L'(T) \neq 0$ certainly exists.

If $m'_4 = l'_3 = 0$, we could take $L = L'$, $M = M'$ in (5.3.11) and then $m'_3 = l'_2$ contradicts (5.3.12 bis), (5.3.30). Hence, by homogeneity, we may suppose that $m'_4 = l'_3 = 1$, and so

$$L'(T) = T^3 + \sigma T^2 + \text{ linear } \in k[T], \tag{5.3.36}$$

$$M'(T) = T^4 + \sigma T^3 + \text{ quadratic } \in k[T], \tag{5.3.37}$$

for some $\sigma \in k$. On squaring (5.3.34) and using (5.3.30) we have

$$L'(\Theta)^2G(\Theta) = M'(\Theta)^2, \tag{5.3.38}$$

and so

$$M'(T)^2 - L'(T)^2G(T) = w'F(T) \tag{5.3.39}$$

for some $w' \in k$, since terms in T^8, T^7 cancel by (5.3.31), (5.3.36),(5.3.37). Here $w' = 0$ is eliminated by the usual argument, and so

(5.3.40) $$w' \in k^*.$$

Following the by now familiar routine, we substitute a_1 for T and obtain

(5.3.41) $$w' \in \left(k(\sqrt{g})^*\right)^2.$$

If $w' \in (k^*)^2$, we may choose v with $w' = v^2$ so that

(5.3.42) $$vY = M'(X)$$

passes through $\mathfrak{a}_1 = (a_1, d_1)$. Since $\mathfrak{a}_2$ is conjugate to $\mathfrak{a}_1$ over k, (5.3.42) also passes through $\mathfrak{a}_2$. Hence the total intersection of (5.3.42) with C is $\mathfrak{A} + 2\mathfrak{C}$, where $\mathfrak{C}$ is the divisor of degree 3 given by $L'(X) = 0$ and (5.3.42). Hence $\mathfrak{A}$ is in the specified kernel.

There remains the possibility

(5.3.43) $$w' = gv^2, \quad v \in k^*.$$

We shall show that this leads to a contradiction. By (5.3.39) we may suppose that

(5.3.44) $$M'(a_1) = vd_1\sqrt{g}.$$

With L, M given by (5.3.7), (5.3.8), we have, by (5.3.9), (5.3.34), that

(5.3.45) $$\{L'(\Theta) + L(\Theta)\}\rho = M'(\Theta) + M(\Theta)$$

and so, by (5.3.30), as usual,

(5.3.46) $$\{M'(T) + M(T)\}^2 - \{L'(T) + L(T)\}^2 G(T) = (pT + q)F(T)$$

for some $p, q, \in k$. Here

(5.3.47) $$p = 2(m_3 - l_2) \neq 0,$$

by (5.3.12 bis). On substituting a_1 for T in (5.3.46) and using (5.3.21), (5.3.44), we get

(5.3.48) $$(u + v)^2 g d_1^2 = (pa_1 + q)d_1^2.$$

Here $d_1 \neq 0$ by "general position," so

(5.3.49) $$a_1 = p^{-1}\{(u+v)^2 g - q\} \in k,$$

contrary to (5.3.19). Hence (5.3.43) cannot occur. This concludes the proof of the theorem for $\mathfrak{A}$ in general position.

The completion of the proof when $\mathfrak{A}$ is not in general position is similar to that for the quintic case at the end of §4.2 but requires some extra twists. Let $\mathfrak{x}_1, \mathfrak{x}_2$ be independent generic points of C and let $\underset{\sim}{k}$, as before, be the subfield of $k(\mathfrak{x}_1, \mathfrak{x}_2)$ fixed under the interchange of $\mathfrak{x}_1$ and $\mathfrak{x}_2$. We denote by $\underset{\sim}{\mathcal{L}}$, $\underset{\sim}{\mathcal{M}}$, $\underset{\sim}{\mu}$, $\underset{\sim}{\mathcal{G}}$ the analogues over $\underset{\sim}{k}$ of $\mathcal{L}$, $\mathcal{M}$, μ, $\mathcal{G}$. Put $\mathfrak{X} = (\mathfrak{x}_1, \mathfrak{x}_2)$. Let $\mu(\mathfrak{A}) = 1$, so

$$\underset{\sim}{\mu}(\mathfrak{A} + 2\mathfrak{X}) = 1$$

and $\mathfrak{A} + 2\mathfrak{X}$ is in general position. Hence, by what has already been proved, either $\mathfrak{A} + 2\mathfrak{X} \in 2\underset{\sim}{\mathcal{G}}$ or $\mathfrak{A} + 2\mathfrak{X}$ is of the type described in Lemma 5.2 (but with $\underset{\sim}{k}$ as ground field).

Arguing as at the end of §4.2, if $\mathfrak{A} + 2\mathfrak{X} \in 2\underset{\sim}{\mathcal{G}}$ then $\mathfrak{A} \in 2\mathcal{G}$.

Suppose now that $\mathfrak{A} + 2\mathfrak{X} \notin 2\underset{\sim}{\mathcal{G}}$ but that there is a $\mathfrak{D}$ of the type (5.2.5), where $\mathfrak{C}$ of degree 3 is defined over k. Then $\mathfrak{D} + \mathfrak{A} + 2\mathfrak{X} \in 2\underset{\sim}{\mathcal{G}}$, so $\mathfrak{D} + \mathfrak{A} \in 2\mathcal{G}$ and the conclusion of the theorem holds.

In fact if $\mathfrak{A} + 2\mathfrak{X} \notin 2\underset{\sim}{\mathcal{G}}$, then such a $\mathfrak{D}$ always exists. It is enough to show that if there is a divisor $\underset{\sim}{\mathfrak{C}}$ of degree 3 defined over $\underset{\sim}{k}$, then there is a divisor $\mathfrak{C}$ of degree 3 defined over k, and this follows by a specialization argument. Let b be any element of k and let $e \in \bar{k}$ satisfy $e^2 = F(b)$. Then $\underset{\sim}{k}$ specializes to k under $\mathfrak{x}_1 \to (b, e)$; $\mathfrak{x}_2 \to (b, -e)$. We can take for $\mathfrak{C}$ any specialization of $\underset{\sim}{\mathfrak{C}}$ over this.

References

[1] J. W. S. Cassels. *Diophantine equations, with special reference to elliptic curves.* J. London Math. Soc. **41** (1966), 193–291. Corrigenda **42** (1967), 183.

[2] L. J. Mordell. *On the rational solutions of the indeterminate equations of the third and fourth degree.* Proc. Camb. Phil. Soc. **21** (1922), 179–192.

[3] D. Mumford. *On the equations defining abelian varieties I.* Invent. Math. **1** (1966), 287–354.

[4] D. Mumford. *On the equations defining abelian varieties II.* Invent. Math. **3** (1967), 75–135.
[5] A. Weil. *L'arithmétique sur les courbes algébriques.* Acta Math. **52** (1928), 281-315.(= Œuvres Scientifiques I, 11–45).
[6] A. Weil. *Sur un théorème de Mordell.* Bull. Sci. Math. (II) **54** (1929), 182–191. (=Œeuvres Scientifiques I, 47–56).

Received April 23, 1982

Professor J. W. S. Cassels
Department of Pure Mathematics and Mathematical Statistics
16 Mill Lane, Cambridge CB2 1SB, England

Number Theoretic Applications of Polynomials with Rational Coefficients Defined by Extremality Conditions

G. V. Chudnovsky

To I.R. Shafarevich

Introduction

It is well known that classes of polynomials in one variable defined by various extremality conditions play an extremely important role in complex analysis. Among these classes we find orthogonal polynomials (especially classical orthogonal polynomials expressed as hypergeometric polynomials) and polynomials least deviating from zero on a given continuum (Chebicheff polynomials). Orthogonal polynomials of the first and second kind appear as denominators and numerators of the Padé approximations to functions of classical analysis and satisfy familiar three-term linear recurrences. These polynomials were used repeatedly to study diophantine approximations of values of functions of classical analysis, especially exponential and logarithmic functions at rational points $x = p/q$ [1], [2], [3], [4], [5]. The methods of Padé approximation in diophantine approximations are quite powerful and convenient to use, since they replace the problem of rational approximations to numbers with the approximations of functions. There are, however, arithmetic restrictions on rational approximations to functions if they are to be used for diophantine approximations. The main restriction on polynomials here is to have rational integer coefficients or rational coefficients with a controllable denominator. Such arithmetic restrictions transform a typical problem of classical analysis into an unusual mixture of arithmetic and analytic difficulties. For example, recurrences defining orthogonal polynomials must be of a special type to guarantee that their solutions will have bounded denominators. In this paper we consider various classes of polynomials generated by imposing arithmetic restrictions on classical approximation theory problems (orthogonal or Chebicheff polynomials).

In the first part of the paper we consider polynomials with rational coefficients arising from rational approximations to logarithmic and algebraic functions. These polynomials are defined through linear recurrences with integer coefficients, and they have denominators growing not faster than in geometric progression. Together with the ordinary three-term linear recurrences, we consider more general matrix linear recurrences that are more convenient for the determination of arithmetic properties of polynomials generated by them. All the recurrences under consideration are representations of the so-called contiguous relations, i.e., linear relations between systems of analytic functions having the same monodromy properties [2], [6], [7]. The transformation from one contiguous system of functions to another, described by linear recurrences, is broadly called the Backlund transformation because its appearance in completely integrable systems as a canonical transformation [6], [7]. In our case, contiguous relations are analyzed in the simplest cases of Fuchsian linear differential equations with regular singularities at $0, 1, \infty$, and several apparent singularities (generalizations of Gauss hypergeometric functions).

The new recurrences are applied to logarithmic and inverse trigonometric functions, and lead to new "dense" sequences of rational approximations to logarithms of rational numbers. Using these "dense" sequences of approximations (see Lemma 1.1), we obtain new improved measures of irrationality of particularly important numbers like $\ln 2, \pi/\sqrt{3}, \pi, \sqrt[3]{2}$. These measures are considerably better than those [1] furnished by the ordinary Padé approximation technique. We present here only those applications of the method of Backlund transformations [8] where all the recurrence relations are given explicitly.

Another series of recurrences arises from Thue polynomials introduced by Thue [9] as a means of approximation of algebraic numbers α of degrees $n \geq 3$. These polynomials $P_r(x), Q_r(x)$ with rational integer coefficients of degree $rn + [\frac{n}{2}]$ are defined by the "Padé-type" relations

$$P_r(x)\alpha - Q_r(x) = (\alpha - x)^{2r+1} S(x)$$

for a polynomial $S(x)$ and arbitrary $r \geq 1$. We pay particular attention to the cubic α, when the three-term recurrence connecting $P_r(x), Q_r(x)$ is given explicitly in terms of invariants and covariants of the cubic polynomial $P(x)$ whose root is α. Applications to measures of diophantine approximations of cubic irrationalities (including cubic roots $\sqrt[3]{D}, D > 0$ and solutions $f(\sqrt{d})$, $d < 0$ of Weber's modular equation) are presented.

While classes of polynomials with rational integer coefficients studied in Chapter I arise from the orthogonality condition and the Padé approximation, polynomials studied in Chapter II can be characterized as Chebicheff polynomials with rational integer coefficients. They can be characterized as polynomials with rational integer coefficients least deviating from zero on a continuum E. Number-theoretic interest in these polynomials was originated in 1936 by the famous Gelfond-Schnirelman approach to the prime number theorem, in which deviations from zero of Chebicheff polynomials with rational integer coefficients were directly related to the distribution of primes. We refer the reader to a recent book by Ferguson [10], where this approach and related results are presented.

We study the original Gelfond-Schnirelman approach in the context of the "integer" transfinite diameter of a continuum E in the cases $E = [0,1]$ and $E = [0,1/4]$, which are important for the prime number theorem. We include an exposition of the results of Fekete and the recent results of Aparicio on upper and lower bounds on the "integer" transfinite diameter.

Though the one-dimensional version of the Gelfond-Schnirelman approach is obviously insufficient, we extend the remarks of Trigub and examine the multidimensional generalization of the Gelfond-Schnirelman method. Analysis of the simplest cases of polynomials with rational integer coefficients least deviating from zero on $[0,1]^d$ shows that this approach works. Interesting developments here (such as an approximation of $\psi(x)$ by solutions of singular integral equations) coming from statistical mechanics and the many-body problems are presented.

It is quite remarkable that the analytic methods employed here are essentially those of the Petersburg school (Chebicheff, Markov, Bernstein), with some arithmetic adjustments.

This work was supported by the U.S. Air Force under Grant AFOSR-81-0190 and the John Simon Guggenheim Memorial Foundation.

References

[1] G. V. Chudnovsky, C. R. Acad. Sci. Paris, v. 288 (1979), A-607–A-609; C. R. Acad. Sci. Paris, v. 288 (1979), A-965–A-967.

[2] G. V. Chudnovsky, in *Bifurcation phenomena in mathematical physics and related topics*, D. Reidel, Boston, USA, 1980, 448–510.

[3] K. Alladi, M. Robinson, Lecture Notes in Mathematics, Springer, 1980, v. 751, 1–9.

[4] F. Beukers, Lecture Notes in Mathematics, Springer, 1981, v. 888, 90–99.
[5] G. V. Chudnovsky, Lecce Lectures, Lecture Notes in Physics, Springer, v. 120, 1980, 103–50.
[6] D. V. Chudnovsky, G. V. Chudnovsky, Lett. Math. Phys., 4 (1980), 373–80.
[7] D. V. Chudnovsky, G. V. Chudnovsky, J. Math. Pures Appl., Paris, France, 61 (1982), 1–16.
[8] D. V. Chudnovsky, Festschrift in honor of F. Gürsey, 1982 (in press).
[9] A. Thue, J. Reine Angew. Math. 135 (1909), 284–305.
[10] Le Baron O. Ferguson, *Approximation by polynomials with integral coefficients*, Mathematical surveys #17, American Mathematical Society, Providence, 1980.

Chapter I. Padé-type Approximtions and Measures of Irrationality of Logarithms and Algebraic Numbers

We study diophantine approximations to numbers that are values of the logarithmic (or inverse trigonometric) or algebraic functions using polynomials with rational coefficients that are denominators and numerators or rational approximations to corresponding functions. The main tool here is given by recurrences and linear ordinary differential equations (l.o.d.e.) with rational function coefficients satisfied by these polynomials. Linear recurrences themselves are reformulations of contiguous relations [1] between systems of functions with the same monodromy group. They were analyzed in the papers of D. V. Chudnovsky and G. V. Chudnovsky, see e.g., [16]–[19], as Backlund transformations.

These contiguous relations between approximants and functions furnish us with "dense" sequences of rational approximations p_n/q_n to numbers α that are values of functions (at rational points). These "dense" sequences of approximations to α allow us to estimate the measure of irrationality of α. The "dense" sequence in this paper means the following:

$$p_n \in \mathbf{Z}, \quad q_n \in \mathbf{Z}; \quad \log|q_n| \sim an, \quad \log|q_n\alpha - p_n| \sim bn$$

with $a > 0$, $b < 0$ as $n \to \infty$. Under these assumptions the exponent of irrationality of α can be expressed in terms of the "density constant"

$1 - a/b$. Namely, (Lemma 1.1) the measure of irrationality is

$$|\alpha - p/q| > |q|^{-1+(a/b)+\epsilon}$$

for all $\epsilon > 0$ and rational integers p, q; $|q| > q_0(\epsilon)$.

"Dense" sequences of rational approximations are established using linear recurrences generated by contiguous relations. Asymptotics of solutions of these recurrences define "density constants" and measures of irrationality. We start with examples of Padé approximants and corresponding orthogonal polynomials. We examine Padé approximations to binomial functions and powers of logarithmic functions. Here the recurrences are consequences of contiguous relations between generalized hypergeometric functions. We describe separately Thue polynomials generating "dense" sequences of rational approximations to algebraic numbers. This way we obtain new effective measures of diophantine approximations to cubic irrationalities. In §4 we study new, rapidly convergent, improved Padé approximations. Improved Padé approximations provide new bounds for measures of irrationality of values of the logarithmic (inverse trigonometric) functions. Particular attention is devoted to two special numbers $\ln 2$ and $\pi/\sqrt{3}$. For these two numbers we present linear recurrences defining new "dense" sequences of rational approximations. Better "density constants" for these sequences considerably improve measures of irrationality for $\ln 2$, $\pi/\sqrt{3}$ given in [2], [3].

§1. Proofs of irrationality of the number θ follow basically the same pattern: one establishes the existence of approximations p/q to θ such that

$$0 < |\theta - p/q| < \frac{1}{\psi(N)} \quad \text{for} \quad p, q \in \mathbf{Z}; \quad |p|, |q| \le N$$

and $\psi(N)/N \to \infty$ for $N \to \infty$. Moreover, if we approximate θ by a "dense" sequence of rational numbers, then we obtain a simple method for finding the measure of irrationality of θ.

We state the quite obvious and repeatedly reproduced

Lemma 1.1. *Let us assume that there exists a sequence of rational integers P_n, Q_n such that*

$$\left.\begin{matrix}\log|P_n|\\ \log|Q_n|\end{matrix}\right\} \sim \overline{a} \cdot n \tag{1.1}$$

as $n \to \infty$ *and*

$$\log|Q_n\theta - P_n| \sim \bar{b}n \tag{1.2}$$

as $n \to \infty$, *where* $\bar{b} < 0$. *Then the number* θ *is irrational and, moreover,*

$$|\theta - \frac{p}{q}| > |q|^{(\bar{a}/\bar{b})-1-\epsilon}$$

for all rational integers p, q *with* $|q| \geq q_0(\epsilon)$.

If some family of rational approximations P_n/Q_n *to* θ *is found, then the determination of the sizes of* P_n *and* Q_n *as in (1.1), is called an "arithmetic" asymptotic, while the error of approximation, as in (1.2), is called an "analytic" asymptototic of rational approximation.*

The essential point in the correct determination of "analytic" and "arithmetic" asymptotics of rational approximations is to establish the (linear) recurrence formulas connecting successive P_n *and* Q_n.

Knowing them, it is rather easy to determine "analytic" asymptotics of P_n, Q_n, using the following Poincaré lemma:

Lemma 1.2. *Let*

$$\sum_{i=1}^{m} a_i(n)X_{n+i} = 0 \tag{1.3}$$

be a linear recurrence with coefficients depending on n *such that*

$$a_i(n) \to a_i \text{ when } n \to \infty.$$

Suppose the roots of the "limit" characteristic equation

$$\sum_{i=0}^{m} a_i\lambda^i = 0$$

are distinct in absolute values:

$$|\lambda_1| > \cdots > |\lambda_m|.$$

Then there are m linearly independent solutions $X_n^{(j)} : j = 1, \ldots, m$ of (1.3) such that

$$\log|X_n^{(j)}| \sim n \log|\lambda_1| : j = 1, \ldots, m \text{ as } n \to \infty$$

and there is only one (up to a scalar multiplier) solution $\overline{X}_n$ of (1.3) such that

$$\log|\overline{X}_n| \sim n \log|\lambda_m|.$$

Though recurrences (1.3) are extremely important for the construction of diophantine approximations, they do not solve the problem of "arithmetic" asymptotics, since for $a_m(n)$ being polynomial in n, the expressions for X_n tend to have complicated denominators. Only in very special cases do these expressions have the denominator growing not faster than geometric progression.

§2. One of the situations, in which we know both the "arithmetic" and "analytic"asymptotics, is the case when the numbers under consideration are values of (generalized) hypergeometric functions.. The corresponding family of rational approximations is a specialization of a system of Padé approximations to (generalized) hypergeometric functions [5], [6]. The "analytic" asymptotic is determined using the Riemann boundary value problem that is associated with the monodromy group of the corresponding differential equation. From the point of view of applications, we present the corresponding results for systems of functions with only the simplest singularities. For completeness we remind readers of the definition of the Padé approximation (cf. [9]):

Definition 2.1. Let $f_1(x), \ldots, f_n(x)$ be functions analytic at $x = 0$ and let $m_1, \ldots, m_n$ be non-negative integers (called weights). Then polynomials $P_1(x), \ldots, P_n(x)$ of degrees at most $m_1, \ldots, m_n$ are called *Padé approximants* to $f_1(x), \ldots, f_n(x)$ if

$$R(x) = \sum_{i=1}^{n} P_i(x) f_i(x)$$

has a zero at $x = 0$ of order $\geq \sum_{i=1}^{n}(m_i + 1) - 1$. The function $R(x)$ is called the *remainder function.*

For systems of functions with the simplest monodromy (like an Abelian group as a monodromy group) we know explicit expressions for the remainder function and the Padé approximants [5]. These expressions are given as simple contour integrals in the complex plane or multiple integrals in the real domain. These explicit expressions can be used in order to obtain an asymptotics of the remainder function and Padé approximants by means of the Laplace (steepest descent) method. Alternatively, asymptotics can be obtained from contiguous relations between solutions of l.o.d.e. satisfied by the remainder function. The corresponding l.o.d.e. and contiguous relations themselves can be obtained as well from integral representations.

We start with the asymptotics of the remainder function and Padé approximants in the Padé approximation problem for binomial, logarithmic and similar hypergeometric functions, see [1], [5], and then present explicit integral representations.

Theorem 2.2. *Let $\omega_1, \dots, \omega_n$ be distinct* (mod Z) *complex numbers. Let $f_1(x), \dots, f_n(x)$ be one of the following systems of functions:*

i) $f_i(x) = \log(1-x)^{i-1} : i = 1, \dots, n;$
ii) $f_i(x) = (1-x)^{\omega_i} : i = 1, \dots, n;$
iii) $f_i(x) = {}_2F_1(1; \omega_i; c; x) : i = 1, \dots, n.$

Let

$$R(x) = \sum_{i=1}^{n} P_i(x) f_i(x)$$

be the remainder function in the Padé approximation to $f_1(x), \dots, f_n(x)$ with weights $m_1, \dots, m_n$ at $x = 0$.

Let

$$m_i = M + m_i^0 \text{ and } \frac{m_i^0}{M} \to 0 \text{ as } M \to \infty : i = 1, \dots, n.$$

Then the asymptotics of $|R(x)|$ and $|P_i(x)|$ is determined everywhere in C *using the following notations:*

$$r_n^-(x) = \min\{|1 - \varsigma_n^j \sqrt[n]{1-x}\,| : j = 0, \dots, n-1\},$$

$$r_n^+(x) = \max\{|1 - \varsigma_n^j \sqrt[n]{1-x}\,| : j = 0, \dots, n-1\}.$$

Then for any $x \neq 0, 1, \infty$ *where* $r_n^-(x) < r_n^+(x)$, *we have*

$$|R(x)| \cong r_n^-(x)^{nM}(1 + 0(\frac{1}{M}));$$
$$|P_i(x)| \cong r_n^+(x)^{nM}(1 + 0(\frac{1}{M})):$$

$i = 1, \ldots, n$ *as* $M \to \infty$. *Here* $\varsigma_n^i = \exp\left(\frac{2\pi\sqrt{-1}\cdot j}{n}\right)$.

Following [5] we present the contour integral representation for the remainder function and Padé approximants in the most general case iii). Moreover, our expressions include the inhomogeneous case of $n+1$ functions $f_0(x), f_1(x), \ldots, f_n(x)$:

$$f_0(x) = 1; \quad f_i(x) = {}_2F_1(1, \omega_i; c; x) : i = 1, \ldots, n.$$

The normalized remainder function $R(x)$ for the Padé approximation problem to the system of functions $f_0(x), \ldots, f_n(x)$ with weights $m_0, \ldots, m_n$ has the following form:

$$R(x) = \frac{m_0!\, m_1! \cdots m_n!}{2\pi i} \oint_C \frac{{}_2F_1(1, s; c; x)\, ds}{F_0(s)} \tag{2.1}$$

for

$$F_0(s) = \prod_{k_0=0}^{m_0} (s + k_0) \cdot \prod_{i=1}^{n} \prod_{k=0}^{m_i} (s + k - \omega_i),$$

and the closed contour C encircling all zeros of $F_0(s)$. Similar integral representations hold for Padé approximants $P_i(x) : i = 0, \ldots, n$, but with the contour C replaced by C_i, where C_i encircles the zeros of $\prod_{k=0}^{m_i}(s + k - \omega_i)$ but none of the other zeros of $F_0(s)$, $i = 0, \ldots, n$ and we formally put $\omega_0 = 0$ and for $m_0 > 0$ assume that $\omega_i - \omega_j \not\equiv 0 \,(\mathrm{mod}\, \mathrm{Z})$ for $i \neq j$.

Explicit expression for Padé approximants and the remainder function in cases i) and ii) can be deduced from Hermite's expression [4] of Padé approximants for systems of exponential functions, see [3]. We present the multiple integral representation for the remainder function in problems i) and ii) following the Hermite method. Similar expressions can be given in problem iii) as well. In problem iii) we put:

$$R(x) = \int_0^x dt_1 \int_0^{t_1} dt_2 \cdots \int_0^{t_{n-2}} dt_{n-1} \cdot R(x \mid t_1, \dots, t_{n-1});$$

$$\begin{aligned}(2.2)\quad R(x \mid t_1, \dots, t_{n-1}) &= (x - t_1)^{m_1}(t_1 - t_2)^{m_2} \cdot \dots \\ &\dots \cdot (t_{n-2} - t_{n-1})^{m_{n-1}} t_{t_{n-1}}^{m_n} \cdot (1 - x)^{\omega_1}(1 - t_1)^{\omega_2 - \omega_1 - m_1 - 1} \cdot \dots \\ &\dots \cdot (1 - t_{n-1})^{\omega_n - \omega_{n-1} - m_{n-1} - 1}.\end{aligned}$$

This integral representation, unlike the contour integral representation (2.1), can be directly used to obtain asymptotics for the remainder function. Formulas similar to (2.2) (see [20]) can be given for the Padé approximants $P_i(x) : i = 1, \dots, n$ if the domain of integration is moved to the complex space. From the integral representation (2.2) one can obtain an integral representation for the Padé approximation problem i), if one formally puts $\omega_i = 0$. Apparently a simpler expression can be obtained if in i) one considers instead the Padé approximation problem for functions $\ln^{i-1} z : i = 1, \dots, n$ at $z = 1$ with equal weights $m_1 = \dots = m_n = m$. The expression for the remainder function $R(z)$ in this problem is the following,

$$R(z) = \int_{D_z} \cdots \int (z - t_1 \cdots t_{n-1})^m \prod_{i=1}^{n-1} \frac{(t_i - 1)^m}{t_i^{m+1}} dt_i,$$

where the domain D_z of integration is (for real $z > 1$) the following one: $D_z = \{(t_1, \dots, t_{n-1}) \in \mathbf{R}^{n-1} : 1 < t_i < z,\ t_1, \dots, t_{n-1} < z\}$. There is an interesting similarity between the expression (2.2) for the Padé approximation to n binomial functions and the Hermite formula for the remainder function in the Padé approximation problem to n exponential functions $e^{\omega_0 x}, \dots, e^{\omega_{n-1} x}$ at $x = 0$ with weights $m_0, \dots, m_{n-1}$ [21]:

$$\begin{aligned} R(x) &= x^{m_0 + \dots + m_{n-1} + n - 1} \int_0^1 dt_1 \dots \int_0^{t_{n-2}} dt_{n-1} \varphi(x \mid \bar{t}), \\ \varphi(x \mid \bar{t}) &= (1 - t_1)^{m_0 - 1}(t_1 - t_2)^{m_1 - 1} \dots t_{n-1}^{m_{n-1}} \\ &\cdot \exp\{\omega_0 x(1 - t_1) + \omega_1(t_1 - t_2) + \dots + \omega_{n-1} x t_{n-1}\}. \end{aligned}$$

Unlike "analytic" asymptotics, denominators of coefficients of Padé approximants $P_i(x)$ cannot be easily determined. To find denominators of Padé approximants and to determine "arithmetic" asymptotics, the recurrences relating Padé approximants with contiguous weights should be analyzed. These recurrences for systems of functions satisfying Fuchsian linear differential equations are called continguous relations, following Riemann [1]. We concentrate here only on the case of logarithmic functions as a particular case of the Gauss hypergeometric function.

The recurrences that are consequences of Gauss contiguous relations between ${}_2F_1$ functions can be presented in the form:

$$F(m+1,l,k;z) = F(m,l,k-1;z) + zF(m,l,k;z);$$
$$F(m,l+1,k;z) = F(m,l,k-1;z) + (z-1)F(m,l,k;z).$$

The specification of initial conditions $F(1,1,k;z)$ gives us $P_n(z)$, $Q_n(z)$, $R_n(z)$ in the Padé approximation problem for $\ln(1-\frac{1}{z})$:

$$P_n(z)\ln(1-\frac{1}{z}) + Q_n(z) = R_n(z)$$

where $R_n(z) = 0(z^{-n-1})$ as $|z| \to \infty$, $P_n(z)$ and $Q_n(z)$ are polynomials of degrees n and $n-1$, respectively.

i) If $F_1(1,1,k;z) = \frac{1}{k-2}\{(-z)^{2-k} - (1-z)^{2-k}\}$ for $k \neq 2$ and $F_1(1,1,2;z) = \ln(1-\frac{1}{z})$, then

$$R_n(z) \stackrel{\text{def}}{=} F_1(n+1,n+1,n+2;z);$$

ii) if $F_2(1,1,k;z) = \delta_{k2}$, then

$$P_n(z) \stackrel{\text{def}}{=} F_2(n+1,n+1,n+2;z);$$

iii) if $F_3(1,1,k;z) = \frac{1}{k-2}\{(-z)^{2-k} - (1-z)^{2-k}\}$, $F_3(1,1,2;z) = 0$, then

$$Q_n(z) \stackrel{\text{def}}{=} F_3(n+1,n+1,n+2;z).$$

These recurrences are usually replaced by a single three-term recurrence

$$(n+1)X_{n+1} - (2n+1)(z-2)X_n + nz^2X_{n-1} = 0 \qquad (2.3)$$

satisfied by $X_n = P_n$, Q_n or R_n. Nevertheless, the recurrence (2.3) conceals the most remarkable arithmetical properties of $P_n(x)$ and $Q_n(x)$:

iv) coefficients of $P_n(z)$ are rational integers; and

v) coefficients of $Q_n(z)$ are rational numbers with the common denominator dividing the least common multiplier of $1, \ldots, n$, denoted by $lcm\{1, \ldots, n\}$ (i.e., growing not faster than $e^{(1+o(1))n}$ as $n \to \infty$).

Statements iv), v) do not follow immediately from (2.3), since one may suspect the denominator to grow like $n!$, because of the division by $n+1$ in order to find X_{n+1}. However, previous matrix recurrences together with (2.3) immediately imply iv), v). Hence one needs *all* the recurrences simultaneously.

Comparing (2.3) with the classical recurrence for Legendre polynomials [7], an immediate relation is established:

$$P_n(z) \cdot z^{-n} = \overline{P}_n, \quad x = 1 - 2z^{-1},$$

where $\overline{P}_n(x)$ is the Legendre polynomial of degree n:

$$\overline{P}_n(x) = 2^{-n}(n!)^{-1} \frac{d^n}{dx^n}\{(x^2-1)^n\}.$$

Similarly,

$$R_n(z) \cdot z^{-n} = \overline{Q}_n(x), \quad x = 1 - 2z^{-1},$$

where $\overline{Q}_n(x)$ is a Legendre function of the second kind and, consequently,

$$R_n(z) = P_n(z) \cdot \frac{1}{2} \ln\left(1 - \frac{1}{z}\right) + Q_n(z).$$

Because of the connection with the classical thoery of orthogonal polynomials one can write

$$Q_n(z) = \int_0^1 \frac{P_n(z) - P_n(x_1)}{z - x_1} \, dx_1.$$

This shows that coefficients of $Q_n(z)$ do have denominators, but they are of the form $\frac{1}{m} : m = 1, \ldots, n$, which proves the properties of P_n, Q_n. The asymptotics of P_n, Q_n follows from (2.3) according to Poincaré's lemma.

The properties i)–v) of Padé approximants and the remainder function in the Padé approximation problem for the logarithmic function enable us to estimate the measure of diophantine approximations of the logarithms of rational numbers. For this we simply use Lemma 1.1. The "arithmetic" and "analytic" asymptotics are given by the properties iv)–v) and the recurrence (2.3), together with Lemma 1.2. We start with the number $\ln 2$, which corresponds in the above-mentioned scheme to $z = -1$. Lemma 1.1 on "dense" sequences of approximations immediately gives us, as in [2], [8], [15]:

$$|q \ln 2 - p| > q^{-3.660137409\cdots - \epsilon}$$

for the rational integers p, q, provided that $|q| \geq q_1(\epsilon)$ for any $\epsilon > 0$.

Similarly, Padé approximations to the logarithmic function at points of the Gaussian field $\mathbf{Q}(i)$ give us a measure of the diophantine approximation to $\pi/\sqrt{3}$ [2], [8], [15].

$$|q\pi/\sqrt{3} - p| > |q|^{-7.30998634\cdots - \epsilon}$$

for rational integers p, q, provided that $|q| \geq q_2(\epsilon)$ for any $\epsilon > 0$. Moreover, results of computer experiments show that [10]:

$$|\pi/\sqrt{3} - p/q| > |q|^{-8.31}$$

for all rational integers p, q, with $|q| \geq 2$. Historically the first explicit published result on the measure of irrationality of $\pi/\sqrt{3}$ belongs to Danilov [11]. Later, similar bounds with different exponents had been obtained independently by several researchers (including Wirsing and Beukers), see [2], [8]. All these results were based on the same system of Padé approximations of the logarithmic function and, hence, on the Hermite technique. The difference in the exponent is explained by different accuracy in the computation of asymptotics. We present now the recurrence leading to the exact asymptotics together with explicit solutions to the recurrence: the exponent $7.309\ldots$ in the measure of irrationality of $\pi/\sqrt{3}$ is connected with the following nice three-term linear recurrence:

$$\begin{aligned} &n(2n+1)(4n-3)X_{n+1} + \{7.16n^3 - n.12n^2 - 6n + 5\}X_n + \\ &(4n+1)(2n-1)(n-1)X_{n-1} = 0. \end{aligned} \tag{2.4}$$

As $n \to \infty$, $x^2 + 7x + 1 = 0$, where $X_n \sim x^n$ as $n \to \infty$.

Then there are two solutions p_n and q_n of this recurrence such that

$$q_n \in \mathbf{Z} \quad \text{for all} \quad n$$

and

$$p_n \cdot lcm\{1, \dots, 2n\} \in \mathbf{Z} \quad \text{for all } n;$$

by the Poincaré theorem

$$|p_n| \sim (2+\sqrt{3})^{2n}, \quad |q_n| \sim (2+\sqrt{3})^{2n}.$$

Then

$$|q_n \frac{\pi}{\sqrt{3}} - p_n| \sim (2-\sqrt{3})^{2n}.$$

The expression for q_n ($\in \mathbf{Z}$) is rather simple:

$$q_n = 2^{1-2n} \sum_{m=0}^{n-1} \binom{2n-1}{m} \binom{4n-2m-2}{2n-1} 3^{n-m}.$$

This number is indeed an integer, and it can be represented in a number of different ways since it is connected with the values of Legendre polynomials. Namely,

$$q_n = \sum_{m=0}^{2n-1} \binom{2n-1}{m} \binom{2n+m-1}{m} \rho^m, \quad \rho = \frac{i\sqrt{3}-1}{2}.$$

Similarly, an expression exists for a "near integer" solution p_n:

$$p_n = \sum_{m=0}^{2n-1} \frac{(2n+m-1)!}{(m!)^2 \cdot (2n-1-m)!} \sigma(m)\rho^m - \sigma(2n-1)q_n$$

for $\sigma(m) = 1 + \frac{1}{2} + \cdots + \frac{1}{m}$. This explains why the denominator of p_n divides $lcm\{1, \dots, 2n\}$.

The asymptotics of p_n, q_n give a weak measure of irrationality of $\pi/\sqrt{3}$:

$$|q \cdot \frac{\pi}{\sqrt{3}} - p| > |q|^{+\chi-\epsilon}$$

for $|q| \geq q_3(\epsilon)$ and $\chi = \frac{\ln(2+\sqrt{3})+1}{\ln(2-\sqrt{3})+1} = -7.3099863$.

Though the possibilities of Padé approximations to the logarithmic function are completely exhausted by the bounds above, the improved rational approximations to the logarithmic function give new measures of irrationality. In §4 we present new recurrences and new bounds of the measure of diophantine approximations to $\pi/\sqrt{3}$ and $\ln 2$. These new measures are remarkable because for the first time they do not make use of the Hermite technique of 1873 [4].

Nevertheless, Hermite's technique of Padé approximations to powers of the logarithmic function provides a nontrivial bound for the measure of irrationality of π. For this we use Padé approximations to the system of functions in i) of Theorem 2.2 for $n = 6$, cf. [3]. The bound obtained so far is the following:

$$|q\pi - p| > |q|^{-18.8899444\ldots}$$

for rational integers p, q with $|q| \geq q_4(\epsilon)$ and any $\epsilon > 0$.

§3. In the study of diophantine approximations to algebraic numbers, we have a class of polynomials with rational coefficients similar to the Padé approximants that were introduced by Thue [22]. Let α be an algebraic number of degree $n \geq 2$ which is a root of an irreducible polynomial $P(x) \in \mathbb{Z}[x]$ of degree n. Then for every integer $r \geq 0$ there exist polynomials $A_{r_n}(x)$, $B_r(x)$ with rational integer coefficients of degree of at most $rn + [\frac{n}{2}]$ such that

$$\alpha A_r(x) - B_r(x) = (x - \alpha)^{2r+1} C_r(x),$$

where $C_r(x)$ is a polynomial. These polynomials provide a "dense" sequence of rational approximations p_r/q_r to α, if one starts with an initial good rational approximation $x = p/q$ to α, and puts $p_r/q_r = B_r(p/q)/A_r(p/q)$. In particular, if the initial approximation p/q is too good, then the exponent of the measure of irrationality of α will be less than n, i.e., we obtain an improvement over the Liouville theorem. It is in this way that A. Thue [22] proved his first ineffective theorem on diophantine approximations to algebraic numbers according to which there are only finitely many rational approximations to α with an exponent of irrationality larger than $n/2$. This method of Thue as well as his theorem was subsequently generalized to the multivariable case by Siegel, Schneider, Gelfond, Dyson, and finally Roth [23]. Roth's theorem states that there are only finitely many approximations to α with an exponent of irrationality larger than $2 + \epsilon$ for every $\epsilon > 0$. However, this theorem as well as the original one of Thue does not give an effective bound on the size of these approximations. Only Baker's theorem gives us a slight effective improvement over the Liouville theorem, and since any effective improvement in the exponent is important, attention to the old Thue method has been revived in the hope that it can give new, powerful results. Typically the case of roots $\alpha = (a/b)^{m/n}$, $(m,n) = 1$, is considered since in this case Thue polynomials are expressed explicitly in terms of Padé approximants $X_r(z)$, $Y_r(z)$ to the function z^ν, $\nu = m/n$, at $z = 1$, [24], [25], [26]. The Padé approximants $X_r(z)$, $Y_r(z)$ are polynomials of degree r and satisfy a simple three-term linear recurrence

$$(r+1-\nu)X_{r+1}(z) - (2r+1)(z+1)X_r(z) + (r+\nu)(z-1)^2 X_{r-1}(z) = 0,$$

(and the same recurrence is satisfied by $Y_r(z)$), $r = 1, \dots$. Polynomials $X_r(z)$, $Y_r(z)$ are, in fact, expressed as hypergeometric polynomials (Jacobi polynomials). We have

$$X_r(z) = {}_2F_1(-r, -r-\nu; 1-\nu; z), \quad Y_r(z) = z^r X_r(z^{-1}). \tag{3.1}$$

In view of the Kummer formula $X_r(z)$, $Y_r(z)$ can be expressed as a hypergeometric polynomial in $1 - z$ as well:

$$X_r(z) = \frac{(2r)!}{r!\,(1-\nu)\cdots(r-\nu)} \cdot {}_2F_1(-r, -r-\nu; -2r; 1-z),$$
$$X_r(z) = \frac{(2r)!}{r!\,(1-\nu)\cdots(r-\nu)} \cdot {}_2F_1(-r, -r+\nu; -2r; 1-z).$$

Using properties of polynomials $X_r(z)$, $Y_r(z)$, many interesting results on the diophantine approximation to roots $\alpha = (a/b)^{m/n}$ were obtained [25] [26], [3] (see specially [3] for new results on denominators of $X_r(z)$). Below we consider another class of algebraic numbers: cubic irrationalities, following [27] (cf. [28]).

We consider cubic irrationalities α that are solutions of the algebraic equation $f(\alpha, 1) = 0$ for the binary cubic form

$$f(x,y) = ax^3 + bx^2y + cxy^2 + dy^3$$

with rational integers a, b, c, d. With this form we associate an invariant D (discriminant): $D = -27a^2d^2 + 18abcd + b^2c^2 - 4ac^3 - 4b^3d$, and two (quadratic and cubic) covariants

$$H(x,y) = -\frac{1}{4}\begin{vmatrix} \frac{\partial^2 f}{\partial x^2} & \frac{\partial^2 f}{\partial x \partial y} \\ \frac{\partial^2 f}{\partial x \partial y} & \frac{\partial^2 f}{\partial y^2} \end{vmatrix}, \quad G(x,y) = \begin{vmatrix} \frac{\partial f}{\partial x} & \frac{\partial f}{\partial y} \\ \frac{\partial H}{\partial x} & \frac{\partial H}{\partial y} \end{vmatrix}.$$

We present recurrences and expressions for Thue polynomials associated with α in the homogeneous form, when an initial approximation to α is x/y (see $y = 1$ above). The approximants $A_r(x,y)$ and $B_r(x,y)$ are homogeneous polynomials in x, y of degree $3r+1$ with rational coefficients and satisfy $\alpha A_r(x,y) - B_r(x,y) = (x-\alpha y)^{2r+1} \cdot C_r(x,y)$ for a polynomial $C_r(x,y)$. The following three-term linear recurrence

$$(3.2) \qquad \left(r+\frac{2}{3}\right) Z_{r+1}(x,y) + (2r+1)G(x,y)Z_r(x,y) \\ - 9(3r+1)\mathcal{D}f(x,y)^2 Z_{r-1}(x,y) = 0$$

has both $A_r(x,y)$ and $B_r(x,y)$ as its solutions. This enables us to express $A_r(x,y)$ and $B_r(x,y)$ in terms of Jacobi polynomials $X_r(z)$ from (3.1) for $\nu = 1/3$. We put for $r \geq 0$

$$X_r(x,y) = y^r X_r(x/y),$$

see (3.1) for $\nu = 1/3$. Then two linerly independent solutions of (3.2) are

$$X_r^{\pm}(x,y) = X_r\left(\frac{1}{2}\{G(x,y) \pm \sqrt{-27\mathcal{D}}\, f(x,y)\}, \frac{1}{2}\{G(x,y) \mp \sqrt{-27\mathcal{D}}\, f(x,y)\}\right).$$

Thue polynomials $A_r(x) = A_r(x,1)$, $B_r(x) = B_r(x,1)$, similar to other classical extremal polynomials, satisfy Fuchsian l.o.d.e. of the second order in addition to the three-term recurrence (3.1). If we put $f(x) = f(x,1)$, then $y(x) = A_r(x), B_r(x)$ are two linerly independent solutions of the l.o.d.e.

$$f(x)y''(x) - 2rf'(x)y'(x) - \frac{r(3r+1)}{2} f''(x)y(x) = 0.$$

It is interesting to note that, though $f(x)$ has three regular singularities, there are no accessory parameters in this equation.

For small $|x - \alpha y|$, polynomials $A_r(x,y)$, $B_r(x,y)$ are approximants to α. It follows from Poincaré's Lemma 1.2 that $\frac{1}{r}\log|A_r(x,y)|$, $\frac{1}{r}\log|B_r(x,y)|$ are asymptotically $\log|\xi_2|$, while $\frac{1}{r}\log|A_r(x,y)\alpha - B_r(x,y)|$ is asymptotically $\log|\xi_1|$ where ξ_1, ξ_2 are roots of $\xi^2 + 2G(x,y)\xi - 27\mathcal{D}f(x,y)^2 = 0$, $|\xi_1| < |\xi_2|$. According to [3] and [27], the common denominator Δ_r of the coefficients of the polynomials $A_r(x,y)$ and $B_r(x,y)$ satisfies

$$\frac{1}{r}\log\Delta_r \to \frac{\pi\sqrt{3}}{6}.$$

Also there are additional divisibility properties [26], [28]: coefficients of the polynomial $X_r(1-3^{3/2}z)$ are divisible by $3^{[3r/2]}$. Hence according to Lemma 1.1 we obtain the following [27]:

Theorem 3.3. *Let x, y be rational integers and x/y be an approximation to α: $|x-\alpha y| = \min\{|x-\beta y|: f(\beta,1)=0\}$ and let $D, G(x,y)=G$, $H(x,y)=H$ be as above. For a rational integer M such that*

$$M^2 \mid (G^2, 27Df^2, H^3)$$

we put $G_1 = |G/M|$, $D_1 = 27Df^2/M^2$. There exists an effective improvement over the Liouville theorem on the diophantine approximations to α, if $3^{3i/2}\cdot(G_1+\sqrt{G_1^2+D_1}) > D_1^2\cdot e^{\pi\sqrt{3}/2}$ where $v_3(D_1)\geq i$ for $i=0,\ldots,3$. The exponent in the measure of irrationality of α is the following. For arbitrary rational integers p and q we have

$$|\alpha - \frac{p}{q}| > |q|^{-\chi-1-\epsilon} \tag{3.4}$$

if $|q| \geq q_\epsilon(\alpha)$ and $q_\epsilon(\alpha)$ depends effectively only on α and an arbitrary $\epsilon>0$. The exponent χ is

$$\chi = \frac{\ln\{(G_1+\sqrt{G_1^2+D_1})\gamma\}}{\ln\{|G_1-\sqrt{G_1^2+D_1}|\gamma\}}, \quad \gamma = e^{\pi\sqrt{3}/6}\cdot 3^{-i} \tag{3.5}$$

where $v_3(D_1)\geq i$ for $0\leq i\leq 3$.

We note that once the bound (3.4) is established for a given cubic irrationality α, the measure of irrationality with the same exponent is true for any other irrational number from the cubic field $\mathbf{Q}(\alpha)$. Namely for an arbitrary irrational $\beta\in\mathbf{Q}(\alpha)$ we have

$$|\beta - \frac{p}{q}| > c_\epsilon\cdot H(\beta)^{-\chi-1-\epsilon}|q|^{-\chi-1-\epsilon}$$

for an effective constant $c_\epsilon>0$ depending only on $\epsilon>0$ and the field $\mathbf{Q}(\alpha)$. Hence in all examples it is sufficient to give only an exponent for a single generator of the field $\mathbf{Q}(\alpha)$. The first series of examples we present gives the measure of diophantine approximations for cubic irrationalities from a given pure cubic extension $\mathbf{Q}(\sqrt[3]{D})$ for a rational integer $D>0$.

Example.

1) $D = 2 : \chi = 1.429709\ldots$; 2) $D = 3 : \chi = 1.692661\ldots$;
3) $D = 6 : \chi = 1.320554\ldots$; 4) $D = 7 : \chi = 1.727503\ldots$;
5) $D = 10 : \chi = 1.413886\ldots$; 6) $D = 12 : \chi = 1.907840\ldots$;
7) $D = 13 : \chi = 1.824735\ldots$; 8) $D = 15 : \chi = 1.493153\ldots$;
9) $D = 17 : \chi = 1.198220\ldots$; 10) $D = 18 : \chi = 1.907841\ldots$;
11) $D = 19 : \chi = 1.269963\ldots$; 12) $D = 20 : \chi = 1.194764\ldots$:

The second class of numbers is given by values of modular functions and is closely connected to the Heegner-Stark studies of one-class discriminants. These numbers are very interesting because they admit a few unusually large partical fractions in the initial part of their continued fraction expansions (Stark [29]). The modular function is $f(z) = q^{-1/48}\prod_{n=1}^{\infty}(1 + q^{n-1/2})$,
$q = e^{2\pi i z}$ for Im $z > 0$. The six appropriate numbers here are $f(\sqrt{d})$ for which the imaginary quadratic field $\mathbf{Q}(\sqrt{d})$, $d < 0$ has a class number $h = 1$ and $|d| \equiv 3 \pmod 8$: $D = -3, -11, -19, -43, -67, -163$. In all these cases we present the corresponding (effective) exponent $\kappa < 3$ of the measure of irrationality of algebrraic numbers α from the field $\mathbf{K}$, $\mathbf{K} = \mathbf{Q}(f(\sqrt{d}))$:

$$|\alpha - \frac{p}{q}| > c \cdot H(\alpha)^{-\kappa}|q|^{-\kappa}$$

for rational integers p, q with an effective constant $c > 0$ depending only on $\mathbf{K}$. Simultaneously we present an equaiion satisfied by $f = f(\sqrt{d})$ [29].

I. Let $d = -3$. Then $f^3 - 2 = 0$, so $\kappa = 2.429709\ldots$.
II. Let $d = -11$. Then $f^3 - 2f^2 + 2f - 2 = 0$ and $\kappa = 2.326120\ldots$.
III. Let $d = -19$. Then $f^3 - 2f - 2 = 0$ and $\kappa = 2.535262\ldots$.
IV. Let $d = -43$. Then $f^3 - 2f^2 - 2 = 0$ and $\kappa = 2.738387\ldots$.
V. Let $d = -67$. Then $f^3 - 2f^2 - 2f - 2 = 0$ and $\kappa = 2.802370\ldots$.
VI. Let $d = -163$. Then $f^3 - 6f^2 + 4f - 2 = 0$ and $\kappa = 2.882945\ldots$.

§4. Here we present recurrences and their solutions that provide "dense" sequences of rational approximations to $\ln 2$, $\pi/\sqrt{3}$ and π with "density constants" better than rational approximations given by the Padé approximation to the logarithmic function. Following Lemma 1.2, one can determine the "analytic" and "arithmetic" asymptotics of solutions of these recurrences. We follow the general scheme of §2 but with matrix and scalar recurcnces of a more complicated form reflecting a different monodromy structure.

The new, better measure of irrationality of ln 2 is based on a new set of contiguous relations, reflecting the presence of apparent singularities. These recurrences are the following:

$$\begin{aligned} G(m+1,n,k;z) &= G(m,n,k-2;z) + (2z-1)G(m,n,k-1;z) \\ &\quad + (z^2-z)G(m,n,k;z); \\ G(m,n+1,k;z) &= G(m,n,k-2;z) + G(m,n,k;z)(z-z^2). \end{aligned} \tag{4.1}$$

Solutions to these recurrences are completely determined by the initial conditions $G(1,1,k;z)$.

These are two kinds of initial conditions that determine the sequences P_n and Q_n:

i) $G_1(1,1,k;z) = \delta_{k2}$; $P_n \stackrel{\text{def}}{=} G_1(N_1,N_2,N_3;-1)$
$N_1 = [0.88n]$, $N_2 = [0.12n]$, $N_3 = n$.

ii) $G_2(1,1,k;z) = \frac{1}{k-2}\{(1-z)^{2-k} - (-z)^{2-k}\}$ for $k \neq 2$,
$G_2(1,1,2;z) = 0$. Then $Q_n \stackrel{\text{def}}{=} G_2(N_1,N_2,N_3;-1)$;
$N_1 = [0.88n]$, $N_2 = [0.12n]$, $N_3 = n$.

The specialization $z = -1$, above, corresponds to the specialization of ln 2 as a value of the logarithmic function $\ln(1-1/z)$. Two solutions P_n and Q_n of (4.1) with initial conditions i) and ii) satisfy the following familiar properties:

iii) P_n are rational integers;

iv) Q_n are rational numbers whose denominators divide $lcm\{1,\dots n\}$.

Now the remainder $R_n = P_n \ln 2 - Q_n$ again arises from the recurrences (4.1) and is very small, when $n \to \infty$ so that Q_n/P_n determines a very good rational approximation to ln 2.

The numbers P_n, Q_n, R_n satisfy a scalar recurrence with coefficients depending on n which is not a three-term recurrence, but a four-term linear recurrence. The asymptotics of its solutions is determined according to

Lemma 1.2 by the roots of the quartic polynomial. Numerically one has

$$\left.\begin{array}{l}\log|P_n|\\ \log|Q_n|\end{array}\right\} \sim 1.5373478\ldots\cdot n$$

and

$$\log|P_n \ln 2 - Q_n| \sim -1.77602924 \cdot n$$

as $n \to \infty$.

Hence one has the following measure of irrationality of $\ln 2$:

$$|q \ln 2 - p| > |q|^{-3.2696549\ldots}.$$

The best exponent in the measure of the diophantine approximation to $\ln 2$ we can achieve this way requires more complicated recurrences than (4.1), correspponding to more apparent singularities. The properties iii), iv) hold, but the asymptotics of P_n, Q_n and R_n are substituted by the following ones:

$$\left.\begin{array}{l}\log|P_n|\\ \log|Q_n|\end{array}\right\} \sim 1.93902189\ldots\cdot n,$$

and

$$\log|P_n \ln 2 - Q_n| \sim 1.93766649 \cdot n$$

as $n \to \infty$. Hence the measure of the diophantine approximation of $\ln 2$ is:

$$|q \ln 2 - p| > |q|^{-3.134400029\ldots}$$

for rational integers p, q with $|q| \geq q'$.

Sometimes new, better measures of irrationality are connected with new three-term linear recurrences rather than with matrix or multi-term recurrences. A new measure of the diophantine approximation for the number $\pi/\sqrt{3}$ arises this way from the Padé-type approximation to the function arctg x/x. We present the corresponding linear recurrences in their general form

$$\begin{aligned}&Y_{n+1} ab(3,0)b(3,1)b(3,-1)d^{-1} + Y_{n-1} a'b(1,-2)b(1,-3)b(1,-1)d'^{-1}\\ &\quad + Y_n\{b(1,0)b(3,-1)\cdot\big(cb(2,1) - eb(3,1)\big)d^{-1}\\ &\qquad + b(3,-2)b(1,-1)\big(c'b(2,-3) - e'b(1,-3)\big)\,d'^{-1} - b(2,-1)\} = 0;\end{aligned}$$

and for $i, j = \pm 1, \ldots, \pm 3$ we denote

$$\begin{aligned} b(1,j) &= b(i; 3n+j, 4n+j); \\ d &= cb(2,1)b(2,0) - cb(1,1)b(3,0)eb(2,0)b(3,1), \\ d' &= c'b(2,-2)\cdot b(2,-3) - c'b(3,-3)b(1,-2) - e'b(2,-2)b(1,-3); \end{aligned}$$

and $b(i;n,m) : i = 1,2,3;\ a, c, e, a', c', e'$ dependent on n and z have the following form:

$$\begin{aligned} b(3;n,m) &= 4(m+1)(4n-2m+7)(4n+5), \\ b(2;n,m) &= (4n-4m+5)(4n+7) + (1-2z)(4n+5)(4n+7)(4n+9), \\ b(1;n,m) &= 4(n'+1)(2n+3)(4n+9); \\ a &= 2(4n+3)(12n+13)\{(12n+13)z - 6(n+1)\}, \\ c &= (12n+13)^2\cdot(4n+4) - (12n+13)12(n+1) \\ &\quad - 6(n+1)(6n+7)(4n+9)z^{-1}(1-z)^{-1}, \\ e &= 6(n+1)(6n+7)\{(12n+13)z - 6(n+1)\}z^{-1}(1-z)^{-1}, \\ a' &= z(12n-7) - (6n-4), \\ c' &= (1-z)z(12n-7) - \frac{(6n-4)(6n-3)(4n+1)}{(12n-7)(4n-1) - 4(3n-2)}; \\ e' &= \frac{(8n+2)(4n-3)\{(12n-7)z - 2(3n-2)\}}{(12n-7)(4n-1) - 4(3n-2)}. \end{aligned}$$

This single recurrence is in fact a consequence of simple matrix relations equivalent to contiguous relations and is similar to one given for $\ln 2$.

There are two linearly independent solutions of this recurrence: one "integer" solution $Y_n^{(in)}$, and another "almost integer" $Y_n^{(nonin)}$ (when the denominators divide $lcm\{1,\ldots,4n\}$).

We present here a formula for an "integer" solution of the recurrence expressed as a sum of products of binomial coefficients:

$$Y_n^{(in)} = \sum_{i_1,i_2=0,\, i_1+i_2<4n}^{3n} \binom{3n}{i_1}\binom{3n}{i_2}\binom{2(4n-i_1-i_2-1)}{4n-i_1-i_2-1}\cdot$$
$$(2(4n-i_1-i_2)-3)^{-1}\cdot z^{-n+i_2}(z-1)^{3n-i_2}(-4)^{-4n+i_1+i_2}.$$

For $\pi/\sqrt{3}$ one should take $z = -3$, and for normalization we put $X_n = 12^{[n/4]}\cdot Y_{[n/4]}$.

The approximants $X_n^{(in)}$, $X_n^{(nonin)}$ possess the following properties:

I. $X_n^{(in)} \in \mathbb{Z}$, denominators of $X_n^{(nonin)}$ divide $lcm\{1,\ldots,n\}$;

II. As $n \to \infty$ one has

$$\left.\begin{array}{l}\log|X_n^{(in)}|\\ \log|X_n^{(nonin)}|\end{array}\right\} \to -1.66439185\ldots \cdot n,$$

and

III. $\log|\frac{\pi}{\sqrt{3}}X_n^{(in)} + X_n^{(nonin)}| \to 2.2006689\ldots \cdot n.$

Applying Lemma 1.1 to this sequence of approximations, we obtain the measure of irrationality for $\pi/\sqrt{3}$, which is considerably better than the previous one:

$$|q\pi/\sqrt{3} - p| > |q|^{-4.817441679\ldots}$$

for rational integers p, q with $|q| \geq q_0$.

One can generalize the recurrences for Y_n in order to obtain the best measure of irrationality for $\pi/\sqrt{3}$ so far. These three-term linear recurrences are very similar to the one given above, and their solutions (denoted by P_n and Q_n) satisfy the following properties.

I. The numbers P_n, Q_n are rational numbers, where P_n are integers and the denominators of Q_n divide $lcm\{1,\ldots,n\}$;

II. We have

$$\left.\begin{array}{l}\log|P_n|\\ \log|Q_n|\end{array}\right\} \sim 2.191056949 \cdot n$$

as $n \to \infty$.

III. The numbers Q_n/P_n approximate $\pi/\sqrt{3}$:

$$\log|P_n \cdot \frac{\pi}{\sqrt{3}} - Q_n| \sim -1.6658281013$$

as $n \to \infty$.

Properties I–III imply, according to Lemma 1.1, the following measure of irrationality of $\pi/\sqrt{3}$:

$$|q\pi/\sqrt{3} - p| > |q|^{-4.792613804\ldots}$$

for all integers p, q with $|q| \geq q_0$.

These and similar measures of the irrationality of $\pi/\sqrt{3}$ arise from three-term linear recurrences similar to the one satisfied by Y_n above. Expressions of P_n (or $Y_n^{(in)}$) are slightly different but are expressed as sums of products of three binomial coefficients, unlike the case of Padé approximations to logarithms, when they are sums of the products of two binomial coefficients only [2], [4], [15].

New dense sequences of approximations are found for π^2, improving on the measures of irrationality for π^2 and π. They have expressions like above but with the "integer" solution being a sum of a large number of binomial coefficients. Also the three-term linear recurrence is substituted by multi-term linear recurrences. We obtained the following measure of irrationality of π^2 (improving that of Apéry)

$$|\pi^2 q - p| > |q|^{-7.5},$$

so that $|\pi q - p| > |q|^{-16}$ for rational integers p, q with $|q| \geq q_1$.

§5. For number theoretical applications, it is important to use simultaneous approximations to several numbers and, in particular, linear forms in $1, \theta, \dots, \theta^{n-1}$ for a given number θ. For example, to study diophantine approximations to a value $f(x_0)$ of a function $f(x)$, one has to apply Padé approximations to a system of functions $1, f(x), \dots, f(x)^{n-1}$ (see 2.1). Theorem 2.2., cases i) and ii), furnish us with analytic tools to examine these Padé approximations. Integral representations (2.1) or (2.2) give us information about the arithmetic properties of coefficients of Padé approximants. This way we can study rational approximations (or, in general, approximations by algebraic numbers of fixed degree) to the numbers $(1-x)^\nu$ or $\ln(1-x)$ for $\nu \in \mathbf{Q}(\nu \notin \mathbf{Z})$ and $x \in \overline{\mathbf{Q}}$, $x \neq 0$. Moreover, since weights in Padé approximations can vary, various similar diophantine approximation problems can be studied using Padé approximations. One of them is the p-adic diophantine approximation to algebraic numbers and algebraic roots, and another is the diophantine approximation by numbers whose denominators have prime factors from a fixed set. Padé approximations to a system of functions lead to matrix (and linear) recurrences generalizing the ones given for $n = 1$. Linear recurrences connecting Padé approximants and the remainder function for equal weights $m_1 = \cdots = m_n = N$ for the Padé approximation to functions in cases i), ii), iii) of

Theorem 2.2 are $n+1$-term linear recurrences with coefficients that are polynomials in N and x. They are deduced from matrix contiguous relations for the generalized hypergeometric functions. The asymptotics of Padé approximants and the remainder functions which follow from lemma 1.2 is presented in Theorem 2.2. Similar recurrences can, in general, be presented for the generalization of Thue polynomials from §3, which are defined as polynomials $P_N(x,y) \in \mathbf{Z}[x,y]$ such that

$$\frac{\partial^{i_1+i_2}}{\partial x^{i_1}\partial y^{i_2}} P_N(x,y)|_{x=y=\alpha} = 0$$

for $i_1\theta_1 + i_2\theta_2 < N$ (with parameters θ_1, θ_2). These Gelfond-Dyson polynomials, as similar Roth polynomials in d variables, lead to better effective measures of irrationality of an algebraic number α, provided one knows $d-1$ initial good approximations to α. In particular cases of algebraic roots α, Gelfond-Dyson polynomials can be determined explicitly through differential equations and the recurrences which they satisfy. A special case of Gelfond-Dyson polynomials is the one considered by Siegel [24], which can be reduced to the Padé approximations of theorem 2.2. case ii) with $m_1 = \cdots = m_n = N$, $\omega_1 = 0$, $\omega_2 = \nu, \ldots, \omega_n = (n-1)\nu$ and $\alpha = (1-x)^\nu$ with $x, \nu \in \mathbf{Q}$; $x \neq 0$. The asymptotics of the remainder function and Padé approximants as $N \to \infty$ is given in 2.2, but for diophantine applications we need to know the common denominator of Padé approximants $P_i(x) : i = 1, \ldots, n$. This denominator was estimated in §3 for $n = 2$ and $\nu = 1/3$. We now present the bounds on denominators in the general case, cf. [27].

For $n = 2$ and an arbitrary rational $\nu = s/r$, $(s,r) = 1$, the common denominator Δ_N of Padé approximants $P_1(x)$, $P_2(x)$, in Padé approximations to $(1-x)^\nu$ at $x = 0$ with weight N, has the following asymptotics

$$\frac{1}{N}\log\Delta_N \to \frac{\pi}{\varphi(r)} \sum_{i=0,(i,r)=1}^{r-1} \operatorname{ctg}\frac{\pi i}{r},$$

as $N \to \infty$. Similarly, for $n > 2$ the asymptotics of the common denominator of coefficients of Padé approximants can be determined explicitly through sums of values of $\frac{d\log\Gamma(z)}{dz}$ with rational z [27]. However, a simple upper bound can be given for the common denominator Δ_N^n of coefficients of Padé approximants $P_1(x), \ldots, P_n(n)$ in the Padé approximation to functions $1, (1-x)^\nu, \ldots, (1-x)^{(n-1)\nu}$ at $x = 0$. For $\nu = s/r$,

$(s,r)=1$ and prime r we have

$$\frac{1}{N}\log\Delta_N \leq (n-1)\left\{\frac{2r}{r-1}\sum_{i=1}^{r} i^{-1} - \frac{r}{r-1}\cdot\log(r-1)\right\}$$

as $N\to\infty$.

This leads to new effective measures of irrationality (or measures of approximations by algebraic numbers of fixed degree) for algebraic roots $\sqrt[r]{D}$ with $D>0$ (see the examples of §3 for $r=3$). For example, estimates of denominators from [27], §6, give uninflated bounds for an effective improvement of the Liouville theorem for numbers $\sqrt[r]{D}$ with a fixed D and $r\geq r_0(D)$. We mention that for $D=2$ one can take $r_0=83$.

References

[1] B. Riemann, *Oeuvres Mathematiques*, Blanchard, Paris, 1968, pp. 353–63.
[2] G. V. Chudnovsky, C. R. Acad. Sci. Paris, 288 (1979), A-607–A-609.
[3] G. V. Chudnovsky, C. R. Acad. Sci. Paris, 288 (1979), A-965–A-967.
[4] Ch. Hermité, C. R. Acad. Sci. Paris, 77 (1873), 18–24, 74–79, 226.
[5] G. V. Chudnovsky, Cargesć lectures in *Bifurcation phenomena in mathematical physics and related topics*, D. Reidel, Boston, 1980, 448–510.
[6] G. V. Chudnovsky, J. Math. Pures Appl. 58 (1979), 445–76.
[7] G. Szegö, *Orthogonal polynomials*, AMS, Providence, 1938.
[8] K. Alladi, M. L. Robinson, Lecture Notes Math., v. 751, Springer, 1979, 1–9.
[9] G. A. Baker, Jr., *Essentials of Padé approximants*, Academic Press, 1975.
[10] G. V. Chudnovsky, Lecture Notes Physics, v. 120, Springer, 1980, 103–150.
[11] V. Danilov, Math. Zametki, 24 (1978), No. 4.
[12] A. Y. Khintchine, *Continued fractions*, University of Chicago Press, 1964.
[13] K. Mahler, Philos. Trans. Roy. Soc. London, 245A (1953), 371–98.

[14] D. V. Chudnovsky, Festschrift in honor of F. Gürsey, 1982 (in press).
[15] F. Beukers, Lecture Notes in Math., Springer, 1981, v. 888, 90–99.
[16] D. V. Chudnovsky, G. V. Chudnovsky, Lett. Math. Phys. 4 (1980), 373.
[17] D. V. Chudnovsky, G. V. Chudnovsky, J. Math. Pures Appl. 61 (1982), 1–16.
[18] D. V. Chudnovsky, G. V. Chudnovsky, Phys. Lett., 82 A (1981), 271.
[19] D. V. Chudnovsky, G. V. Chudnovsky, Phys. Lett., 87 A (1981), 325.
[20] K. Mahler, Math. Ann. 105 (1931), 267–76.
[21] K. Mahler, Math. Ann. 168 (1967), 200–27.
[22] A. Thue, J. Reine. Angew. Math. 135 (1909), 284–305.
[23] W. J. LeVeque, *Topics in number theory*, Addison-Wesley, v. 2, 1957.
[24] C. L. Siegel, Abh. Preus. Akad. Wiss. No. 1 (1929).
[25] C. L. Siegel, Akad. d. Wiss. in Göttingen II, Mat.-Phys. l., 1970, no. 8, Gött. (1970), 169–95.
[26] A. Baker, Quart. J. Math. Oxford 15 (1964), 375–83.
[27] G. V. Chudnovsky, Ann. of Math. 117 (1983), N° 3.
[28] D. K. Faddeev, *Studies in number theory*, ed. by A. V. Malyshev, N.Y. 1968, 51–56.
[29] H. M. Stark, in *Applications of computers in algebra*, Academic Press, 1970, 21–35.

Chapter II. On the Notion of the "Integer" Transfinite Diameter and the Gelfond-Schnirelman Method in the Prime Number Theorem

§1. We remind the reader of the definition of the transfinite diameter of a continuum E defined by Fekete [1]. We put

$$d_n(E) = \max_{z_1,\ldots,z_n \in E} \prod_{i<j} |z_i - z_j|^{2/n(n-1)}.$$

Then $d(E) = \lim_{n\to\infty} d_n(E)$ is called the transfinite diamter of E. Alternatively, the transfinite diameter can be described using the maximum of the absolute value of a deviation on E of polynomials of growing degree. Among all polynomials $P_n(z) = z^n + \cdots + a_0$ which have the leading coefficient one, we choose the polynomial with the minimal value of its norm $|P_n|_E = \max_{z\in E}|P_n(z)|$. The value of this norm is denoted by ρ_n^n. Then $d(E) = \lim_{n\to\infty} \rho_n$. For an interval of length l in the plane, the transfinite diamter is $l/4$.

Polynomials least deviating from zero were proposed as an instrument for the elementary proof of the prime number theorem by Gelfond and Schnirelman in 1936 in their analysis of Chebicheff's proof of the Bertrand postulate. An account of this extraordinary attempt can be found in [2] and [3]. The modern reader can find an account of the Gelfond-Schnirelman method for the prime number theorem in the historical section of the book by Le Baron Ferguson [4]. Less readily available, but not less interesting are the works of other representatives of the Russian school in constructive function theory (see Trigub [5] and Aparicio [11]).

We repeat briefly the essence of the original simple attempt of Gelfond and Schnirelman. The main number theoretical function under consideration is the Chebicheff function $\psi(x)$:

$$\psi(x) = \sum_{p^m \leq x} \ln p,$$

where the summation is extended over primes p sich that $p^m \leq x$. The prime number theorem itself is equivalent to the statement $\psi(x) \sim x$ for large x[6]. The function $\psi(x)$ is a logarithm of a more familiar arithmetic function: $e^{\psi(x)} = lcm\{1, \ldots, x\}$ for an integer x. The idea of Gelfond-Schnirelman is to estimate $\psi(x)$ from below by taking an integral of a polynomial with rational integer coefficients least deviating from zero on the interval [0,1]. Namely, let $P(x) = \sum_{i=0}^n p_i x^i$ be a non-zero polynomial of degree n with integer rational coefficients such that

$$\max_{x\in[0,1]} |P(x)| \leq \delta^n.$$

Then

$$\alpha = \int_0^1 p^{2N}(x)\,dx > 0,$$

and α is a rational number whose denominator obviously divides

$$lcm\{1, \ldots, 2nN + 1\} = e^{\psi(2nN+1)}.$$

Hence $\alpha \cdot e^{\psi(2nN+1)}$ is a non-zero rational integer and $\alpha \leq \delta^{2nN}$, so that $\delta^{-2nN} \leq e^{\psi(2nN+1)}$ or $-2nN \cdot \log \delta \leq \psi(2nN+1)$ for an integer $N \geq 1$. This lower bound for $\psi(x)$ would lead to the prime number theorem, if as Gelfond-Schnireleman had hoped, $\delta \rightarrow 1/e$ as $n \rightarrow \infty$. Unfortunately, as Gelfond found much later, $\delta > 1/e$ and even $\delta > (1+\sqrt{2})^{-1} \neq 0.414213\ldots$ (more precise results on values of δ are presented below).

Despite the failure of the straightforward application of the original Gelfond-Schnirelman idea, it was soon realized that the same analytic method could still be used for the proof of the prime number theorem if one considered only polynomials with several variables with integer rational coefficients least deviating from zero on a d-dimensional cube $[0,1]^d$.

The multidimensional approach has essentially been proposed by Trigub [5] and Ferguson [4]. Moreover, Trigub [5] has suggested that, perhaps, an appropriate polynomial $P(x_1,\ldots,x_d)$ can be constructed using an analog of the Roth method by demanding that $P(x_1,\ldots,x_d)$ have zeros of different orders at given points inside the cube $[0,1]^d$. This avenue has not yet been properly pursued, and instead the examples of polynomials $P(x_1,\ldots,x_d)$ considered here are given explicitly, rather than proved to exist by the Dirichlet box principle. Results for polynomials arising from different discriminants and connected with the Selberg integrals are presented below. Though we do not prove the prime number theorem in its full generality, our constants seem to be close to the expected 1.

§2. We start with the notion of the integer transfinite diameter of a continuum E. For a given $n \geq 1$ we put

$$\rho_n(E) = \min_{P_n(x)\in \mathbf{Z}[x]} \max_{x\in E} |P(x)|^{1/n},$$

where the min is taken over the set of all polynomials $P_n(x)$ with integer rational coefficients and of degree of at most n. Then the quantity

$$\delta(E) = \lim_{n\rightarrow\infty} \rho_n(E)$$

is called the integer transfinite diameter of E. The only obvious lower bound on $\delta(E)$ is $d(E)$, the transfinite diameter of E difined in §1. There is a lower bound on $\delta(E)$ for $E = [a,b]$ given by the Fekete theorem [7], see Gelfond [8]:

Proposition 2.1. *For $b > a$, $b - a < 1$ and $E = [a, b]$ we have $\delta(E) \le \sqrt{(b-a)/4}$.*

The precise value of $\delta(E)$ is unknown and it essentially depends on the arithmetic properties of a and b. It is of utmost interest for us to consider the case of $E = [0, 1]$.

Lemma 2.2 [5], [9]. *We have $\delta([0,1]) = \sqrt{\delta([0,1/4])}$.*

This lemma is simply established using the quadratic map $x \to x(1-x)$. There is a convenient method to improve both the upper and lower bounds of $\delta(E)$, using studies of the class H of polynomials with rational integer coefficients, all the roots of which are lying in the interval E. In H we take a subclass H_1 of polynomials irreducible over $\mathbf{Z}$. Another important quantity connected with the integer transfinite diameter of E is expressed in terms of the leading coefficients $a(P)$ of polynomials $P(x) \in H_1$. Namely, we define

$$l(E) = \lim_{n \to \infty} \{ \min_{P_n(x) \in H_1} |a(P_n)|^{-1/n} \},$$

where min is taken over polynomials $P_n(x) \in H_1$ of degree of at most n and with the leading coefficient $a(P_n)$. Apparently $l(E)$ is a much better lower bound for $\delta(E)$ than that given by an ordinary transfinite diameter $d(E)$. This remark is a consequence of the following simple lemma:

Lemma 2.3. *Let $P_i(x) \in \mathbf{Z}[x]$ be a system of polynomials having no common zeros such that each polynomial $P_i(x)$ has degree n_i, all roots in the continuum E, and the leading coefficient a_i. Then for an arbitrary polynomial $P(x) \in \mathbf{Z}[x]$ of degree n, we have*

$$\max_{x \in E} |P(x)|^{1/n} \ge \overline{\lim_{n \to \infty}} \{ |a_i|^{-1/n_i} \}.$$

Corollary 2.4. *For an artibrary continuum E, $\delta(E) \ge l(E)$.*

The lower bound from Corollary 2.4 is in many cases better than that given by $d(E)$. One may even conjecture that $\delta(E) = l(E)$ at least for E being an interval with rational ends, for example for $E = [0, 1]$ or $E = [0, 1/4]$. The construction of a polynomial $P_i(x)$ satisfing the conditions of Lemma 2.3 and such that $|a_i|^{1/n_i}$ is decreasing can be achieved using different methods, one of which is based on the transformation of cyclotomic polynomials [9], and another of which is based on rational transformations of an interval into itself [10].

The first bound for $l(E)$ for $E = [0, 1/4]$ is from polynomials $P_m(x)$ with rational integer coefficients, whose zeros form a complete system of conjugate algebraic numbers:

$$x_j = (6 + e^{2\pi ij/(2m+1)} + e^{-2\pi ij/(2m+1)})^{-1}$$

for $j = 0, \ldots, m-1$ and $i = \sqrt{-1}$. The leading coefficient a_m of polynomial $P_m(x) \in \mathbf{Z}[x]$ that has roots x_j can be determined explicitly as $a_m = \prod_{j=0}^{m-1} |x_j^{-1}|$. Hence

$$a_m = \prod_{j=0}^{m-1} |(e^{2\pi ij/(2m+1)})^2 + 6(e^{2\pi ij/(2m+1)}) + 1|.$$

If we denote the root, largest in absolute value, of $x^2 + 6x + 1 = 0$, $-3 - 2\sqrt{2}$ by α, then $a_m^{1/m} = |\alpha| \cdot |(\alpha^{-(2m+1)} + 1)/(\alpha^{-1} + 1)|^{1/m}$, or $\lim_{m\to\infty} a_m^{1/m} = |\alpha| = (1 + \sqrt{2})^2$. This gives an immediate lower bound for $\delta(E)$:

$$\delta(E) \geq (1 + \sqrt{2})^{-1}$$

for $E = [0, 1]$. This bound is stronger than the one given by the value of the transfinite diamter of E, and it is this bound that shows the failure of the initial attempt by Gelfond-Schnirelman (because $\delta > e^{-1}$). This bound was obtained independently in [5] and [9]. Another transformation, this time rational, leads to a conjectural value of $l(E)$ and $\delta(E)$ for $E = [0, 1]$. The rational transformation for $E = [0, 1/4]$ proposed by Aparacio [10] is the following one: $t \to t(1 - 4t)/(1 - 3t)^2$. This transformation can be used to define, starting from a given polynomial $Q(x)$ belonging to the class H, another polynomial $Q_1(x)$ from the same class H according to the following rule:

$$Q_1(t) = (1 - 3t)^{2d} Q\left(\frac{t(1 - 4t)}{(1 - 3t)^2}\right),$$

where $\deg(Q) = d$. In this way one defines polynomials $Q_n(x)$ of degrees $2n$ starting from the initial polynomial $Q_0(x) = 1 - 5x$. This sequence of polynomials can be used via Lemma 2.3 to get a lower bound for $\delta(E)$, $E = [0, 1/4]$. For this, one has to evaluate the lower bound of $l(E)$ given by a limit of $|q_n|^{2^{-n}}$ with q_n being the leading coefficient of $Q_n(x)$. Using an iterative formula for the computations of $Q_n(x)$, we obtain a product expansion [10] for $\lambda = \lim_{n\to\infty} |q_n|^{-2^{-n}}$. This formula, following [10], has the form $\lambda = \prod_{k=0}^{\infty} \theta_k^{2^{-k}}$ where $\theta_0 = 1$, $\theta_{k+1} = \theta_k + \theta_k^{-1}$. This gives the following estimate of the constant λ:

$$\lambda = 5.64937647\ldots.$$

The constant λ^{-1} gives us the lower bound for $l(E)$ and, hence, for $\delta(E)$ with $E = [0, 1/4]$. One may assume that λ^{-1} is, in fact, a correct value of $\delta(E)$ for $E = [0, 1/4]$.

The upper bounds on $\delta(E)$ that are close to the value λ^{-1} are achieved by considering polynomials of the following form

$$P(x) = x^a(1-4x)^b \prod_{j=0}^{J} Q_j(x)^{v_j},$$

where a, b are positive and v_j are non-zero rational integers. Examples with $J = 0, 1, 2$ are the following:

$$P_0(x) = x^4(1-4x),$$
$$P_1(x) = x^5(1-4x)(1-5x),$$
$$P_2(x) = x^{11}(1-4x)^2(1-5x)^2(29x^2-11x+1),$$

where $Q_0(x) = 1-5x$, $Q_1(x) = 29x^2-11x+1$. This gives the following upper bounds for $\delta(E)$ for $E = [0, 1/4]$; $\delta(E) \le 0.2$; $\delta(E) \le 0.186165509\ldots$; $\delta(E) \le 0.1840868\ldots$ as compared with the lower bound given by $\lambda^{-1} : \delta(E) \ge 0.17701068$. Better upper bounds are achieved if $J \ge 2$.

§3. The multidimensional generalization of the Gelfond-Schnirelm method gives the possibility of an alternative analytic proof of the prime number theorem with the possibility of improving the error term. According to the method outlined in §1, the main problem is to find polynomials $P(x_1, \ldots, x_d)$ with rational integer coefficients least deviating from zero in $[0, 1]^d$. For one matter, however, this generalization cannot be too much different from the one-dimensional case.

Remark 3.1. Let $E = I^d$ where I is an interval. There exists a polynomial $P(x_1, \ldots, x_d)$ with rational integer coefficients in absolute value less than 1 in E, if and only if the length of I is less than 4.

We consider polynomials $P(x_1, \ldots, x_d) \in \mathbb{Z}[x_1, \ldots, x_d]$ having different degrees in different variables in each monomial in $P(x_1, \ldots, x_d)$. This approach seems more reasonable since it involves sums of values of the ψ-function such as $\int_1^x \psi(y)\,dy$, which are typically easier to evaluate than $\psi(x)$. We remind [6] that the prime number theorem is equivalent to $\int_1^x \psi(y)\,dy \sim x^2/2$ for large x. The relationship between the values of the ψ-function and polynomials least deviating from zero is based on the following consequence of the Gelfond-Schnirelman method. Let

$$P(x_1, \ldots, x_d) \in \mathbb{Z}[x_1, \ldots, x_d] \text{ and } \max_{0 \le x_i \le 1} |P(x_1, \ldots, x_d)| \le \delta.$$

Then if

$$\int_0^1 \cdots \int_0^1 dx_1 \ldots dx_d P(x_1, \ldots, x_d) \ne 0,$$

$$\delta^{-1} \le lcm\{(i_1+1)^{-1} \ldots (i_d+1)^{-1}\}, \tag{3.1}$$

where lcm is taken over all d-plets $(i_1, \ldots, i_d)$ such that $x_1^{i_1} \ldots x_d^{i_d}$ is a monomial in $P(x_1, \ldots, x_d)$. Formula (3.1) is here the basis for obtaining lower bounds for values of the ψ-function. Another auxiliary statement deals with the interval $[0, 1/4]$. If $P(x_1, \ldots, x_d) \in \mathbb{Z}[x_1, \ldots, x_d]$ and $\max_{0 \le x_i \le 1/4} |P(x_1, \ldots, x_d)| \le \epsilon < 1$, then

$$\epsilon^{-1} \le lcm\{j_1^{-1} \ldots j_d^{-1}\}, \tag{3.2}$$

where lcm is taken over all d-plets $(j_1, \ldots, j_d)$ such that $j_k \le 2i_k + 1$ for $k = 1, \ldots, d$ and $x_1^{i_1} \ldots x_d^{i_d}$ is a monomial in $P(x_1, \ldots, x_d)$. In bounds (3.1) and (3.2) it is essential that the corresponding integrals are non-zero. Usually it can be achieved by taking an arbitrary polynomial $P_0(x_1, \ldots, x_d)$ and taking its powers P_0^N.

The simplest polynomial with integer coefficients least deviating from zero on I^d is suggested by the definition of the transfinite diameter of $I : \Delta(x_1, \ldots, x_d) = \prod_{i<j}(x_i - x_j)^2$. Here

$$\max_{x_i \in I} |\Delta(x_1, \ldots, x_d)|^N \le \nu_d^{d(d-1)N}$$

for an arbitrary N and $\lim_{d\to\infty}\nu_d = \text{length}(I)/4\,(= d(I))$. In order to apply the bound (3.1) for the lower bound of the values of the ψ-function, we express the right-hand side of (3.1) for the polynomial $\Delta(x_1,\ldots,x_d)^N$ in terms of values of the ψ-function:

$$\log lcm\{(i_1+1)\ldots(i_d+1)\} \le \psi(2N(d-1)+1)+\psi((2d-3)N+1) \\ +\psi(2N(d-2)+1)+\cdots+\psi((N(d-1)+1).$$

Hence with the help of inequality (3.1) we obtain an obvious, lower bound:

$$\sum_{j=0}^{d-1}\psi(N(d-1+j)+1) \ge d(d-1)N\log\nu_d \tag{3.3}$$

with $\nu_d \to 4$ as $d\to\infty$. This gives an inflated lower bound for the integral of the ψ-function:

$$\int_1^x \psi(y)\,dy \ge \left\{\frac{\log 16}{3}+o(1)\right\}\frac{X^2}{2}.$$

The quantity $\log 16/3 = 0.924196241\ldots$ is still below 1 but is better than the Chebicheff value of $0.921\ldots$ that used ingenious arithmetic considerations.

Obviously the polynomial $\Delta(x_1,\ldots,x_d)$ is not a polynomial least deviating from zero on I^d. For $I = [0,1]$ it is more natural to multiply $\Delta(x_1,\ldots,x_d)^N$ by $\prod_{i=1}^d x_i^M(1-x_i)^M$ for a certain $M > 1$. It so happens that in this particular case there is the famous Selberg explicit expression for the corresponding integral [12]:

$$\int_0^1\cdots\int_0^1 \Delta(x_1,\ldots,x_d)^z \prod_{i=1}^d x_i^{u-1}(1-x_i)^{y-1}\,dx_i \\ = \prod_{j=0}^{d-1}\frac{\Gamma(u+jz)\Gamma(y+jz)\Gamma(1+(j+1)z)}{\Gamma(u+y+(n+j-1)z)\Gamma(1+z)} \tag{3.4}$$

for $\mathrm{Re}\,u > 0$, $\mathrm{Re}\,y > 0$, $\mathrm{Re}\,z \ge 0$. However, since we want to consider polynomials more complicated than in (3.4), where an explicit expression in terms of Γ-factors does not exist, we ignore the explicit formula and take a purely analytical approach. This approach is based on inequality

(3.1) as its only arithmetic part. Since we are interested only in the asymptotic law of the distribution of primes, instead of estimating the integral $\int_0^1 \cdots \int_0^1 P(x_1, \ldots, x_d)\, dx_1 \ldots dx_d$ for $P(x_1, \ldots, x_d) \in \mathbf{Z}[x_1, \ldots, x_d]$ least deviating from zero on $[0.1]^d$, we consider integrals

$$\int_0^1 \cdots \int_0^1 P(x_1, \ldots, x_d)^N dx_1 \ldots dx_d$$

for a sufficiently large integer N. Hence, according to the Laplace method, the asymptotics of this integral is given by the contribution of critical points of $P(x_1, \ldots, x_d)$ in the cube $[0,1]^d$. Namely, let $\overline{x}_j = (x_1^j, \ldots, x_d^j)$ be all solutions of the equations

$$\frac{\partial}{\partial x_i} \log P(x_1, \ldots, x_d) = 0; \quad i = 1, \ldots, d$$

contained in $[0,1]^d$. Then the asymptotics of

$$\int_0^1 \cdots \int_0^1 P(x_1, \ldots, x_d)^N dx_1 \ldots dx_d$$

is given by a linear combination of $P(\overline{x}_j)^N$. We use this approach in the case of $d \to \infty$ when the equations determining the critical points turn into an integral equation.

We start with a polynomial $\Delta(x_1, \ldots, x_d) \prod_{i=1}^d x_i^a (1 - x_i)^b$. From the symmetry considerations (despite arithmetic objections) it is better to consider the symmetric interval $[-1, 1]$. In this case the asymptotics of the integral is determined by the critical points of

$$\prod_{1 \le i < j \le d} (x_i - x_j)^2 \cdot \prod_{i=1}^{b} (1 - x_i)^{2p} \cdot \prod_{i=1}^{d} (1 + x_i)^{2q}.$$

The critical points (maximums) of this expression are determined by Stieltjes' theorem [13] as equilibrium configurations of one-dimensional Coulomb particles with fixed particles at ± 1 with charges $p \ge 0$ and $q \ge 0$.

Lemma 3.5. *Let $p \ge 0$, $q \ge 0$ and let $x_1, \ldots, x_d$ be numbers from $[-1, 1]$ such that the function*

$$T(x_1, \ldots, x_d) = \prod_{i=1}^{d} (x_i + 1)^q (1 - x_i)^p \cdot \prod_{1 \le i < j \le n} |x_i - x_j|$$

attains its maximal value in $[-1,1]^d$. *Then* $x_1, \dots, x_d$ *are all zeros of the Jacobi polynomial* $P_d^{(\alpha,\beta)}(x)$ *for* $\alpha = 2p-1$, $\beta = 2q-1$. *This maximal value* T *is:*

$$T = \{2^{-d(d-1)} \prod_{\nu=1}^{d} \nu^{\nu-2d+2}(\nu+\alpha)^{\nu-1}(\nu+\beta)^{\nu-1}(d+\nu+\alpha+\beta)^{d-\nu}\}^{1/2}$$
$$\times \left\{2^d \binom{2d+\alpha+\beta}{d}^{-1}\right\}^{(d-1)} \cdot \binom{d+\alpha}{d}^p \binom{d+\beta}{d}^q.$$

We take $E = [0,1/4]$ *(see (3.2)), and in this case the polynomial* $P(x_1,\dots,x_d) \in \mathbf{Z}[x_1,\dots,x_d]$ *is*

$$\prod_{i<j}(x_i-x_j)(x_i+x_j-1) \cdot \prod_{i=1}^{d}\{x_i(1-x_i)\}^p$$
$$\cdot \prod_{i=1}^{d}\{1-2x_i\}^{2q} = \prod_{i<j}(y_i-y_j) \cdot \prod_{i=1}^{d} y_i^p(1-4y_i)^q$$

for $y_i = x_i(1-x_i) \in [0,1/4]$ *with* $x_i \in [0,1]$.

The asymptotics for T as $d \to \infty$ is achieved in the case when p and q grow linearly with d or $\alpha = ad$, $\beta = bd$, for constants $a \geq 0$ and $b \geq 0$ with $d \to \infty$. From (3.2) we obtain:

$$\int_0^{(d-1)N} \psi((a+b)dN + (d-1)N + y)\,dy$$
$$\geq \{-\log T_{p,q} + \frac{d(d-1)}{2}\log 8 + pd\log 8 + qd\log 2\}N = AN$$

for $2p-1 = ad$, $2q-1 = bd$ and $T_{p,q}$ being the maximum given by Lemma 3.5. We apply this formula when d is sufficiently large ($d \to \infty$), and hence an asymptotics for A has the following form

$$\begin{aligned} A \cong \bar{A}d^2 \overset{\text{def}}{=} d^2\{&\log(1+a)\cdot\frac{(1+a)^2}{4} + \log(1+b)\cdot\frac{(1+b)^2}{4} \\ &- \frac{a^2}{4}\log a - \frac{b^2}{4}\log b - \frac{1}{4}(2+a+b)^2\log(2+a+b) \\ &+ \frac{1}{4}(1+a+b)^2\log(1+a+b) - (1+a)\log 2\}. \end{aligned} \tag{3.6}$$

Hence for $N \to \infty$ (sufficiently large with respect to d) we obtain

$$(3.7) \qquad \int_{dN}^{2dN} \psi((a+b)dN + y)\,dy \geq \tilde{A}d^2 N.$$

The maximal value of $\overline{A}/(1.5+a+b)$ for $a \geq 0$ and $b \geq 0$ is achieved for $b = 0$, and $a \cong 0.78$, when

$$\frac{\overline{A}}{(1.5+a)} \cong 0.990358\ldots.$$

Hence, in this case the value of "Chebicheff's" constant is $0.990358\ldots$. It is, however, unexpected that $b = 0$ would be an optimal choice in view of §2, where the corresponding polynomial $1-4t$ for $E = [0, 1/4]$ plays an important role. This phenomenon can be properly understood only by analyzing integral equations determining the asymptotics of the integral (critical point of $P(x_1, \ldots, x_d)$) as $d \to \infty$.

We consider, as above, the polynomial

$$Q(x_1, \ldots, x_d) = \prod_{1 \leq i < j \leq d} (x_i - x_j)^2 \cdot \prod_{i=1}^{d} (x_i + 1)^{V_1} \cdot (x_i - 1)^{V_{-1}}$$

for $V_1 \geq 0$, $V_{-1} \geq 0$; $V_1 = v_1 d$, $V_{-1} = v_{-1} d$ for $v_1 \geq 0$, $v_{-1} \geq 0$. The system of algebraic equations determining the critical points of

$$Q(x_1, \ldots, x_d) : \frac{\partial}{\partial x_i} \log Q(x_1, \ldots, x_d) = 0, \quad i = 1, \ldots, d$$

takes the following form

$$2 \cdot \sum_{j \neq i} (x_i - x_j)^{-1} + v_1 d(x_i + 1)^{-1} + v_{-1} d(x_i - 1)^{-1} = 0 :$$

$i = 1, \ldots, d$ and $x_i \in [-1, +1]$. As $d \to \infty$ we can apply classical methods of statistical mechanics, and for $d \to \infty$ we can consider a distribution function $\rho(\alpha)$ of x_i on $[-1, +1]$. Classical methods [14] suggest that x_i fill the subinterval $[-a, a] \subset [-1, 1]$ and that the density $\rho(\alpha)$ is characterized as $\frac{1}{d}\sum_{i=1}^{d} f(x_i) = \int_{-a}^{a} f(\alpha)\rho(\alpha)d\alpha + o(1)$ for $d \to \infty$ and an arbitrary integrable function $f(\alpha)$. Hence the equations determining the critical points of $Q(x_1,\dots,x_d)$ turn into a singular integral equation:

$$2\int_{-a}^{a}\frac{\rho(\alpha)\,d\alpha}{x-\alpha}+\frac{v_1}{x+1}+\frac{v_{-1}}{x-1}=0 \tag{3.8}$$

for $x\in[-a,a]$. Following the Laplace method, we have

$$\begin{aligned}\frac{1}{N}\log\Big|\int_{-1}^{+1}\cdots\int_{-1}^{+1}Q(x_1,\dots,x_d)^N\,dx_1\dots dx_d\Big|\\ =d^2\Big\{\int_{-a}^{a}\int_{-a}^{a}\rho(\alpha)\rho(\beta)\log|\alpha-\beta|\,d\alpha\,d\beta\\ +v_1\int_{-a}^{a}\rho(\alpha)\log|\alpha+1|\,d\alpha+v_{-1}\int_{-a}^{a}\rho(\alpha)\log|\alpha-1|\,d\alpha\Big\},\end{aligned} \tag{3.9}$$

as $d\to\infty$ and $N\to\infty$.

We return now to the solution of the singular linear integral equation which is a particular case of the Prandtl equation (see [15]). This equation is of considerable interest for quantum field theory and plasma physics and the same equation appears in the determination of the asymptotics of Padé approximations [16]. The solution is given by the following explicit formula:

$$\begin{aligned}\rho(x)&=\frac{1}{\pi\sqrt{a^2-x^2}}\left[P-\frac{1}{\pi}\int_{-a}^{a}\frac{f'(t)\sqrt{a^2-t^2}}{t-x}\,dt\right],\\ P&=\frac{1}{\pi\ln(2/a)}\int_{-a}^{a}\frac{f(t)\,dt}{\sqrt{a^2-t^2}},\end{aligned} \tag{3.10}$$

where $f(x)=-\frac{1}{2}\{v_1\ln\frac{1}{|1-x|}+v_{-1}\ln\frac{1}{|1+x|}\}+c_0$, and where the constant c_0 is determined from the equation $\int_{-a}^{a}\rho(t)\ln\frac{1}{|t-x|}dt=f(x)$. We present an explicit expression for $\rho(x)$ in the following form:

$$\rho(x)=\frac{-\sqrt{1-a^2}}{2\pi\sqrt{a^2-x^2}}\cdot\left\{\frac{v_1}{x-1}-\frac{v_1}{a-1}-\frac{v_{-1}}{x+1}+\frac{v_{-1}}{a+1}\right\}.$$

Let us consider a particularly important case when $v_1=v_{-1}$, so that

$$\rho(x)=\frac{v_1}{\pi}\cdot\frac{\sqrt{a^2-x^2}}{\sqrt{1-a^2}\cdot(1-x^2)}$$

for $x \in [-a, a]$ and $0 < a < 1$, and

$$\frac{v_1\{1-\sqrt{1-a^2}\}}{\sqrt{1-a^2}} = 1, \quad \text{or } a^2 = \frac{2v_1+1}{(v_1+1)^2}. \tag{3.11}$$

An example above (with $b = 0$) shows that the direct transfer of the one-dimensional considerations of §2 to multidimensional integrals of §3 is impossible. We present instead a different general scheme for the improvement of the "Chebicheff constant" and the proof of the prime number theorem. This scheme is based on the solution of singular integral equations similar to (3.8) in terms of hyperelliptic integrals. For this we start with polynomials $Q_1(x), \ldots, Q_k(x)$ with rational integer coefficients having all their zeros in the interval $E = [0, 1/4]$. Then the d-dimensional integral that can be used for improvement of the lower bound of the ψ-function as in (3.2) is the following one

$$J = \int_0^1 \cdots \int_0^1 dx_1 \ldots dx_d \{\prod_{i<j}(t_i - t_j)^2 \prod_{l=1}^{k} \prod_{i=1}^{d} Q_l(t_i)^{v_l}\}^N \tag{3.12}$$

for $t_i = x_i(1-x_i) \in [0, 1/4]$ and $x_i \in [0, 1]$ and $v_1 \geq 0, \ldots, v_k \geq 0$. Let $\{\alpha_1, \ldots, \alpha_n\}$ be a complete set of (distinct) zeros of $Q_1(t), \ldots, Q_k(t)$. As $d \to \infty$, this system of algebraic equations determining the asymptotics of (3.12) turns into a single singular integral equation for the distribution function $\rho(\alpha)$. Let $0 \leq \alpha_1 < \alpha_2 < \cdots < \alpha_n \leq 1/4$. Then there are $n+1$ intervals $E_s = [a_s, b_s]$ such that $a_0 = 0$, $b_0 < \alpha_1$; $\alpha_n < a_n$, $b_n = 1/4$; $\alpha_s < a_s \leq b_s < \alpha_{s+1}$ ($s = 1, \ldots, n$). The singular integral equation takes the following form

$$\sum_{s=0}^{n} \int_{a_s}^{b_s} \frac{\rho(\alpha)\, d\alpha}{\alpha - x} = \frac{1}{2} \sum_{l=1}^{n} \frac{w_l}{x - \alpha_l} \tag{3.13}$$

for $x \in U_{s=0}^{n}[a_s, b_s]$, together with the normalization condition

$$\sum_{s=0}^{n} \int_{a_s}^{b_s} \rho(\alpha)\, d\alpha = 1,$$

and the regularity conditions $\rho(a_s) = \rho(b_s) = 0$, if $b_s - a_s \neq 0$ and $w_{s-1} > 0$, $w_s > 0$.

An explicit expression is the following one

$$\rho(x) = \frac{(-1)^{n-s+1}}{\pi^2\sqrt{\prod_{s=0}^{n}|(x-a_s)(x-b_s)|}} \times \tag{3.14}$$
$$\left[P_n(x) + \sum_{s=0}^{n}(-1)^{n-s}\int_{a_s}^{b_s} \boxed{\mathrm{A}}\frac{dt}{t-x}\right]$$

where

$$\boxed{\mathrm{A}} = \sqrt{\prod_{s_1=0}^{n}|(t-a_{s_1})t-b_{s_1})|}\cdot\left\{\sum_{l=1}^{n}\frac{w_l}{2(t-\alpha_l)}\right\}$$

and $x \in [a_s, b_s] : s = 0, \dots, n$, and $P_n(x)$ is a polynomial of degree n (see [17]).

§4. Simultaneously with a lower bound for values of the ψ-function, the Gelfond-Schnirelman method gives an upper bound for values of the ψ-function, and hence for the number of primes. For this we again use integrals over the polynomials with integer rational coefficients, but this time taken over a different path (domain) of integration. The essence of the method can be illustated by an example of the integral $\int_0^1 x^n(1-x)^n dx$, which is an equivalent of P. Erdös' method of obtaining upper and lower bounds for $\pi(x)$ used widely in all books on number theory (see [18]). In this estimate we look at $\binom{2n}{n}$ written as

$$\log\binom{2n}{n} = \psi(2n) - \psi(n) + \psi\left(\frac{2n}{3}\right) - \left(\frac{n}{2}\right) + \cdots. \tag{4.1}$$

From this we derive simultaneously upper and lower bounds for values of the ψ-function:

$$n\cdot\log 4 + o(n) \le \psi(2n);$$
$$\psi(2n) - \psi(n) \le n\log 4 + o(n).$$

Thus we obtain

$$\log 2\cdot n \le \psi(n) \le \log 4\cdot n \tag{4.2}$$

for large n. Identities like (4.1) form a basis for the original Chebicheff-Sylvester method [19], and they all follow from the representation of a factorial as a sum of the values of the ψ-function. For an arbitrary $x \geq 1$ we, following Chebicheff, put $T(x) = \log\{[x]!\}$. Then

$$T(x) = \sum_{m \leq x} \psi\left(\frac{x}{m}\right).$$

Upper and lower bounds (4.2) for $\psi(x)$ can be deduced simultaneously from the integral representaion, since $\binom{2n}{n}$ and $\binom{2n}{n}^{-1}$ both can be written as B-function integrals. Namely,

$$n^{-1}\binom{2n}{n}^{-1} = \int_0^1 t^n(1-t)^{n-1}\,dt$$

and

$$\binom{2n}{n} = -\frac{i}{2\pi}\int_1^{(0+)} (-t)^{-n-2}(1-t)^{2n}\,dt.$$

This example shows how an integral of a polynomial with integer rational coefficients over the interval $[-1,1]$ can be inverted and represented as an integral over a unit circle of a rational function with a singularity inside a circle. Namely, as in §§ 1–2, looking on the term-by-term integration of $P(z)\cdot(1+z)^n z^{-k}$, with $d(P) \leq 2k-n$, we get:

$$\prod_{k<p\leq n} p \quad \text{divides} \quad \int_{|z|=1} P(z)(1+z)^n \frac{dz}{z^k},$$

if $P(z) \in \mathbf{Z}[z]$. Hence

$$\prod_{k<p\leq n} p \leq \max_{|z|=1} |(1+z)^n P(z)|$$

for $P(z) \in \mathbf{Z}[z]$, $d(P) \leq 2k-n$. In this formula, as above in §2, it is assumed that the integral $\int_{|z|=1} P(z)(1+z)^n \frac{dz}{z^k}$ is non-zero. Since to bound this integral we are using the critical points method, we can always insure the nonvanishing of the integral by considering $P(z)^N$ instead of $P(z)$, for a large integer N. We can instead consider integrals

$$\frac{1}{\pi}\int_{-2}^{2} (t+2)^n P(t) \frac{dt}{\sqrt{4-t^2}}$$

for $P(t) \in \mathbb{Z}[t]$, and obtain

$$\prod_{k<p\leq n} p \leq \max_{t\in[-2,2]} |(t+2)^{n/2} P(t)|,$$

$P(t) \in \mathbb{Z}[t]$, $d(P) \leq \frac{2k-n}{n}$. A similar consideration exists in the multi-dimensional situation. This way we achieve

$$(4.3) \qquad \prod_{k<p\leq n} p^d \leq \max_{-2\leq t_i\leq 2} |P(t_1,\ldots,t_d) \prod_{i=1}^{d} (2+t_i)^{n/2}|,$$

where $P(t_1,\ldots,t_d) \in \mathbb{Z}[t_1,\ldots,t_d]$ and $d_{t_i}(P) \leq k - \frac{n}{2}$. Hence we have a situation completely parallel to that of (3.1)–(3.2) but now with upper bounds for values of the ψ-function instead of lower bounds. Again, the natural choice of the polynomial $P(t_1,\ldots,t_d)$ is the following one

$$P(t_1,\ldots,t_d) = \prod_{i=1}^{d} (t_i - 2)^w \cdot \prod_{i<j} (t_i - t_j)^{2v}$$

for appropriate $w > 0$, $v > 0$. This way we obtain an upper bound for $\psi_1(x) = \int_1^x \psi(y)dy$ of the same degree of accuracy as its lower bound from §3.

It is useful to remark that the methods of both §3 and §4 can be brought to a uniform form of integration over the same interval (say [0,1]), if one allows a polynomial with rational coefficients to have denominators which are powers of 2. In particular, this can provide us with a "mock" proof of the prime number theorem using multiple integrals of polynomials. The form of a polynomial and the proof itself are, unfortunately, dependent on an identity which is known to be equivalent to the prime number theorem itself. This is the identity (see [6])

$$(4.4) \qquad \sum_{n=1}^{\infty} \frac{\mu(n)}{n} = 0,$$

where $\mu(n)$ is the Möbius function (cf. Landau [20] for this and other methods of derivation of the prime number theorem). We use now the original Chebicheff method of (4.1), (4.2). Namely, we apply the Möbius inversion formula to the function $T(x) = \log\{[x]!\}$. We thus have $\psi(x) =$

$\sum_{n\leq x}\mu(n)T(\frac{x}{n})$. Similar to (4.1) we put

$$\psi(2x)-2\psi(x)=\sum_{n\leq 2x}\mu(n)\Big\{T\Big(\frac{2x}{n}\Big)-\Big(\frac{x}{n}\Big)\Big\}.$$

Also we have

$$T\Big(\frac{2x}{n}\Big)-2T\Big(\frac{x}{n}\Big)=\log\{[\frac{2x}{n}]!/[\frac{x}{n}]!^2\}.$$

Hence, using the identity (4.4) (see the method of [20]), we obtain

$$\psi(2x)-2\psi(x)=o(x)$$

for a large x. This method (due essentially to Landau) clearly shows the "mock" proof based on a consideration of the following product of rational numbers that are either binomial coefficients or their inverses:

$$\prod_{n\leq x}\binom{2[x/n]}{[x/n]}^{\mu(n)}.$$

It remains only to represent each of the binomial coefficients $\binom{2[x/n]}{[x/n]}$ as one-dimensional integrals over $[0,1]$, if $\mu(n)=-1$, or over the unit circle (or $[-2,2]$) if $\mu(n)=+1$.

When the work on Chapter II was completed, the author found that the Gelfond-Schnirelman approach (apprently without knowledge of the original attempt) was successfully pursued by M. Nair [21], [22]. The results of M. Nair [22] are essentially those of section 3 of Chapter II.

Also recent work of H. G. Diamond, K. McCurley [23] and P. Erdös revives interest in the Sylvester approach [19] similar to that of §4.

More work in this direction (approximation by polynomials with integer coefficients, integral estimates and the prime number theorem etc.) has to be mentioned, see D. Cantor [24]. The elementary approach to the prime number theorem, similar to that of Gelfond-Schnirelmen, was rediscovered recently by several researchers, cf. the brief discussion of this approach among others in the new review of H. Diamond, Bull. Amer. Math. Soc., 7, (1982), 553 (added in the proof).

Added in proof. Recently the author, following the discussion of §5, Ch. I, proved the following **Theorem**. Let $f(x)$ be an algebraic function with the Taylor expansion $f(x)=\sum_{n=0}^{\infty}a_nx^n$, $a_n\in\mathbf{Q}$. Then for every $\epsilon > 0$ and rational integers a, b with $|b|^\epsilon > c_1|a|^{2+\epsilon}$, $|f(a/b) - P/Q| > |Q|^{-2-\epsilon}$ where P and Q are arbitrary rational integers with $|Q| > Q_0(\epsilon, a, b)$. Here $c_1 = c_1(\epsilon) > 0$ and Q_0 are effective constants.

References

[1] L. V. Ahlfors, *Conformal invariants*, McGraw-Hill, 1973.

[2] A. O. Gelfond, L. G. Schnirelman, Uspekhi Math., Nauk, No. 2, (1936) (Russian).

[3] A. O. Gelfond, Comment "On determining the number of primes not exceeding the given values" and "On primes" in P. L. Chebicheff, *Complete Works*, v. 1., Academy, Moscow, 1946, 2nd printing (Russian).

[4] Le Baron O. Ferguson, *Approximation by polynomials with integral coefficients*, Mathematical surveys #17, American Mathematical Society, Providence, 1980.

[5] R. M. Trigub, *Metric questions of the theory of functions and mappings*, No. 2., pp. 267-333, Naukova Dumka Publishing House, Kiev, 1971 (Russian).

[6] A. E. Ingham, *The distribution of prime numbers*, Cambridge Tracts, #30, Cambridge Univ. Press, 1932.

[7] M. Fekete, C. R. Acad. Sci. Paris. 1923, 17.

[8] A. O. Gelfond, Uspekhi Math. Nauk. 10 (1955), No. 1, 41–56 (Russian).

[9] E. Aparicio, Revista Matematica Hispano-Americana, 4 Serie, t. XXXVIII, No. 6, 259–70 Madrid, 1978.

[10] E. Aparicio, Revista de la Universidad de Santader, Numero 2, Parte 1 (1979), 289–91.

[11] E. Aparicio, Revista Matematica Hispano-Americana, 4 Serie, v.XXXVI (1976), 105–24.

[12] A. Selberg, Norsk Mathematisk Tidsskrift, 26 (1944), 71–78.

[13] G. Szegö, *Orthogonal polynomials*, American Mathematical Society Colloquium Publication #23, Providence, Rhode Island, 1939.

[14] L. Hulthén, Arkiv for Mat. Astron. Fysik, 26A, Hafte 3, No. 11 (1938), 1–106.

[15] D. V. Chudnovsky, G. V. Chudnovsky, Letters Nuovo Cimento, 19, No. 8 (1977), 300–02.
[16] J. Nuttall, *Bifurcation phenomena in mathematical physics and related topics*, D. Reidel Publishing Company, Boston, 1980, 185–202.
[17] N. I. Muskhelishvili, *Singular integral equations*, P. Noordhoff, N. V. Groningen, Holland, 1953.
[18] W. J. LeVeque, *Topics in number theory*, v. 1, Addison-Wesley, 1956.
[19] J. J. Sylvester, Messenger. Math. (2), 21 (1891), p. 120.
[20] E. Landau, Sitz. Akad. Wissen. Wien, Math-Nat. Klasse, bd. CXVII, Abt. IIa, 1908.
[21] M. Nair, The American Mathematical Monthly, 89, No. 2, (1982), 126–29.
[22] M. Nair, J. London Math. Soc. (2), 25 (1982), 385–91.
[23] H. G. Diamond, K. S. McCurley, Lecture Notes Math., v. 899, Springer, 1981, 239–53.
[24] D. G. Cantor, J. Reine Angew. Math., 316 (1980), 160–207.

Received August 6, 1982

Dr. G. V. Chudnovsky
Mathematics Department
Columbia University
New York, New York 10027

Infinite Descent on Elliptic Curves with Complex Multiplication

John Coates

To I.R. Shafarevic

1. Introduction

It is a pleasure to dedicate this paper to I. R. Safarevic, in recognition of his important work on the arithmetic of elliptic curves. As anyone who has worked on the arithmetic of elliptic curves is acutely aware, it is still dominated today, despite its long and rich history, by a wealth of tantilizing conjectures, which are convincingly supported by numerical evidence. The most important amongst these conjectures, at least from the point of view of diophantine equations, is the conjecture of Birch and Swinnerton-Dyer, which grew out of the attempt to apply to elliptic curves the quantitative local to global principles employed by Siegel in his celebrated work on quadratic forms. This conjecture is so well known that there is no need to repeat its precise statement here. However, we do wish to point out that no-one has yet found a direct and natural link between the existence of rational points of infinite order on an elliptic curve defined over a number field and the behaviour of its Hasse-Weil L-Series at the point $s = 1$ in the complex plane, as is predicted by the conjecture of Birch and Swinnerton-Dyer. Guided by Artin and Tate's [15] success with the geometric analogue, most recent work has attempted to establish such a connexion indirectly by p-adic techniques which combine the classical infinite descent of Mordell and Weil with ideas from Iwasawa's theory of $\mathbf{Z}_p$-extensions of number fields. The first results in this direction were found by Mazur [10], who studied descent theory on abelian varieties over an arbitrary $\mathbf{Z}_p$-extension of the base field, assuming only that the abelian variety has good ordinary reduction at every ramified prime for the $\mathbf{Z}_p$-extension (we shall describe this by simply saying that the abelian variety is *ordinary* for the $\mathbf{Z}_p$-extension). The present paper, which was motivated by earlier work of Wiles and myself [3], pursues the study of this descent theory in a very special case, which nevertheless contains important families of elliptic curves, whose arithmetic properties are presumably typical of all elliptic curves. Namely, we consider elliptic curves with complex multiplication, and a certain non-cyclotomic $\mathbf{Z}_p$-extension of the base field, whose existence is closely associated with the hypothesis of complex multiplication, and for which the elliptic curve is ordinary (see §2 for the precise definition). Much of the material in this article is not recent work and was presented in the Hermann Weyl Lectures at the Institute for Advanced Study, Princeton, in 1979, and will eventually form part of a more detailed set of notes on these lectures. However, the crucial results of §3, which are due to Bernadette Perrin-Riou [12], were only proven after these lectures took place. Finally, there have been two important recent developments on the problems discussed in this paper. Firstly, P. Schneider (see [13], [14], and a manuscript in preparation) has now established the equality of the analytic height and algebraic height for arbitrary abelian varieties and $\mathbf{Z}_p$-extensions of the base field, for which the abelian variety is ordinary. Secondly, Mazur and Tate (see their article in this volume) have found some striking new descriptions of the analytic height attached to an elliptic curve over a $\mathbf{Z}_p$-extension, subject always to the hypothesis that the elliptic curve is ordinary for the $\mathbf{Z}_p$-extension.

Notation. Let K be an imaginary quadratic field, $\mathcal{O}$ the ring of integers of K, and F a finite extension of K. Throughout this paper, E will denote an elliptic curve defind over F which has complex multiplication by K in the sense that its ring of F-endomorphisms is isomorhic to $\mathcal{O}$. As usual, we fix an isomorphism of $\mathcal{O}$ with $\mathrm{End}_F(E)$ such that the embedding of K in F induced by the action of $\mathrm{End}_F(E)$ on the tangent space to E/F at the origin is the given embedding of K in F. Note also that the hypotheses that $\mathrm{End}_F(E)$ is the maximal order of K involves no real loss of generality since every elliptic curve E over F with $\mathbf{Q} \otimes_{\mathbf{Z}} \mathrm{End}_F(E)$ isomorphic to K is isogenous over F to one with this property. If N is an extension field of F, we write $E(N)$ for the group of N-rational points on E. If L/N is a Galois extension of fields, we write $G(L/N)$ for the Galois group of L over N. Let $\overline{N}$ denote a fixed algebraic closure of N, and A a discrete module for $G(\overline{N}/N)$. The Galois cohomology groups of $G(\overline{N}/N)$ acting on A will be denoted by $H^i(N, A)$; when $A = E(\overline{N})$, we simply write $H^i(N, E)$. For each $\mathcal{O}$-module D, we put D_α for the kernel on D of a non-zero element α of $\mathcal{O}$. If $\mathfrak{h}$ is an integral ideal of K, let $D_{\mathfrak{h}} = \bigcap_{\alpha \in \mathfrak{h}} D_\alpha$. As usual, if $D = E(\overline{F})$, we simply write E_α, and $E_{\mathfrak{h}}$. Finally, if $\beta \in K$, β^* will denote its conjugate over $\mathbf{Q}$.

1. Descent Theory over the Base Field F

We begin by recalling that the Tate-Safarevic group $Ш_F$ of E over F is defined by the exactness of the sequence

$$(1) \qquad 0 \to Ш_F \to H^1(F,E) \to \prod_v H^1(F_v,E),$$

where the product on the right is taken over all finite places v of F, and F_v denotes the completion of F at v. Now take α to be a non-zero element of $\mathcal{O} = \mathrm{End}_F(E)$, which is not an automorphism. Taking Galois cohomology of the exact sequence of $G(\overline{F}/F)$-modules

$$0 \to E_\alpha \to E(\overline{F}) \xrightarrow{\alpha} E(\overline{F}) \to 0,$$

we obtain the exact sequence

$$(2) \qquad 0 \to E(F)/\alpha E(F) \to H^1(F,E_\alpha) \to \left(H^1(F,E)\right)_\alpha \to 0.$$

By definition, $(Ш_F)_\alpha$ is a subgroup of $\left(H^1(F,E)\right)_\alpha$. The Selmer group $S_F^{(\alpha)}$ of E over F relative to α is defined to be the inverse image of $(Ш_F)_\alpha$ in $H^1(F,E_\alpha)$, under the map at the right hand end of (2). Thus we have the fundamental exact sequence

$$(3) \qquad 0 \to E(F)/\alpha E(F) \to S_F^{(\alpha)} \to (Ш_F)_\alpha \to 0.$$

We now introduce two modified forms of the Selmer group, whose definitions are motivated by the Iwasawa theory of §2. Indeed, in both cases, we modify the local conditions in precisely the primes of F which are ramified in the $\mathbf{Z}_p$-extension introduced in §2. We first specify the conditions that will be imposed on the prime number p for the rest of this paper. Henceforth, we assume that p satisfies:–

Hypothesis on p. (i) *p splits in K into two distinct primes, say* $(p) = \mathfrak{p}\mathfrak{p}^*$; (ii) *$E$ has good reduction at every prime of F above p*; (iii) $p \neq 2$.

As is well known, we note that (i) and (ii) are equivalent to the assertion that E has good ordinary reduction at each prime above p. Also the condition $p \neq 2$ is not needed in §1, but it is essential for the later parts of the paper. Write $K_{\mathfrak{p}}(= \mathbf{Q}_{\mathfrak{p}})$ and $\mathcal{O}_{\mathfrak{p}}(= \mathbf{Z}_{\mathfrak{p}})$ for the completion of K at $\mathfrak{p}$ and its ring of integers, and put $D_{\mathfrak{p}} = K_{\mathfrak{p}}/\mathcal{O}_{\mathfrak{p}}$. Now take $\alpha \neq 0$ to be any element of $\mathcal{O} = \mathrm{End}_F(E)$, which is not an automorphism. The first modification of the Selmer group is to neglect to impose local conditions at the primes above $\mathfrak{p}$, as follows. Define $Ш'_F$ by the exactness of the sequence

$$(4) \qquad 0 \to Ш'_F \to H^1(F,E) \to \prod_{v \nmid \mathfrak{p}} H^1(F_v,E),$$

where the product is now taken over all finite primes v of F, which do not divide $\mathfrak{p}$. We then define $S'^{(\alpha)}_F$ to be the inverse image of $(Ш'_F)_\alpha$ in $H^1(F,E_\alpha)$ under the map on the right of (2). The second modification of the Selmer group is to impose more stringent local conditions at the primes above $\mathfrak{p}$. Indeed, if v is any place of F, we have the local analogue of (2), namely

$$(5) \qquad 0 \to E(F_v)/\alpha E(F_v) \to H^1(F_v,E_\alpha) \to \left(H^1(F_v,E)\right)_\alpha \to 0.$$

Writing $r_{v,\alpha}$ for the restriction map from $H^1(F,E_\alpha)$ to $H^1(F_v,E_\alpha)$, it is plain from the definition of the Selmer group that the image of $S^{(\alpha)}_F$ under $r_{v,\alpha}$ is contained in the image of $E(F_v)/\alpha E(F_v)$ under the map on the left of (5). Hence the restriction map defines an application

$$s_{v\alpha}\colon S^{(\alpha)}_F \longrightarrow E(F_v)/\alpha E(F_v).$$

We now define $R^{(\alpha)}_F$ by the exactness of the sequence

$$(6) \qquad 0 \to R^{(\alpha)}_F \to S^{(\alpha)}_F \xrightarrow{h_\alpha} \prod_{v \mid \mathfrak{p}} E(F_v)/\alpha E(F_v),$$

where h_α is given by the product of the $s_{v,\alpha}$ over all v dividing $\mathfrak{p}$.

These modified forms of the Selmer group are only interesting for suitable choices of the endomorphism α, which we now explain. For the rest of the paper, we fix an element π of $\mathcal{O}$ satisfying

$$(7) \qquad (\pi) = \mathfrak{p}^k,$$

for some integers $k \geq 1$. Put $q = \pi\pi^* = p^k$. For each positive integer n, we have the natural maps

$$(8) \qquad E_{\pi^n} \hookrightarrow E_{\pi^{n+1}},$$

$$E_{\pi^{*n+1}} \xrightarrow{\pi^*} E_{\pi^{*n}}, \tag{9}$$

the first being the natural inclusion, and the second being given by multiplication by π^*. We now pass to the limit and define

$$S_F = \varinjlim S_F^{(\pi^n)}, \quad S'_F = \varinjlim S_F'^{(\pi^n)}, \tag{10}$$

$$\mathfrak{R}_F = \varprojlim R_F^{(\pi^{*n})}, \quad \mathfrak{S}_F = \varprojlim S_F^{(\pi^{*n})}. \tag{11}$$

Here the inductive limits are taken with respect to the maps induced form (8), and the projective limits with respect to the maps induced from (9).

We introduce more notation. If X is a finite set, $\#(X)$ will denote its cardinality. Let A be an $\mathcal{O}$-module. We write

$$A(p) = \bigcup_{n\geq 1} A_{p^n}, \quad A(\mathfrak{p}) = \bigcup_{n\geq 1} A_{\pi^n}.$$

We also put $T_{\pi^*}(A) = \varprojlim A_{\pi^{*n}}$, where the projective limit is taken with respect to the maps induced by multiplication by π^*.

If v is a place of F, write k_v for the residue field of v, and Nv for the cardinality of k_v. When E has good reduction at v, we denote the reduction of E modulo v by $\tilde{E}_v/k_v$. Let ψ_F denote the Grossencharacter of E over F in the sense of Deuring-Weil. Thus ψ_F is a Grossencharacter of F, which is unramified precisely at the places where E has good reduction, and which has the property that, for each v where E has good reduction, $\psi_F(v)$ is the unique element of $\mathcal{O} = \mathrm{End}_F(E)$, whose reduction modulo v is the Frobenius endomorphism of $\tilde{E}_v$ relative to k_v. If a, b are elements of $K_{\mathfrak{p}}$, we write $a \sim b$ if $a = \epsilon b$, where ϵ is a unit in $K_{\mathfrak{p}}$; we use the same notation for elements of K, viewing them in $K_{\mathfrak{p}}$ via the canonical inclusion.

Lemma 1. *For each prime v of F dividing $\mathfrak{p}$, we have*

(i) $\#(\tilde{E}_v(k_v)) \sim 1 - \psi_F(v)/Nv$;

(ii) *Reduction modulo v induces an isomorphism from*

$$\varprojlim E(F_v)/\pi^{*n} E(F_v)$$

onto the p-primary subgroup of $\tilde{E}_v(k_v)$.

Proof. The definition of ψ_F implies easily that

$$\#(\tilde{E}_v(k_v)) = (1 - \psi_F(v))(1 - \psi_F(v)^*).$$

By the theory of complex multiplication, $\mathfrak{p}$ divides $\psi_F(v)$ because v divides $\mathfrak{p}$, and $\psi_F(v)\psi_F(v)^* = Nv$. Hence the right-hand side of the displayed expression is equal to $1 - \psi_F(v)/Nv$, up to multiplication by a unit in $K_\mathfrak{p}$. This proves (i). Let $E_{1,v}(F_v)$ denote the group of F_v-rational points in the kernel of reduction modulo v. Since v lies above $\mathfrak{p}$, (7) implies that π^{*^n} is an automorphism of $E_{1,v}(F_v)$, and so reduction modulo v induces an isomorphism

$$E(F_v)/\pi^{*^n}E(F_v) \xrightarrow{\sim} \tilde{E}_v(k_v)/\pi^{*^n}\tilde{E}_v(k_v).$$

Assertion (ii) of the lemma now follows, since one sees easily that the p-primary subgroup of $\tilde{E}_v(k_v)$ is killed by π^{*^n} for all sufficiently large n.

Putting $\alpha = \pi^{*^n}$ in (6), and passing to the projective limit, we conclude from (ii) of Lemma 1 that we have an exact sequence

$$0 \to \mathfrak{R}_F \to \mathfrak{S}_F \to \prod_{v|\mathfrak{p}} \tilde{E}_v(k_v)(p). \tag{12}$$

Proposition 2. *We have*

$$[\mathfrak{S}_F : \mathfrak{R}_F].[\text{Ш}'_F(\mathfrak{p}) : \text{Ш}_F(\mathfrak{p})] = \#\Big(\prod_{v|\mathfrak{p}} \tilde{E}_v(k_v)(p)\Big) \sim \prod_{v|\mathfrak{p}}(1 - \psi_F(v)/Nv).$$

Proof. We first recall the definition of the Weil pairing, suitably normalized for the proof of Proposition 2. If $u \in E_{\pi^n}$, let $\tilde{u}$ be the unique element of E_{π^n} satisfying $\pi^{*^n}\tilde{u} = u$. Write $\mathcal{F}'_n = F(E_{\pi^n})$. There exist rational functions $f_{n,u}$, $g_{n,u}$ on E, which are defined over $\mathcal{F}'_n$, and whose divisors are given respectively by, in an obvious notation,

$$q^n((u) - (0)), \qquad \sum_{\mu \in E_{\pi^n}} ((\tilde{u} + \mu) - (\mu)),$$

where 0 denotes the zero point of E. Since $f_{n,u} \circ \pi^{*^n}$ and $g_{n,u}^{q^n}$ have the same divisors, we can multiply $f_{n,u}$ by a non-zero element of $\mathcal{F}'_n$ so that

$$f_{n,u} \circ \pi^{*^n} = g_{n,u}^{q^n}.$$

If v is any element of $E_{\pi^{*n}}$, we define

$$w_n(u,v) = g_{n,u}(P+v)/g_{n,u}(P).$$

Plainly $w_n(u,v)$ belongs to the group μ_{q^n} of q^n-th roots of unity, and it is well known that the map

$$\rho_n\colon E_{\pi^{*n}} \longrightarrow \mathrm{Hom}(E_{\pi^n}, \mu_{q^n}) \tag{13}$$

given by $\rho_n(v)(u) = w_n(u,v)$ is an isomorphism of $G(\overline{F}/F)$-modules. Moreover, it is easy to verify that, for all $n \geq 1$, we have

$$\rho_{n+1}(v)(u) = \rho_n(\pi^* v)(u) \qquad u \in E_{\pi^n}, v \in E_{\pi^{*n+1}}. \tag{14}$$

We now begin the proof proper of Proposition 2. Noting that the image on the right hand side of (1) actually lies in the direct sum rather than the product, we define B_n by the exactness of the sequence

$$0 \to (\text{Ш}_F)_{\pi^n} \to \big(H^1(F,E)\big)_{\pi^n} \to \oplus_v \big(H^1(F_v,E)\big)_{\pi^n} \to B_n \to 0,$$

where v runs over all finite places of F. Simple diagram chasing shows that there exits a map

$$g_n\colon \prod_{v|\mathfrak{p}} \big(H^1(F_v,E)\big)_{\pi^n} \longrightarrow B_n$$

such that the sequence

$$0 \to (\text{Ш}_F)_{\pi^n} \to (\text{Ш}'_F)_{\pi^n} \to \prod_{v|\mathfrak{p}} \big(H^1(F_v,E)\big)_{\pi^n} \xrightarrow{g_n} B_n \tag{15}$$

is exact. We can now pass to the inductive limit in the obvious sense. It follows that the index of $\text{Ш}_F(\mathfrak{p})$ in $\text{Ш}'_F(\mathfrak{p})$ is the order of the kernel of g, where $g = \varinjlim g_n$ is the induced map between $\prod_{v|\mathfrak{p}} H^1(F_v,E)(\mathfrak{p})$ and $B = \varinjlim B_n$. To calculate the order of this kernel, we dualize the map g_n. In general, if A is a discrete, we write $\hat{A}$ for the Pontrjagin dual of A, and $\hat{f}$ for the dual of a homomorphism between two such groups. By Tate local duality, the dual of the discrete group $H^1(F_v,E)$ is $E(F_v)$, and this in turn induces a duality between $\big(H^1(F_v,E)\big)_{\pi^n}$ and $E(F_v)/\pi^{*n}E(F_v)$. Moreover, a theorem of Cassels ([1], p. 198) shows that the dual of B_n is canonically isomorphic to $S_F^{(\pi^{*n})}$. Further, an analysis of the proof in [1] shows that : – (i). When $E_{\pi^{*n}}$ is identified with $\mathrm{Hom}(E_{\pi^n}, \mu_{q^n})$ via the map ρ_n above, then $\hat{g}_n$ is precisely the map $h_{\pi^{*n}}$ of (6); (ii). By virtue of (14), the dual of the natural map from B_n to B_{n+1} is the map from $S_F^{(\pi^{*n+1})}$ to $S^{(\pi^{*n})}$ induced by multiplication by π^*. Hence $\hat{g} = \varprojlim \hat{g}_n$ is none other than the map on the right of the exact sequence (12). Thus

$$\#(\mathrm{Ker}\, g) = \#(\mathrm{Coker}\, \hat{g}) = \#\Big(\prod_{v|\mathfrak{p}} \tilde{E}_v(k_v)(p)\Big)/[\mathfrak{S}_F : \mathfrak{R}_F].$$

This completes the proof of Proposition 2.

To end this section, we analyse the natural exact sequences associated with the groups $\mathfrak{S}_F$ and $\mathfrak{R}_F$, which are defined via passage to the projective limit. Putting $\alpha = \pi^{*^n}$ in (3), and passing to the projective limit, we obtain the exact sequence

$$0 \to E(F) \oplus_{\mathfrak{O}} \mathfrak{O}_{\mathfrak{p}^*} \to \mathfrak{S}_F \to T_{\pi^*}(\text{Ш}_F) \to 0. \tag{1.6}$$

Define the subgroup $E_1(F)$ of $E(F)$ via the exactness of

$$0 \to E_1(F) \to E(F) \to \prod_{v|\mathfrak{p}} \tilde{E}_v(k_v), \tag{17}$$

where the map on the right is the product of the reduction homomorphisms for all v dividing $\mathfrak{p}$. Taking the tensor product over $\mathfrak{O}$ with $\mathfrak{O}_{\mathfrak{p}^*}$, and recalling that we can identify $\tilde{E}_v(k_v) \otimes_{\mathfrak{O}} \mathfrak{O}_{\mathfrak{p}^*}$ with the p-primary subgroup of $\tilde{E}_v(k_v)$, we obtain the exact sequence

$$0 \to E_1(F) \otimes_{\mathfrak{O}} \mathfrak{O}_{\mathfrak{p}^*} \to E(F) \otimes_{\mathfrak{O}} \mathfrak{O}_{\mathfrak{p}^*} \to \prod_{v|\mathfrak{p}} \tilde{E}_v(k_v)(p). \tag{18}$$

It follows from (12), (16) and this last sequence that

$$(E(F) \otimes_{\mathfrak{O}} \mathfrak{O}_{\mathfrak{p}^*}) \cap \mathfrak{R}_F = E_1(F) \otimes_{\mathfrak{O}} \mathfrak{O}_{\mathfrak{p}^*}, \tag{19}$$

where the two groups on the left are regarded as subgroups of $\mathfrak{S}_F$. Define Θ_F via the exact sequence

$$0 \to E_1(F) \oplus_{\mathfrak{O}} \mathfrak{O}_{\mathfrak{p}^*} \to \mathfrak{R}_F \to \Theta_F \to 0. \tag{20}$$

It follows from (16) and (19) that there is a natural injection $\Theta_F \hookrightarrow T_{\pi^*}(\text{Ш}_F)$, and the following lemma is then clear.

Lemma 3.

$$[\mathfrak{S}_F : \mathfrak{R}_F] = [E(F) \otimes_{\mathfrak{O}} \mathfrak{O}_{\mathfrak{p}^*} : E_1(F) \otimes_{\mathfrak{O}} \mathfrak{O}_{\mathfrak{p}^*}].[T_{\pi^*}(\text{Ш}_F) : \Theta_F].$$

Example 1. Take $F = K = \mathbf{Q}(\sqrt{-3})$, and E the elliptic curve defined over K by the equation $y^2 = x^3 - 2$. The prime $p = 7$ satisfies the conditions imposed on p. A simple calculation shows that $\#(\tilde{E}_{\mathfrak{p}}(k_{\mathfrak{p}})) = 7$. Moreover, $E(K)$ contains the point $P = (3, 5)$, which is certainly not in $E_1(K)$ because its coordinates are integral. We conclude from the above index calculations that, in this case, we have

$$[E(F) \otimes_{\mathfrak{O}} \mathfrak{O}_{\mathfrak{p}^*} : E_1(F) \otimes_{\mathfrak{O}} \mathfrak{O}_{\mathfrak{p}^*}] = 7,$$

and

$$[\text{Ш}'_F(\mathfrak{p}) : \text{Ш}_F(\mathfrak{p})] = [T_{\pi^*}(\text{Ш}_F) : \Theta_F] = 1.$$

2. Introduction of Iwasawa Theory

We now introduce a certain non-cyclotomic $\mathbf{Z}_p$-extension of the base field F, which will be the basic tool used in the rest of the paper to study the arithmetic of E over F. The most natural way to define this $\mathbf{Z}_p$-extension is via points of finite order on E. Let $E_{\mathfrak{p}^\infty} = \bigcup_{n \geq 1} E_{\mathfrak{p}^n}$, and define $\mathcal{F}_\infty = F(E_{\mathfrak{p}^\infty})$. The action of the Galois group of $\mathcal{F}_\infty$ over F on $E_{\mathfrak{p}^\infty}$ defines a canonical injection

$$\chi_\infty : G(\mathcal{F}_\infty/F) \hookrightarrow \mathbf{Z}_p^\times, \tag{21}$$

whose image is of finite index in $\mathbf{Z}_p^\times$. The decomposition

$$\mathbf{Z}_p^\times = \mu_{p-1} \times (1 + p\,\mathbf{Z}_p)$$

(in general, if m is an integer ≥ 1, μ_m will denote the group of m-th roots of unity) gives rise via χ_∞ to a corresponding decomposition of the Galois group $G(\mathcal{F}_\infty/F) = \Delta \times \Gamma$, where $\chi_\infty(\Delta) \subset \mu_{p-1}$ and $\chi_\infty(\Gamma) \subset (1 + p\mathbf{Z}_p)$. Thus the fixed field of Δ is a $\mathbf{Z}_p$-extension F_∞ of F, whose Galois group can be identified with Γ. It is obvious that the fixed field of Γ is the field $\mathcal{F} = F(E_{\mathfrak{p}})$. For simplicity, we write χ (respectively, κ) for the restriction of χ_∞ to Δ (respectively, Γ).

By class field theory, there is a unique $\mathbf{Z}_p$-extension of the imaginary quadratic field K, which is unramified outside $\mathfrak{p}$. We denote this $\mathbf{Z}_p$-extension by K_∞. The classical theory of complex multiplication shows immediately that the field F_∞ defined in the previous paragraph is in fact the compositum of F and K_∞. It follows that F_∞/F is unramified outside the set of primes above $\mathfrak{p}$, and that each time above $\mathfrak{p}$ is ramified in F_∞/F. Indeed, both assertions are plainly true for K_∞/K, and so they remain true for the translated $\mathbf{Z}_p$-extension F_∞/F.

We next give two important technical lemmas. If v is a prime of F_∞, we define $F_{\infty,v}$ to be the union of the completions at v of all finite extensions of $\mathbf{Q}$ contained in F_∞. Note that $F_{\infty,v}$ will not, in general, be complete.

Lemma 4. *For each prime v of F_∞ lying above $\mathfrak{p}$, the group*

$$E_{\mathfrak{p}^{*\infty}} \cap E(F_{\infty,v})$$

is finite.

Proof. We assume that $E_{\mathfrak{p}} \subset E(F)$, so that $F_\infty = F(E_{\mathfrak{p}^\infty})$, since it certainly suffices to prove the lemma in this case. As F_∞/F is a $\mathbf{Z}_p$-extension, the fact that v is ramified in F_∞/F implies that v is totally ramified in $F_\infty/F(E_{\mathfrak{p}^n})$ for all sufficiently large n. Hence the residue field $k_{\infty,v}$ of v is a finite field. Recall that, by our hypothesis on p, E has good reduction at v. Writing $\tilde{E}_v$ for the reduction of E modulo v, it follows that $\tilde{E}_v(k_{\infty,v})$ is a finite abelian group. But, as v does not divide $\mathfrak{p}^*$, it is well known that reduction modulo v induces an injection of $E_{\mathfrak{p}^{*\infty}} \cap E(F_{\infty,v})$ into $\tilde{E}_v(k_{\infty,v})$, thereby proving the lemma.

Remark. Lemma 4, together with the Weil pairing (13), shows also that $\mu_{p^\infty} \cap F^\times_{\infty,v}$ is a finite group for all v dividing $\mathfrak{p}$.

Lemma 5. *Under our hypotheses on p (i.e., p splits in K, E has good reduction at every prime above p, and $p \neq 2$), E has good reduction everywhere over the field $\mathcal{F} = F(E_{\mathfrak{p}})$.*

We omit the proof of this lemma, which is based on a slight variant of the criterion of Néron-Ogg-Safarevic (see [3], or [2] for a more detailed proof).

We now begin the study of the arithmetic of E over F_∞. Recall that $E_\mathfrak{p} \subset E(F)$ if and only if $E_{\mathfrak{p}^\infty} \subset E(F_\infty)$. Also, as always, $\Gamma = G(F_\infty/F)$.

Lemma 6. *Assume that $E_\mathfrak{p} \subset E(F)$. Then, for all $i \geq 1$, we have*

$$H^i(\Gamma, E_{\mathfrak{p}^\infty}) = 0.$$

Proof. We only sketch a proof of this well-known fact. Putting $G_n = G(F(E_{\mathfrak{p}^n})/F)$, we have

$$H^i(\Gamma, E_{\mathfrak{p}^\infty}) = \varinjlim H^i(G_n, E_{\mathfrak{p}^n}).$$

Now $E_{\mathfrak{p}^n}$ is a cyclic group of order p^n with $p \neq 2$, and, as G_n is a group of automorphisms of $E_{\mathfrak{p}^n}$, it follows that G_n is also cyclic. A simple direct calculation then shows that $H^i(G_n, E_{\mathfrak{p}^n}) = 0$ for all $i \geq 1$.

Proposition 7. *The group $E(F_\infty)$, modulo its torsion subgroup, is a free abelian group.*

Proof. We use an entirely analogous argument to Iwasawa ([9], p. 270), who first proved the cyclotomic analogue. Note also that the proposition holds in greater generality ([11], p. 404). We can suppose that $E_\mathfrak{p} \subset E(F)$, since a subgroup of a free abelian group is itself free. Let $\Gamma_n = \Gamma^{p^n}$, let F_n be the fixed field of Γ_n, W_n the torsion subgroup of $E(F_n)$, and $V_n = E(F_n)/W_n$. Define W_∞, V_∞ in a similar manner for F_∞. Taking Γ_n-cohomology of the exact sequence

$$0 \to W_\infty \to E(F_\infty) \to V_\infty \to 0,$$

we obtain the exact sequence

$$0 \to V_n \to V_\infty^{\Gamma_n} \to H^1(\Gamma_n, W_\infty).$$

Now Lemmas 4 and 6 show that $H^1(\Gamma_n, W_\infty)$ is a finite group. Since the Mordell-Weil theorem shows that V_n is finitely generated, it follows that

$V_\infty^{\Gamma_n}$ is a finitely generated free abelian group for all $n \geq 0$. If $m \geq n$, a simple argument (see [11], p. 403) proves that $V_\infty^{\Gamma_m}/V_\infty^{\Gamma_n}$ is torsion free. Hence, if $m \geq n$, $V_\infty^{\Gamma_n}$ is in fact a direct summand of $V_\infty^{\Gamma_m}$, and the assertion of the proposition follows easily.

We can now repeat the definitions of the Tate-Safarevic and Selmer groups given in §1, simply replacing F in the earlier definitions by the field F_∞. In particular, we now write

$$(22) \qquad S_{F_\infty} = \varinjlim S_{F_\infty}^{(\pi^n)}, \quad S'_{F_\infty} = \varinjlim S'^{(\pi^n)}_{F_\infty}.$$

In fact, unlike the situation over the base field F, it is immaterial over F_∞ whether we impose local conditions at the primes above $\mathfrak{p}$.

Lemma 8. *We have* $\text{Ш}_{F_\infty}(\mathfrak{p}) = \text{Ш}'_{F_\infty}(\mathfrak{p})$, *and* $S_{F_\infty} = S'_{F_\infty}$.

Proof. It suffices to establish the first assertion, since it clearly implies the second. We must show that $\left(H^1(F_{\infty,v}, E)\right)_\pi = 0$ for all primes v of F_∞ above $\mathfrak{p}$. As before, let F_n be the fixed field of Γ_n. Since, by definition, $F_{\infty,v}$ is the union of the $F_{n,v}$ for all $n \geq 0$, we see easily that

$$H^1(F_{\infty,v}, E) = \varinjlim H^1(F_{n,v}, E),$$

where the inductive limit is taken relative to the restriction maps. By Tate local duality, $\left(H^1(F_{n,v}, E)\right)_\pi$ is dual to $D_n = E(F_{n,v})/\pi^* E(F_{n,v})$, and the restriction maps are dual to the norm maps. Now, as remarked in the proof of Lemma 1, reduction modulo v defines an isomorphism $D_n \xrightarrow{\sim} \tilde{E}_v(k_{n,v})/\pi^* \tilde{E}_v(k_{n,v})$, where $k_{n,v}$ is the residue field of the restriction of v to F_n. As v is totally ramified in F_∞/F_n for all large n, we conclude that, for all large n, the order of D_n is fixed and Γ_n operates trivially on D_n. Hence the projective limit of the D_n with respect to the norm maps is 0, and the proof of the lemma is complete.

We next show how one can recover S'_F from $S'_{F_\infty} = S_{F_\infty}$. Since taking cohomology commutes with direct limits, we have

$$S'_F \hookrightarrow H^1(F, E_{\mathfrak{p}^\infty}), \quad S'_{F_\infty} \hookrightarrow H^1(F_\infty, E_{\mathfrak{p}^\infty}).$$

Theorem 9. *The restriction map on cohomology induces an isomorphism*

$$S'_F \xrightarrow{\sim} (S'_{F_\infty})^\Gamma.$$

Proof. We first remark that it suffices to prove the result when $E_{\mathfrak{p}} \subset E(F)$. Indeed, if this is not the case, the restriction maps induce isomorphisms

$$S'_F \xrightarrow{\sim} (S'_{\mathcal{F}})^{\Delta}, \quad S'_{F_\infty} \xrightarrow{\sim} (S'_{\mathcal{F}_\infty})^{\Delta},$$

because the order of Δ is prime to p, and so, for any Δ-module A, $H^i(\Delta, A)$ has no p-torsion for all $i \geq 1$.

We now assume that $E_{\mathfrak{p}} \subset E(F)$. It follows from Lemma 6, and the usual restriction-inflation exact sequence, that the restriction map yields an isomorphism

$$H^1(F, E_{\mathfrak{p}^\infty}) \xrightarrow{\sim} H^1(F_\infty, E_{\mathfrak{p}^\infty})^{\Gamma}. \tag{23}$$

Let v be any place of F_∞, and w the restriction of v to F. We plainly have a commutative diagram

$$\begin{array}{ccc} H^1(F_\infty, E_{\mathfrak{p}^\infty}) & \xrightarrow{\lambda_{\infty,v}} & H^1(F_{\infty,v}E)(\mathfrak{p}) \\ i\uparrow & & \uparrow i_v \\ H^1(F, E_{\mathfrak{p}^\infty}) & \xrightarrow{\lambda_w} & H^1(F_w, E)(\mathfrak{p}), \end{array}$$

where the vertical maps are the restriction maps, and where the horizontal maps are the appropriate localisation homomorphisms (e.g., λ_w is the restriction homomorphism to $H^1(F_w, E_{\mathfrak{p}^\infty})$, followed by the canonical map from this group to $H^1(F_w, E)(\mathfrak{p})$). Since

$$S'_F = \bigcap_{w \nmid \mathfrak{p}} \operatorname{Ker} \lambda_w, \quad S_{F_\infty} = \bigcap_{v \nmid \mathfrak{p}} \operatorname{Ker} \lambda_{\infty,v},$$

it is clear from this diagram that i defines an injection of S'_F into $(S'_{F_\infty})^{\Gamma}$. To prove that this map is surjective, it plainly suffices, in view of (23), to show that i_v is injective for all places v of F_∞ not lying above $\mathfrak{p}$. Let v be such a place of F_∞. Since we have assumed $E_{\mathfrak{p}} \subset E(F)$, Lemma 5 shows that E has good reduction at v. Now the kernel of i_v is equal to $H^1(G(F_{\infty,v}/F_w), E(F_{\infty,v}))(\mathfrak{p})$, and this cohomology group is zero when v does not divide $\mathfrak{p}$, because $F_{\infty,v}/F_w$ is unramified and E has good reduction at w (see Corollary 4.4 on p. 204 of [10]). This completes the proof of the theorem.

Corollary 10. *$E(F_\infty)$ is a torsion group if and only if both $E(F)$ is torsion and the restriction map from $\text{Ш}'_F(\mathfrak{p})$ to $\text{Ш}'_{F_\infty}(\mathfrak{p})$ is injective.*

Proof. Passing to the inductive limit, relative to the maps induced from (8), in the standard exact sequence

$$0 \to E(F)/\pi^n E(F) \to S_F'^{(\pi^n)} \to (\text{Ш}'_F)_{\pi^n} \to 0,$$

we obtain the exact sequence

$$(24) \qquad 0 \to E(F) \otimes_{\mathfrak{O}} D_{\mathfrak{p}} \to S'_F \to \text{Ш}'_F(\mathfrak{p}) \to 0,$$

where, as before $D_{\mathfrak{p}} = K_{\mathfrak{p}}/\mathfrak{O}_{\mathfrak{p}}$. The same sequence is valid if we replace F by F_∞, and taking invariants under $\Gamma = G(F_\infty/F)$ gives the exact sequence

$$0 \to (E(F_\infty) \otimes_{\mathfrak{O}} D_{\mathfrak{p}})^\Gamma \to (S'_{F_\infty})^\Gamma \to (\text{Ш}'_{F_\infty}(\mathfrak{p}))^\Gamma.$$

Letting ρ_F denote the restriction map from $\text{Ш}'_F(\mathfrak{p})$ to $\text{Ш}'_{F_\infty}(\mathfrak{p})$, we conclude from Theorem 9 that we have an exact sequence

$$0 \to E(F) \otimes_{\mathfrak{O}} D_{\mathfrak{p}} \to (E(F_\infty) \otimes_{\mathfrak{O}} D_{\mathfrak{p}})^\Gamma \to \operatorname{Ker} \rho_F \to 0.$$

Now if B is any discrete p-primary Γ-module, it is a standard elementary fact that $B = 0$ if and only if $B^\Gamma = 0$. Hence Corollary 10 follows from the above exact sequence, on recalling that $E(F) \otimes_{\mathfrak{O}} D_{\mathfrak{p}}$ (respectively, $E(F_\infty) \otimes_{\mathfrak{O}} D_{\mathfrak{p}}$) is zero if and only if $E(F)$ (respectively, $E(F_\infty)$) is a torsion group.

We note the following description of the kernel of the restriction map

$$\rho_F \colon \text{Ш}'_F(\mathfrak{p}) \longrightarrow \text{Ш}'_{F_\infty}(\mathfrak{p}).$$

Lemma 11. *We have a canonical isomorphism*

$$\operatorname{Ker} \rho_F \xrightarrow{\sim} H^1(\Gamma, E(F_\infty))(\mathfrak{p}).$$

Proof. Since the extension $F(E_{\mathfrak{p}})/F$ has degree prime to p, one sees easily that it suffices to prove the lemma when $E_{\mathfrak{p}} \subset E(F)$. Assuming this to be the case, Lemma 5 shows that E has good reduction everywhere over F. The assertion of the lemma now follows immediately from the definition of the modified Tate-Safarevic groups, the standard restriction-inflation exact sequence, and the fact (see the proof of Theorem 9) that, for each place v of F_∞ not lying above $\mathfrak{p}$, we have $H^1(G(F_{\infty,v}/F_v), E(F_{\infty,v})) = 0$.

We now give an alternative description of the Selmer group $S_{F_\infty} = S'_{F_\infty}$, which is often more useful. As before, let $\mathcal{F}_\infty = F(E_{\mathfrak{p}^\infty})$, and write $\mathfrak{M}_\infty$ for the maximal abelian p-extension of $\mathcal{F}_\infty$, which is unramified outside the primes above $\mathfrak{p}$. Write

$$(25) \qquad X_\infty = G(\mathfrak{M}_\infty/F_\infty).$$

Since $\mathfrak{M}_\infty$ is plainly Galois over F, the Galois group $G(\mathcal{F}_\infty/F)$ operates on X_∞ via inner automorphisms in the standard manner (i.e. if $\sigma \in G(\mathcal{F}_\infty/F)$, pick $\tau \in G(\mathfrak{M}_\infty/F)$ whose restriction to $\mathcal{F}_\infty$ is σ, and define $\sigma \cdot x = \tau x \tau^{-1}$ for $x \in X_\infty$). Recall that $G(\mathcal{F}_\infty/F) = \Delta \times \Gamma$, and that χ denotes the restriction of χ_∞ to Δ. As the order of Δ, which we denote by d, is prime to p, each $\mathbf{Z}_p[\Delta]$-modulo A decomposes as a direct sum

$$A = \bigoplus_{i \bmod d} A^{(\chi^i)},$$

where $A^{(\chi^i)}$ denotes the subspace of A on which Δ acts via χ^i. Finally, if A, B are $G(\mathcal{F}_\infty/F)$-modules, we always endow $\mathrm{Hom}(A, B)$ with the $G(\mathcal{F}_\infty/F)$-structure given by $(\sigma f)(a) = \sigma f(\sigma^{-1}a)$ for $\sigma \in G(\mathcal{F}_\infty/F)$.

Theorem 12. *There are canonical $G(F_\infty/F)$-isomorphisms*

$$S_{F_\infty} \xrightarrow{\sim} \mathrm{Hom}_\Delta(X_\infty, E_{\mathfrak{p}^\infty}) \xrightarrow{\sim} \mathrm{Hom}(X_\infty^{(\chi)}, E_{\mathfrak{p}^\infty}).$$

Proof. Note that the second iomorphism is obvious. Also, by an entirely similar reasoning to that given in the proof of Theorem 9, we can assume that $E_\mathfrak{p} \subset E(F)$, so that $E_{\mathfrak{p}^\infty}$ is rational over F_∞. Put $V_\infty = \mathrm{Hom}(X_\infty, E_{\mathfrak{p}^\infty})$. Hence both V_∞ and $S_{\mathcal{F}_\infty}$ are subgroups of

$$\mathrm{Hom}(G(\overline{F}_\infty/F_\infty), E_{\mathfrak{p}^\infty}).$$

We first show that $V_\infty \subset S_{\mathcal{F}_\infty}$. Since $E_\mathfrak{p} \subset E(F)$, Lemma 5 shows that E has good reduction everywhere over F. Let v be any finite place of F_∞, let $H_{\infty,v}$ denote the maximal unramified extension of $F_{\infty,v}$, and let $D_v = G(H_{\infty,v}/F_{\infty,v})$. Since E has good reduction at v, we have

$$H^1(D_v, E(H_{\infty,v})) = 0 \tag{26}$$

(see p. 204 of [10]). Suppose now that $x \in V_\infty$. By the definition of X_∞, for all places v of F_∞ not lying above $\mathfrak{p}$, the restriction of x to the decomposition group of v factors through the Galois group D_v. It then follows easily from (26) that the image of x in $H^1(F_{\infty,v}, E)$ is necessarily 0 for all v which do not divide $\mathfrak{p}$, whence $x \in S_{\mathcal{F}_\infty}$.

To prove that $S_{F_\infty} \subset V_\infty$, let n be an integer ≥ 1, and take $x_n \in S_{F_\infty}^{(\pi^n)} \subset \mathrm{Hom}(G(\overline{F}_\infty/F_\infty), E_{\pi^n})$. By the definition of the Selmer group, for each finite place v of F_∞, the restriction $x_{n,v}$ of x_n to the decomposition group of v is contained in the image of $E(F_{\infty,v})/\pi^n E(F_{\infty,v})$ in $\mathrm{Hom}(G(\overline{F}_{\infty,v}/F_{\infty,v}), E_{\pi^n})$. Suppose now that v does not lie above $\mathfrak{p}$. Since E has good reduction at v and $(v,\pi) = 1$, E_{π^n} is mapped injectively under reduction modulo v, whence it follows easily that, if $x_{n,v} = P_{n,v}$ modulo $\pi^n E(F_{\infty,v})$, the extension $F_{\infty,v}(Q_{n,v})/F_{\infty,v}$ is unramified, where $Q_{n,v}$ satisfies $\pi^n Q_{n,v} = P_{n,v}$. In other words, $x_{n,v}$ must factor through the Galois group D_v of the maximal unramified extension of $F_{\infty,v}$. Hence the global element x_n factors through $\mathrm{Hom}(X_\infty, E_{\pi^n})$. Thus $S_{F_\infty}^{(\pi^n)} \subset \mathrm{Hom}(X_\infty, E_{\pi^n})$, and, passing to the direct limit, we obtain $S_{F_\infty} \subset V_\infty$. This completes the proof of Theorem 12.

We can now use global class field theory to study the $G(\mathcal{F}_\infty/F)$-module X_∞. Let $\Lambda = \mathbf{Z}_p[[T]]$ be the ring of formal series in an indeterminate T with coefficients in $\mathbf{Z}_p$. As usual in the theory of $\mathbf{Z}_p$-extensions (see [9]), it is best to interpret the Γ-action on the compact $\mathbf{Z}_p$-module X_∞ as a Λ-action, in the following manner. Fix once and for all a topological generator γ_o of Γ, and then defne $T.x = (\gamma_o - 1)x$ for x in X_∞. This extends to an action of $\mathbf{Z}_p[T]$ on X_∞ in the obvious fashion, and finally to an action of Λ by continuity.

Lemma 13. *The Galois group X_∞ is a finitely generated Λ-module.*

Proof. The argument being standard (see [9], p. 255), we only give it very briefly. Write $\mathcal{F}_n$ for the n-th layer of the $\mathbf{Z}_p$-extension $\mathcal{F}_\infty/\mathcal{F}$, and $\mathfrak{M}_n$ for the maximal abelian p-extension of $\mathcal{F}_n$, which is unramified outside $\mathfrak{p}$. Since $\mathfrak{M}_n$ is also plainly the maximal abelian extension of $\mathcal{F}_n$ contained in $\mathfrak{M}_\infty$, a simple calculation with commutators shows that

$$X_\infty/\omega_n X_\infty = G(\mathfrak{M}_n/\mathcal{F}_\infty), \tag{27}$$

where $\omega_n = (1+T)^{p^n} - 1$. On the other hand, global class field theory gives the following explicit description of $G(\mathfrak{M}_n/\mathcal{F}_n)$ as a $G(\mathcal{F}_n/F)$-module. If v is a place of $\mathcal{F}_n$ above $\mathfrak{p}$, let $U_{n,v}$ denote the local units $\equiv 1 \bmod v$ of the completion of $\mathcal{F}_n$ at v. Write $\mathcal{E}_n$ for the group of global units of $\mathcal{F}_n$, which are $\equiv 1 \bmod v$ for each place v of $\mathcal{F}_n$ above $\mathfrak{p}$. Let i_n be the diagonal embedding of $\mathcal{E}_n$ in $\prod_{v|\mathfrak{p}} U_{n,v}$. Then Artin's reciprocity law gives the exact sequence of $G(\mathcal{F}_n/F)$-modules

$$0 \to \left(\prod_{v|\mathfrak{p}} U_{n,v}\right)/\overline{i_n(\mathcal{E}_n)} \to G(\mathfrak{M}_n/\mathcal{F}_n) \to A_n \to 0, \tag{28}$$

where A_n denotes the p-primary subgroup of the ideal class group of $\mathcal{F}_n$, and where the bar over $i_n(\mathcal{E}_n)$ signifies its closure in the $\mathfrak{p}$-adic topology. Taking $n = 0$ in particular, we conclude that X_∞/TX_∞ is a finitely generated $\mathbf{Z}_p$-module, whence a standard argument shows that X_∞ is a finitely generated Λ-module. This completes the proof of the lemma.

Remark. It follows easily from Theorem 9, Theorem 12, and Lemma 13 that the Pontrjagin dual of $\text{Ш}_F(p)$ is a finitely generated $\mathbf{Z}_p$-module. In fact, this is well known to be true for arbitrary abelian varieties defined over a number field.

We write r_n for the $\mathbf{Z}_p$-rank of $\overline{i_n(\mathcal{E}_n)}$ modulo torsion, so that

$$r_n \leq [\mathcal{F}_n : K] - 1$$

by Dirichlet's theorem. Let $\delta_n = [\mathcal{F}_n : K] - 1 - r_n$.

Lemma 14. *(i) X_∞ is Λ-torsion if and only if δ_n is bounded as $n \to \infty$; (ii) If $\delta_o = 0$, then X_∞ is Λ-torsion.*

Proof. Let ρ_n be the $\mathbf{Z}_p$-rank of $X_\infty/\omega_n X_\infty$ modulo torsion. It follows easily from the structure theory of finitely generated Λ-modules that X_∞ is Λ-torsion if and only if ρ_n is bounded as $n \to \infty$. Assertion (i) is now clear, because (27) and (28) show that $\rho_n = \delta_n$ for all $n \geq 0$. It is also plain from the structure theory that X_∞ must be Λ-torsion if X_∞/TX_∞ is finite (i.e., $\delta_o = 0$). This gives (ii), and completes the proof of the lemma.

In view of Lemma 14, it is natural to introduce the following weak form of Leopoldt's conjecture on the p-adic independence of global units in number fields.

Conjecture A (Weak $\mathfrak{p}$-adic Leopoldt conjecture for $\mathcal{F}_\infty$). *δ_n is bounded as $n \to \infty$.*

We remark that Iwasawa [9] has unconditionally proven the analogue of Conjecture A for the cyclotomic $\mathbf{Z}_p$-extension of an arbitrary totally real number field (even though Leopoldt's conjecture remains unproven in this case). While we have not succeeded in establishing Conjecture A in general, we emphasize that it is valid for a fairly wide class of elliptic curves with complex multiplication.

Proposition 15. *Conjecture A is valid if $F(E_\mathfrak{p})$ is abelian over K. In particular, Conjecture A holds when $F(E_{tor})$ is abelian over K, where E_{tor} denotes the torsion subgroup of $E(\overline{F})$.*.

For example, it follows from Proposition 15 that Conjecture A is valid if the base field F coincides with the field of complex multiplication K. Although we do not give details here, two proofs of Proposition 15 are known. Firstly, when $F(E_\mathfrak{p})$ is abelian over K, the p-adic analogue of Baker's theorem on linear forms in the p-adic logarithms of algebraic numbers can be used to prove that $\delta_n = 0$ for all $n \geq 0$. Secondly, when $F(E_{tor})$ is abelian over K, a generalization of the techniques of [4] can also be used to prove Conjecture A (but not the stronger assertion that $\delta_n = 0$ for all $n \geq 0$).

We now derive on important consequence of Conjecture A.

Theorem 16. *Conjecture A implies that $E(F_\infty)$ modulo its torsion subgroup is a finitely generated abelian group.*

Proof. Let H denote $E(F_\infty)$ modulo its torsion subgroup, and put $B = H \otimes_\mathfrak{O} \mathfrak{O}_\mathfrak{p}$. Since Proposition 7 shows that H is a free abelian group, one sees easily that, in order to prove that H is finitely generated, it suffices to show that B is a finitely generated $\mathfrak{O}_\mathfrak{p}$-module. Recall that, if N is a discrete abelian group, $\hat{N}$ denotes the Pontrjagin dual of N. By definition, the Pontrjagin dual of $E(F_\infty) \otimes_\mathfrak{O} D_\mathfrak{p}$ is equal to

$$\text{(29)} \qquad \mathrm{Hom}_{\mathfrak{O}_\mathfrak{p}}(E(F_\infty) \otimes_\mathfrak{O} D_\mathfrak{p}, D_\mathfrak{p}) \xrightarrow{\sim} \mathrm{Hom}_\mathfrak{O}(H, \mathfrak{O}_\mathfrak{p}) \xrightarrow{\sim} \mathrm{Hom}_{\mathfrak{O}_\mathfrak{p}}(B, \mathfrak{O}_\mathfrak{p}).$$

Dualizing the canonical inclusion $E(F_\infty) \otimes_\mathfrak{O} D_\mathfrak{p} \hookrightarrow S_{F_\infty}$, we obtain a surjection

$$\hat{S}_{F_\infty} \to \mathrm{Hom}_{\mathfrak{O}_\mathfrak{p}}(B, \mathfrak{O}_\mathfrak{p}).$$

Now, assuming Conjecture A, X_∞ is Λ-torsion, and so Theorem 12 shows that $\hat{S}_{F_\infty}$ is Λ-torsion. It then follows from the structure theory for Noetherian torsion Λ-modules that the $\mathbf{Z}_p$-rank of $\hat{S}_{F_\infty}$ modulo torsion is finite, whence also $\mathrm{Hom}_{\mathfrak{O}_\mathfrak{p}}(B, \mathfrak{O}_\mathfrak{p})$ has finite $\mathbf{Z}_p$-rank. Since H is a free abelian group, we deduce that B itself must be a finitely generated $\mathfrak{O}_\mathfrak{p}$-module, as required. This completes the proof of Theorem 16.

3. Canonical p-adic Height and its Connexion with Kummer Theory

The remarkable result given in this section (Theorem 17) is due to Bernadette Perrin-Riou [12]. Using quite different methods, P. Schneider has subsequently established an analogous result for arbitrary abelian varieties and for $\mathbf{Z}_p$-extensions of the base field for which the abelian variety is ordinary. We simply state the result, and refer the reader to [12] for a detailed account of the proof.

We begin by rapidly describing the p-adic analogue of the Néron-Tate canonical height, which is attached to E and our $\mathbf{Z}_p$-extension F_∞/F. Take a generalized Weierstrass model for E over F

$$y^2 + a_1xy + a_3y = x^3 + a_2x^2 + a_4x + a_6.$$

If v is a finite place of F, we write ord_v for the order valuation of v, normalized so that $\mathrm{ord}_v(F_v^\times) = \mathbf{Z}$. We always suppose that $E(F_v)$ is endowed with its natural v-adic topology. It is a well known and elementary fact, due to Néron and Tate, that there exists a unique function

$$\lambda_v : E(F_v) \setminus \{0\} \longrightarrow \mathbf{Q},$$

which is continuous when $\mathbf{Q}$ is endowed with the real topology, and which satisfies the following two properties :– (i) the limit as P tends to 0 of the expression $\lambda_v(P) - \mathrm{ord}_v(t(P))$ exists for every local parameter t at 0; (ii) for all points $P \neq 0$, $Q \neq 0$ in $E(F_v)$ with $P \pm Q \neq 0$, we have

$$\lambda_v(P+Q) + \lambda_v(P-Q) - 2\lambda_v(P) - 2\lambda_v(Q) = \frac{1}{6}\mathrm{ord}_v\Big(\big(x(P)-x(Q)\big)^6/\Delta\Big),$$

where Δ denotes the discriminant of our Weierstrass equation. Next suppose that v is a place of F dividing $\mathfrak{p}$, and let $E_{1,v}(F_v)$ denote the kernel of reduction modulo v on E. As always, $K_{\mathfrak{p}}$ will denote the completion of K at $\mathfrak{p}$, and we normalize the p-adic logarithm by setting $\log_p(p) = 0$. It is shown in [12] that there exists a unique function

$$\eta_v \colon E_{1,v}(F_v) \setminus \{0\} \longrightarrow K_{\mathfrak{p}},$$

which is continuous when $K_{\mathfrak{p}}$ is endowed with the $\mathfrak{p}$-adic topology, and which satisfies the following two properties:- (i) for all points $P \neq 0$, $Q \neq 0$ in $E_{1,v}(F_v)$ with $P \pm Q \neq 0$, we have

$$\eta_v(P+Q)+\eta_v(P-Q)-2\eta_v(P)-2\eta_v(Q) = \frac{1}{6}\log_p\Big(N_{F_v|K_{\mathfrak{p}}}\big(x(P)-x(Q)\big)^6/\Delta\Big),$$

where $N_{F_v|K_{\mathfrak{p}}}$ denotes the local norm from F_v to $K_{\mathfrak{p}}$; (ii) for all $\alpha \neq 0$ in $\mathfrak{o}$ and all $P \neq 0$ in $E_{1,v}(F_v)$ with $\alpha P \neq 0$, we have

$$\eta_v(\alpha P) = (\deg\alpha)\eta_v(P) + \frac{1}{12}\log(N_{F_v|K_{\mathfrak{p}}}\xi_\alpha(P)),$$

where $\deg\alpha$ denotes the number of elements in the kernel of α, and where

$$\xi_\alpha(P) = \alpha^{12}\prod_{Q\in E_\alpha\setminus\{0\}}\big(x(P)-x(Q)\big)^6/\Delta.$$

As in the case of the Néron-Tate height, we can combine these local factors to define a global height height function. Since the theory of complex multiplication shows that F necessarily contains the Hilbert class field of K, it follows that the norm from F to K of each finite place v of F is necessarily principal, say $N_{F/K}v = (\tau_v)$, where the generator τ_v is well defined up to multiplication by a root of unity in K. Recall that $E_1(F)$ is the subgroup of finite index in $E(F)$ consisting of those points which belong to the kernel of reduction modulo v for each place v of F dividing $\mathfrak{p}$. We define

$$h_{\mathfrak{p}}: E_1(F) \longrightarrow K_{\mathfrak{p}}$$

by setting $h_{\mathfrak{p}}(0) = 0$, and otherwise

$$h_{\mathfrak{p}}(P) = \sum_{v} \lambda_v(P) \log_p(\tau_v) - \sum_{v|\mathfrak{p}} \eta_v(P).$$

This definition does not depend on the choice of the τ_v, and one sees easily that $h_{\mathfrak{p}}$ is a quadratic function on $E_1(F)$, which also satisfies

$$h_{\mathfrak{p}}(\alpha P) = (\deg \alpha) h_{\mathfrak{p}}(P) \qquad \text{for all } \alpha \in \mathfrak{o}. \tag{30}$$

Plainly $h_{\mathfrak{p}}$ has a unique extension to a quadratic function on $E(F)$, which we also denote by $h_{\mathfrak{p}}$. It is clear that (30) remains valid for this extension. Finally, it is an exercise in linear algebra to prove that there is a unique bilinear form

$$\langle\, ,\, \rangle_{\mathfrak{p}}: E(F) \times E(F) \longrightarrow K_{\mathfrak{p}} \tag{31}$$

such that $h_{\mathfrak{p}}(P) = \langle P, P \rangle_{\mathfrak{p}}$ for all $P \in E(F)$, and whose behaviour with respect to endomorphisms $\alpha \in \mathfrak{o}$ is given by

$$\langle \alpha P, Q \rangle_{\mathfrak{p}} = \alpha \langle P, Q \rangle_{\mathfrak{p}}, \quad \langle P, \alpha Q \rangle_{\mathfrak{p}} = \alpha^* \langle P, Q \rangle_{\mathfrak{p}}.$$

This biliner form $\langle\, ,\, \rangle_{\mathfrak{p}}$ is our desired analogue of the Néron-Tate height. In particular, it can be used to define the $\mathfrak{p}$-adic volume $\mathrm{Vol}_{\mathfrak{p}}(E(F))$ of $E(F)$ by

$$\mathrm{Vol}_{\mathfrak{p}}(E(F)) = \frac{\det \langle z_i, z_j \rangle_{\mathfrak{p}}}{[E(F) : \sum_{i=1}^{n_F} \mathfrak{o} z_i]}, \tag{32}$$

where n_F denotes the $\mathfrak{o}$-rank of $E(F)$, and $z_1, \ldots, z_{n_F}$ is any maximal $\mathfrak{o}$-linearly independent set in $E(F)$. Although we do not go into details here, it is theoretically quite straightforward to calculate $\mathrm{Vol}_{\mathfrak{p}}(E(F))$ once the Mordell-Weil group itself is known. The calculations made so far support the conjecture that $\mathrm{Vol}_{\mathfrak{p}}(E(F))$ is always non-zero.

We now explain the connection between this bilinear form $\langle\, , \rangle_{\mathfrak{p}}$, and the theory of infinite descent, which was discovered by Bernadette Perrin-Riou [12]. Let

$$i_F \colon E_1(F) \otimes_{\mathfrak{O}} \mathfrak{o}_{\mathfrak{p}}. \hookrightarrow \mathfrak{R}_F = \varprojlim R_F^{(\pi^{*n})}$$

denote the natural inclusion in the exact sequence (20). On the other hand, the composition of the natural inclusion of $E(F) \otimes_{\mathfrak{O}} D_{\mathfrak{p}}$ in S'_F with the injection of S'_F into S_{F_∞} given by restriction (see Theorem 9) gives a canonical injection of $E(F) \otimes_{\mathfrak{O}} D_{\mathfrak{p}}$ in S_{F_∞}. Dualizing this map, we get (cf. (29)) a canonical surjection

$$\hat{S}_{F_\infty} \hookrightarrow \widehat{E(F) \otimes_{\mathfrak{O}} D_{\mathfrak{p}}} = \mathrm{Hom}_{\mathfrak{o}_{\mathfrak{p}}}(E(F) \otimes_{\mathfrak{O}} \mathfrak{o}_{\mathfrak{p}}, \mathfrak{o}_{\mathfrak{p}}).$$

Finally, taking the map induced between the Γ-invariants, we obtain a canonical map

$$j_F \colon (\hat{S}_{F_\infty})^{\Gamma} \longrightarrow \mathrm{Hom}_{\mathfrak{o}_{\mathfrak{p}}}(E(F) \otimes_{\mathfrak{O}} \mathfrak{o}_{\mathfrak{p}}, \mathfrak{o}_{\mathfrak{p}}),$$

which is no longer, in general, surjective. If H is an extension field of F, we write $E(H)(\mathfrak{p})$ for the subgroup of the torsion subgroup of $E(H)$ which is annihilated by all sufficiently large powers of π. Define $m_{\mathfrak{p}}$ by the formula

$$m_{\mathfrak{p}} = \frac{\#(E(\mathcal{F})(\mathfrak{p}))}{[\mathcal{F} : F]}, \tag{33}$$

where, as always, $\mathcal{F} = F(E_{\mathfrak{p}})$.

Theorem 17 (B. Perrin-Riou [12]). *There exists a canonical map*

$$\phi_F \colon \mathfrak{R}_F \longrightarrow (\hat{S}_{F_\infty})^{\Gamma} \tag{34}$$

with the following property: if we define the map

$$\theta_F \colon E_1(F) \otimes_{\mathfrak{O}} \mathfrak{o}_{\mathfrak{p}}. \longrightarrow \mathrm{Hom}_{\mathfrak{o}_{\mathfrak{p}}}(E(F) \otimes_{\mathfrak{O}} \mathfrak{o}_{\mathfrak{p}}, \mathfrak{o}_{\mathfrak{p}}) \tag{35}$$

by $\theta_F = j_F \circ \phi_F \circ i_F$, then, for all $x \in E(F)$ and $y \in E_1(F)$, we have

$$(\theta_F(y \otimes 1))(x \otimes 1) = m_{\mathfrak{p}}^{-1} \langle x, y \rangle_{\mathfrak{p}}.$$

Granted Theorem 17, it is an easy exercise to establish the following result. Recall that n_F dentes the $\mathfrak{O}$-rank of $E(F)$ modulo torsion.

Lemma 18. (i) Coker θ_F *is finite if and only if* $\mathrm{Vol}_{\mathfrak{p}}(E(F)) \neq 0$*;* (ii) *If* Coker θ_F *is finite, then*

$$\#(\mathrm{Coker}\theta_F) \sim m_{\mathfrak{p}}^{-n_F}.\mathrm{Vol}_{\mathfrak{p}}(E(F)).\#(E(F)(\mathfrak{p})).[E(F)\otimes_{\mathfrak{O}}\mathfrak{O}_{\mathfrak{p}^*} : E_1(F)\otimes_{\mathfrak{O}}\mathfrak{O}_{\mathfrak{p}^*}].$$

Finally, we shall also need the following important result, which is proven in [12].

Theorem 19 (B. Perrin-Riou [12]). *If Conjecture A is valid, then the map* ϕ_F *of Theorem 17 is an isomorphism.*

4. Analogy with the Birch-Swinnerton-Dyer Conjecture

The ideas underlying this section were first exploited by Artin and Tate [15] in their work on the geometric analogue. It seems fair to say that they lie at the heart of the study of the arithmetic of elliptic curves by the techniques of Iwasawa theory and p-adic analysis. We go a little further than previous work, in that we have avoided imposing the hypothesis that $Ш_F(\mathfrak{p})$ is finite.

If G is a p-primary abelian group, G_{div} will denote the maximal divisible subgroup of G.

Lemma 20. (i) $Ш_F(\mathfrak{p})_{\mathrm{div}} = Ш'_F(\mathfrak{p})_{\mathrm{div}}$*;* (ii) $\#(Ш'_F(\mathfrak{p})/Ш'_F(\mathfrak{p})_{\mathrm{div}}) = \#(Ш_F(\mathfrak{p})/Ш_F(\mathfrak{p})_{\mathrm{div}}).[Ш'_F(\mathfrak{p}) : Ш_F(\mathfrak{p})]$; (iii) *If* W_F *denotes the maximal divisible subgroup of* S'_F*, we have an exact sequence*

$$0 \to E(F)\otimes_{\mathfrak{O}} D_{\mathfrak{p}} \to W_F \to Ш_F(\mathfrak{p})_{\mathrm{div}} \to 0. \tag{36}$$

Proof. (i) and (ii) follow from the fact that $Ш'_F(\mathfrak{p})/Ш_F(\mathfrak{p})$ is finite by Proposition 2. Assertion (iii) is obtained by taking divisible subgroups in the exact sequence (24), and nothing that $E(F)\otimes_{\mathfrak{O}} D_{\mathfrak{p}}$ is divisible.

As before, let $\Lambda = \mathbf{Z}_p[[T]]$ be the ring of formal power series in T with coefficients in $\mathbf{Z}_p$. If A is a finitely generated Λ-torsion Λ-module, the structure theory shows that there is a Λ-homomorphism, with finite kernel and cokernel, from A to a Λ-module of the form

$$\Lambda/(f_1) \oplus \cdots \oplus \Lambda/(f_r),$$

where r is an integer ≥ 1, and $f_1, \ldots, f_r$ are non-zero elements of Λ. The characteristic power series of A is defined to be $f_A(T) = f_1 \ldots f_r$. It is uniquely determined up to multiplication by a unit in Λ. We omit the proof of the following elementary lemma on the characteristic power series of Λ-modules. We write $(A)_\Gamma = A/TA$.

Lemma 21. (i) *If* $0 \to A \to B \to C \to 0$ *is an exact sequence of finitely generated Λ-torsion Λ-modules, then* $f_B(T) = f_A(T).f_C(T)$, *up to multiplication by a unit in Λ;* (ii) *If A is a finitely generated Λ-torsion Λ-module, then $f_A(0) \neq 0$ if and ony if $(A)_\Gamma$ is finite, or equivalently A^Γ is finite. When $f_A(0) \neq 0$, we have*

$$f_A(0) \sim \#((A)_\Gamma)/\#(A^\Gamma). \tag{37}$$

Throughout the rest of the paper, we make the following assumption:–

Assumption. Conjecture A is valid (i.e. the weak $\mathfrak{p}$-adic Leopoldt conjecture holds for $\mathcal{F}_\infty = F(E_{\mathfrak{p}^\infty})$).

We now collect together in a single proposition the consequences of this assumption, which will be used later.

Proposition 22. *Conjecture A implies that* (i) $\hat{S}_{F_\infty}$ *is Λ-torsion,* (ii) $\hat{S}_{F_\infty}$ *has no non-zero finite Λ-submodules,* (iii) *the map ϕ_F of Theorem 17 induces an isomorphism* $\phi_F\colon \mathfrak{R}_F \xrightarrow{\sim} (\hat{S}_{F_\infty})^\Gamma$, *and* (iv) $\text{Ш}_F(\mathfrak{p})_{\text{div}}$ *and* $\text{Ш}_F(\mathfrak{p}^*)_{\text{div}}$ *have the same rank as $\mathbf{Z}_p$-modules.*

Proof. (i) is clear from Lemma 14 and Theorem 12. (ii) follows from Theorem 12 and R. Greenberg's [6] result that Conjecture A implies that X_∞ has no non-zero finite Λ-submodule. (iii) was already noted in Theorem 19, and is due to B. Perrin-Riou [12]. To prove (iv), we note that, as $\hat{S}_{F_\infty}$ is Λ-torsion, $(\hat{S}_{F_\infty})^\Gamma$ and $(\hat{S}_{F_\infty})_\Gamma$ have the same rank as $\mathbf{Z}_p$-modules. Assertion (iv) now follows easily from (iii), the exact sequences (20) and (24), and Theorem 9. This completes the proof of Proposition 22.

We now study the behaviour of the characteristic power series of $\hat{S}_{F_\infty}$ at the point $T = 0$. Recall that n_F denotes the $\mathfrak{o}$-rank of $E(F)$ modulo torsion. We define t_F to be the $\mathbf{Z}_p$-rank of the Pontrjagin dual of $\text{Ш}_F(\mathfrak{p})_{\text{div}}$ (thus $\text{Ш}_F(\mathfrak{p})$ is finite if and only if $t_F = 0$). Put

$$r_F = n_F + t_F. \tag{38}$$

It follows from the exact sequence (36) that $\hat{W}_F$, the dual of the divisible subgroup of S'_F, is a free $\mathbf{Z}_p$-module of rank r_F. On composing the inclusion of W_F in S'_F with the restriction map from S'_F to S_{F_∞} (see Theorem 9), we obtain a canonical injection of W_F in S_{F_∞}. Dualizing this map, we obtain a canonical surjection of $\hat{S}_{F_\infty}$ onto $\hat{W}_F$. We then define the Λ-module Y by the exactness of the sequence

$$0 \to Y \to \hat{S}_{F_\infty} \to \hat{W}_F \to 0. \tag{39}$$

Let $f(T)$ denote the characteristic power series of $\hat{S}_{F_\infty}$, and $g(T)$ the characteristic power series of Y. We claim that, after multiplying $f(T)$ by a unit if necessary, we have

$$f(T) = T^{r_F} g(T). \tag{40}$$

This is clear from the exact sequence (39) and (i) of Lemma 21, because the characteristic power series of $\hat{W}_F$ is plainly T^{r_F}.

Taking Γ-invariants of (39), we obtain the long exact sequence

$$0 \to Y^\Gamma \to (\hat{S}_{F_\infty})^\Gamma \to \hat{W}_F \to (Y)_\Gamma \to (\hat{S}_{F_\infty})_\Gamma \to \hat{W}_F \to 0. \tag{41}$$

We can simplify this sequence in two ways. Firstly, recalling that Theorem 9 shows that $(\hat{S}_{F_\infty})_\Gamma$ is canonically isomorphic to S'_F, and nothing the exact sequence

$$0 \to N_F \to S'_F \to \hat{W}_F \to 0,$$

where N_F is the dual of $\text{Ш}'_F(\mathfrak{p})/\text{Ш}'_F(\mathfrak{p})_{\text{div}}$, we see that we can replace the two terms on the extreme right of (41) by the finite group N_F. Secondly, assertion (iii) of Proposition 22 shows that we can identify $(\hat{S}_{F_\infty})^\Gamma$ with $\mathfrak{R}_F$ via the map ϕ_F. After these modifications, we obtain the exact sequence

$$0 \to Y^\Gamma \to \mathfrak{R}_F \xrightarrow{\alpha_F} \hat{W}_F \to (Y)_\Gamma \to N_F \to 0, \tag{42}$$

where α_F denotes the canonical map indicated.

Lemma 23. *The following three assertions are equivalent:-*(i) $g(0) \neq 0$, (ii) $\operatorname{Coker} \alpha_F$ *is finite, and* (iii) α_F *is injective. When these hold, we have*

$$g(0) \sim \#(N_F) \,.\, \#(\operatorname{Coker} \alpha_F).$$

Proof. By lemma 20, $g(0) \neq 0$ if and only if $(Y)_\Gamma$ is finite, and (42) shows that the finiteness of $(Y)_\Gamma$ is equivalent to the finiteness of $\operatorname{Coker} \alpha_F$, because N_F is finite. Also $g(0) \neq 0$ if and only if Y^Γ is finite. But Y^Γ is finite if and only if $Y^\Gamma = 0$ because Y has no finite non-zero Γ-submodule. The rest of Lemma 23 is clear from Lemma 20 and the exact sequence (42).

We assume from now on that $\operatorname{Vol}_{\mathfrak{p}}(E(F)) \neq 0$, i.e. that the determinant of the biliner from $\langle\ ,\ \rangle_{\mathfrak{p}}$ is not 0. Recall that we have the exact sequence

$$0 \to E_1(F) \otimes_{\mathfrak{O}} \mathfrak{O}_{\mathfrak{p}^*} \to \mathfrak{R}_F \to \Theta_F \to 0,$$

where Θ_F is a submodule of finite index in $T_{\pi^*}(\text{Ш}_F)$. Also dualizing the exact sequence (39), we obtain the exact sequence

$$0 \to \widehat{\text{Ш}_F(\mathfrak{p})_{\mathrm{div}}} \to \hat{W}_F \to \operatorname{Hom}_{\mathfrak{O}_{\mathfrak{p}}}(E(F) \otimes_{\mathfrak{O}} \mathfrak{O}_{\mathfrak{p}}, \mathfrak{O}_{\mathfrak{p}}) \to 0.$$

Put $V_F = \alpha_F(E_1(F) \otimes_{\mathfrak{O}} \mathfrak{O}_{p^*})$, so that $V_F \subset \hat{W}_F$. Our hypothesis that $\operatorname{Vol}_{\mathfrak{p}}(E(F)) \neq 0$ implies that the map θ_F of Theorem 17 is injective. We conclude both that α_F is injective on $E_1(F) \otimes_{\mathfrak{O}} \mathfrak{O}_{\mathfrak{p}^*}$, and that

$$V_F \cap \widehat{\text{Ш}_F(\mathfrak{p})_{\mathrm{div}}} = 0. \tag{43}$$

Hence we have an exact sequence

$$0 \to \widehat{\text{Ш}_F(\mathfrak{p})_{\mathrm{div}}} \to \hat{W}_F / V_F \to \operatorname{Coker} \theta_F \to 0, \tag{44}$$

where $\operatorname{Coker}\theta_F$ is finite because $\operatorname{Vol}_{\mathfrak{p}}(E(F)) \neq 0$, and the order of $\operatorname{Coker} \theta_F$ is given by Lemma 18. It is plain from these remarks that α_F induces a homomorphism

$$\beta_F : \Theta_F = \mathfrak{R}_F / (E(F) \otimes_{\mathfrak{O}} \mathfrak{O}_{\mathfrak{p}^*}) \to \hat{W}_F / V_F,$$

with $\operatorname{Ker} \alpha_F = \operatorname{Ker} \beta_F$, and $\operatorname{Coker} \alpha_F = \operatorname{Coker} \beta_F$. There seems no reason why the image of β_F is necessarily contained in the subgroup $\widehat{\text{Ш}_F(\mathfrak{p})_{\mathrm{div}}}$ of

$\hat{W}_F/V_F$. For this reason, we define the subgroup $\sum_F$ of Θ_F by

$$\sum\nolimits_F = \beta_F^{-1}\big(\beta_F(\Theta_F) \cap \widehat{\text{Ш}_F(\mathfrak{p})_{\text{div}}}\big). \tag{45}$$

Hence the restriction of β_F to $\sum_F$ defines a canonical homorphism

$$\gamma_F\colon \sum\nolimits_F \longrightarrow \widehat{\text{Ш}_F(\mathfrak{p})_{\text{div}}}, \tag{46}$$

with $\operatorname{Ker}\gamma_F = \operatorname{Ker}\beta_F$. Also, we clearly have a commutative diagram

$$\begin{array}{ccccccc} 0 \to & \sum_F & \to & \Theta_F & \to & \Theta_F/\sum_F \to 0 \\ & \downarrow \gamma_F & & \downarrow \beta_F & & \downarrow \\ 0 \to & \widehat{\text{Ш}_F(\mathfrak{p})_{\text{div}}} & \to & \hat{W}_F/V_F & \to & \operatorname{Coker}\theta_F \to 0, \end{array}$$

where the vertical map on the right is injective. Since $\operatorname{Coker}\theta_F$ is finite, it follows that $\sum_F$ is also of finite index in $T_{\pi^*}(\text{Ш}_F)$, that $\operatorname{Coker}\gamma_F$ is a subgroup of finite index in $\operatorname{Coker}\beta_F$, and that

$$[\operatorname{Coker}\beta_F : \operatorname{Coker}\gamma_F] = \#(\operatorname{Coker}\theta_F)\Big/\Big[\Theta_F : \sum\nolimits_F\Big]. \tag{47}$$

Finally, we note that to give the canonical map γ_F is equivalent to giving a canonical pairing

$$\{\ ,\ \}_{\mathfrak{p}} : \sum\nolimits_F \times \text{Ш}_F(\mathfrak{p})_{\text{div}} \to D_{\mathfrak{p}} = \mathbf{Q}_p/\mathbf{Z}_p, \tag{48}$$

where $\{a,b\}_{\mathfrak{p}} = \big(\gamma_F(a)\big)(b)$. We want to stress that this canonical pairing $\{\ ,\ \}_{\mathfrak{p}}$, whose existence we have just proven under the two relatively mild assumptions (which can be verified for many E and many primes p) that Conjecture Λ is valid and $\operatorname{Vol}_{\mathfrak{p}}\big(E(F)\big) \neq 0$, is very mysterious. At present, no example is known in which $\sum_F$ and $\text{Ш}_F(\mathfrak{p})_{\text{div}}$ are not both 0. On the other hand, there is little hard theoretical evidence to indicate that this is always true. We attach to $\{\ ,\ \}_{\mathfrak{p}}$ the invariant

$$\rho_{\mathfrak{p}}(\text{Ш}_F) = \#(\operatorname{Coker}\gamma_F)\Big/\Big[T_{\pi^*}(\text{Ш}_F) : \sum\nolimits_F\Big], \tag{49}$$

it being understood that $\rho_{\mathfrak{p}}(\text{Ш}_F)$ is infinite if $\operatorname{Coker}\gamma_F$ is infinite. One can immediately pose the following question. Does there exist a dual pairing

$$T_{\pi^*}(\text{Ш}_F) \times \text{Ш}_F(\mathfrak{p})_{\text{div}} \to D_{\mathfrak{p}},$$

whose restriction on the left to $\sum_F$ is the pairing $\{\ ,\ \}_{\mathfrak{p}}$? Of course, an affirmative answer to this question would not only show that $\rho_{\mathfrak{p}}(\text{Ш}_F)$ is finite, but would also prove that $\rho_{\mathfrak{p}}(\text{Ш}_F) = 1$.

We can now state our final result. Recall that γ_0 denotes our fixed topological generator of Γ. Let $u = \chi_\infty(\gamma_0)$ be the image of γ_0 under the canonical character giving the action of Γ on $E_{\mathfrak{p}^\infty}$. We then define, for $s \neq 0$ in $\mathbf{Z}_p$,

$$L_{\mathfrak{p}}(E/F, s) = \begin{cases} f(u^{s-1} - 1) & \text{if } E_{\mathfrak{p}} \not\subset E(F) \\ \dfrac{f(u^{s-1} - 1)}{(u^s - 1)} & \text{if } E_{\mathfrak{p}} \subset E(F). \end{cases}$$

The conjectural link of $L_{\mathfrak{p}}(E/F, s)$ with p-adic L-functions (see [4], [5]), which we do not discuss in the present paper, provides the reason for introducing the pole term when $E_{\mathfrak{p}} \subset E(F)$.

Theorem 24. *Assume Conjecture A is valid, and that* $\mathrm{Vol}_{\mathfrak{p}}(E(F)) \neq 0$. *Then* (i) $L_{\mathfrak{p}}(E/F, s)$ *has a zero at* $s = 1$ *of multiplicity* $\geq r_F = n_F + t_F$; (ii) *the multiplicity of this zero is exactly* r_F *if and only if the canonical pairing* $\{\ ,\ \}_{\mathfrak{p}}$ *given by (48) has a finite kernel on the right, or equivalently a trivial kernel on the left;* (iii) *If the multiplicity of this zero is exactly* r_F, *then*

$$\lim_{s \to 1} \frac{L_{\mathfrak{p}}(E/F, s)}{(s-1)^{r_F}} \sim \mathrm{Vol}_{\mathfrak{p}}(E(F)).\#(\text{Ш}_F(\mathfrak{p})/\text{Ш}_F(\mathfrak{p})_{\mathrm{div}}).m_{\mathfrak{p}}^{t_F}\rho_{\mathfrak{p}}(\text{Ш}_F).\prod_{v|\mathfrak{p}}\left(1 - \frac{\psi_F(v)}{Nv}\right),$$

where $m_{\mathfrak{p}}$ *and* $\rho_{\mathfrak{p}}(\text{Ш}_F)$ *are given by (33) and (49), and* ψ_F *denotes the Grossencharacter of* E *over* F.

Proof. (i) is simply assertion (40). (ii) follows from Lemma 23 and the fact explained above that $\mathrm{Ker}\,\alpha_F = \mathrm{Ker}\,\gamma_F$, and $\mathrm{Coker}\,\alpha_F$ is finite if and only if $\mathrm{Coker}\,\gamma_F$ is finite. Finally, to establish (iii), one combines Proposition 2, Lemma 3, Lemma 18, Lemma 23, and the formula (47), and simplifies the resulting expression, noting that

$$m_{\mathfrak{p}} \sim u - 1 \sim \log_{\mathfrak{p}}(u).$$

This completes the proof.

Concluding Remarks

1. When $t_F = 0$, the analogy of Theorem 24 with the refinement of the Birch-Swinnerton-Dyer conjecture for the complex L-function $L(\overline{\psi}_F, s)$ given by Gross in [8], is evident. As is explained in [8], our hypotheses that $p \neq 2$ and that p splits in K guarantees that the Tamagawa factors at the bad primes in the complex conjecture are prime to p, and so do not specifically appear in the formula given in Theorem 24, which is necessarily only valid up to multiplication by a p-adic unit.

2. Allowing the possibility that $t_F > 0$ (i.e. that $\text{Ш}_F(\mathfrak{p})$ is infinite), it is natural to ask whether t_F is independent of the choice of the ordinary prime p. Certainly, a mystical belief in the close analogy between the behaviour of both the complex and p-adic L-functions attached to E/F at the points $s = 1$ suggests that t_F should be independent of the ordinary prime p, and that a modified form of the complex Birch-Swinnterton-Dyer conjecture should hold even when $t_F > 0$. Some highly important recent work of R. Greenberg [7] provides fragmentary theoretical evidence for this belief.

3. Although we do not go into details here (see [5]), a combination of Theorem 24 and the calculation of the values of the complex L-functions $L(\overline{\psi}_F^k, k)(k = 1, 2, \ldots)$ can often be used to completely determine the arithmetic of E relative to small ordinary primes p. We simply mention, without proof, several examples, which it does not seem possible to treat by classical methods. The calculations of the complex values $L(\overline{\psi}_F^k, k)$ used in the proof of these results were made by N. Stephens.

Example 2. Take $F = K = Q(\sqrt{-1})$, and E the elliptic curve defined over K by the equation $y^2 = x^3 - x$. It is classical that $n_K = 0$, and one can now show that $\text{Ш}_K(p) = 0$ for $p = 5, 13, 17, 29, 37, 41, 53$.

Example 3. Take $F = K = Q(\sqrt{-1})$, and E the elliptic curve defined over K by the equation $y^2 = x^3 - 2x$. It is classical that $n_K = 1$, and one can now show that $\text{Ш}_K(p) = 0$ for $p = 5, 17, 29$. Moreover, we have $\text{Vol}_{F_p}(E(K)) \sim 1$ if $p = 5$, and $\text{Vol}_{F_p}(E(K)) \sim p$ if $p = 17, 29$.

References

[1] Cassels, J., Arithmetic on curves of genus 1 (VIII), Crelle 217 (1965), 180–199.

[2] Coates, J., Arithmetic on elliptic curves with complex multiplication, Hermann Weyl Lectures, I.A.S. Princeton, 1979, to appear.

[3] Coates, J., Wiles, A., On the conjecture of Birch and Swinnerton-Dyer, Invent. Math., 39 (1977), 223–251.

[4] Coates, J., Wiles, A., On p-adic L-functions and elliptic units, J. Australian Math. Soc., 26 (1978), 1–25.

[5] Coates, J., Goldstein, C., Some remarks on the main conjecture for elliptic curves with complex multiplication, to appear in American J. Math.

[6] Greenberg, R., On the structure of certain Galois groups, Invent. Math., 47 (1978), 85–99.

[7] Greenberg, R., On the Conjecture Birch and Swinnerton-Dyer, to appear in Invent. Math.

[8] Gross, B., On the conjecture of Birch and Swinnerton-Dyer for elliptic curves with complex multiplication, Progress in Math. 26 (Birkhäuser), 219–236.

[9] Iwasawa, K., On $\mathbf{Z}_l$-extensions of algebraic number fields, Ann. of Math., 98 (1973), 246–326.

[10] Mazur, B., Rational points of abelian varieties with values in towers of number fields, Invent. Math., 18 (1972), 183–266.

[11] Perrin-Riou, B., Groupe de Selmer d'une courbe elliptique à multiplication complexe, Compositio Math., 43 (1981), 387–417.

[12] Perrin-Riou, B., Descente infinie et hauteur p-adique sur les courbes elliptiques à multiplication complexe, Invent. Math., 70 (1983), 369–398.

[13] Schneider, P., Iwasawa L-functions of varieties over algebraic number fields. A first approach., to appear in Invent. Math.

[14] Schneider, P., p-adic height pairings I, Invent. Math., 69 (1982), 401–409.

[15] Tate, J., On the conjecture of Birch and Swinnerton-Dyer and a geometric analogue, Séminaire Bourbaki, No. 306, Février, 1966.

Received September 19, 1982

Professor John Coates
Mathématiques, Bat. 425
Université de Paris-Sud
91405 Orsay, France

On the Ubiquity of "Pathology" in Products

Nicholas M. Katz

Dedicated to I. R. Shafarevic on his 60th birthday

Introduction

Fix a prime number p, and an algebraically closed field k of characteristic p. For any proper smooth k-scheme X, we denote by

$$W\Omega^{\cdot}_X = \varprojlim_n W_n\Omega^{\cdot}_X$$

its DeRham-Witt complex. One knows (c.f. [Ill, 1, 2]) that $W\Omega^{\cdot}_X$ calculates the crystalline cohomology of X, i.e., one has a canonical isomorphism

$$H^*_{\text{cris}}(X) \simeq \mathbf{H}^*(X, W\Omega^{\cdot}_X),$$

which is the inverse limit of canonical isomorphisms

$$H^*_{\text{cris}}(X/W_n(k)) \simeq \mathbf{H}^*(X, W_n\Omega^*_X).$$

One also knows that $H^*_{\text{cris}}(X)$ is a finitely generated W-module, and that for every n, the W_n-modles

$$H^*_{\text{cris}}(X/W_n(k)), \quad H^i(X, W_n\Omega^j), \quad (\text{all } (i,j))$$

are finitely generated. However, there is no reason in general that for given (i,j), the W-module

$$H^i(X, W\Omega^j_X) = \varprojlim_n H^i(X, W_n\Omega^j_X)$$

should be finitely generated. Indeed, in 1958 (c.f. [Se]) Serre discovered that for E a supersingular elliptic curve over k, the W-module (i.e., $i=2$, $j=0$, $X=E\times E$)

$$H^2(E\times E, WO)$$

is *not* finitely generated.

Following Illusie-Raynaud and Bloch-Gabber-Kato, we say that a proper smooth k-scheme X is "Hodge-Witt" if, for *all* (i,j), the W-module $H^i(X, W\Omega^j_X)$ *is* finitely generated, and we say that X is "ordinary" if

$$H^i(X, d\Omega^j_{X/k}) = 0 \qquad \text{for all } i,j.$$

One knows (c.f. [Ill-Ray]) that "ordinary" $\Rightarrow$ "Hodge-Witt."

In this note, we will show that Serre's $E\times E$ example of a variety which is not Hodge-Witt is archetypical. One precise result is

Theorem. *Let X and Y be proper smooth connected non-empty k-schemes, such that both $H^*_{\mathrm{cris}}(X)$ and $H^*_{\mathrm{cris}}(Y)$ are torsion-free. Then the product $X\times Y$ is Hodge-Witt if and only if at least one of X or Y is ordinary, and the other is Hodge-Witt.*

Taking $X=Y$, we obtain

Corollary. *Let X be a proper smooth connected non-empty k-scheme with $H^*_{\mathrm{cris}}(X)$ torsion-free. Then the product $X\times X$ is Hodge-Witt if and only if X is ordinary.*

As will be clear to the reader, the work presented here is based in an essential fashion upon the seminal work of Bloch, Illusie-Raynaud, and Ekedahl. I would like to thank R. Crew for explaining to me the "duality" between the approach to Newton and Hodge polygons taken in [Ka], and that taken by Demazure [De] and by Berthelot-Ogus [B-O].

I. Reduction to a Question about Polygons

Let (M,F) be a "σ-F-crystal on k" (cf. [Ka]), i.e., M is a free, finitely generated W-module, and F is a σ-linear endomorphism of M which induces an automorphism of $M\otimes\mathbf{Q}$. We say that (M,F) is polygonally ordinary if its Newton and Hodge polygons coincide. We say that (M,F)

is polygonally Hodge-Witt if the maximal Newton-Hodge decomposition (c.f. [Ka]) of (M, F) (i.e., using *all* those break-points of the Newton polygon which lie *on* the Hodge polygon) is a decomposition into F-crystals of the following types a) or b):

a) There is a single Hodge slope:

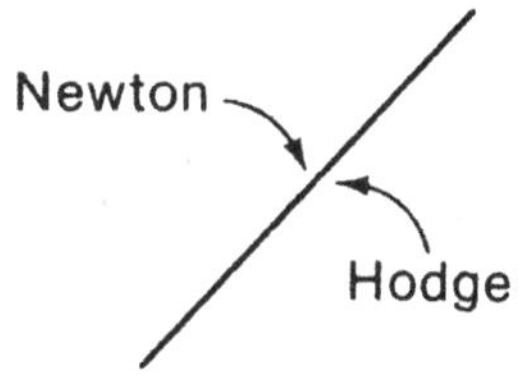

b) There are precisely two Hodge slopes, they are successive integers, say i and $i+1$, and all Newton slopes lie in the open interval $]i, i+1[$, so that the Newton polygon is strictly above the Hodge polygon except at the endpoints, e.g.,

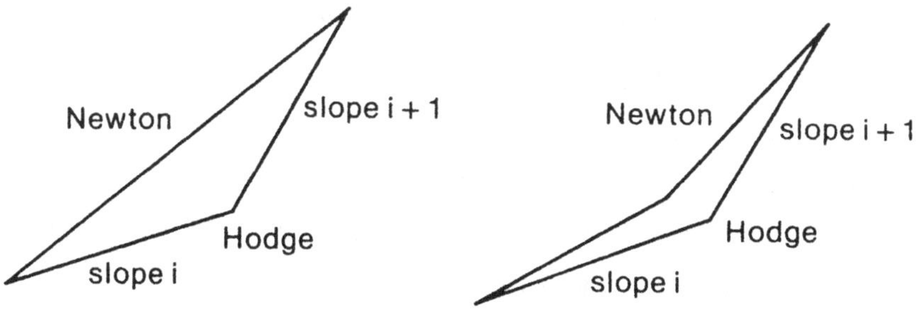

Lemma 1. *Suppse that an F-crystal (M, F) is expressible as a direct sum of F-crystals,*

$$(M, F) = \bigoplus (M_i, F_i).$$

Then (M, F) is polygonally ordinary (respectively, polygonally Hodge-Witt) if and only if each summand (M_i, F_i) is.

Proof. For any subset S of $\mathbf{Q}_{\geq 0}$, let us denote by

$$\pi(S) = \pi_M(S) \in \mathrm{End}_W(M \otimes \mathbf{Q})$$

the projection onto the direct sum over all $\lambda \in S$ of all the Newton-slope= λ isotypical pieces of $M \otimes \mathbf{Q}$. By its functoriality, $\pi_M(S)$ commutes with all W-linear endomorphisms of $M \otimes \mathbf{Q}$ which themselves commute with F. Clearly (M, F) is polygonally Hodge-Witt if and ony if both of the following conditions hold:

A. For every integer $i \geq 0$ the projector $\pi(\{i\})$ lies in $\mathrm{End}_W(M)$, and $\mathrm{Image}\big(\pi(\{i\}) \mid M\big)$, if non-zero, has all its Hodge slopes equal to i.

B. For every integer $i \geq 0$, the projector $\pi(]i, i+1[)$ lies in $\mathrm{End}_W(M)$, and $Im\big(\pi(]i, i+1[) \mid M\big)$, if non-zero, has each of its Hodge slopes equal to either i or $i+1$.

In a similar vein, (M, F) is polygonally ordinary if and only if:

$$A \text{ holds } \textit{and} \text{ if } \pi(]i, i+1[) = 0 \text{ for every integer } i.$$

By functoriality, we have, for any S,

$$\pi_M(S) = \sum pr_i \circ \pi_M(S)$$
$$\pi_{M_i}(S) = pr_i \circ \pi_M(S) \circ incl_i$$

where we have denoted by

$$M \underset{incl_i}{\overset{pr_i}{\rightleftarrows}} M_i$$

the i'th projection and inclusion corresponding to the given decomposition

$$(M, F) = \bigoplus (M_i, F_i).$$

Therefore $\pi_M(S)$ lies in $\mathrm{End}_W(M)$ if and only if the individual $\pi_{M_i}(S)$ do, and if this is the case, then

$$\mathrm{Image}\big(\pi_M(S) \mid M\big) = \bigoplus \mathrm{Image}\big(\pi_{M_i}(S) \mid M_i\big).$$

The set of Hodge slopes (with multiplicities) of a direct sum is simply the union of the sets of Hodge slopes of the summands.

Applying these remarks to $S = \{i\}$ and to $S =]i, i+1[$, we see that the conditions A and B (resp. A and $\pi(]i, i+1[) = 0 \forall i$) hold for a direct sum if and only if they hold for each summand.

Q.E.D.

We will make essential use of the following

Key Theorem (Illusie-Raynaud, Ekedahl). *Let X be a proper smooth connected k-scheme, whose $H^*_{\text{cris}}(X)$ is torsion-free. Consider the following conditions:*

(HDR) *The Hodge $\Rightarrow$ DeRham spectral sequence for X/k*

$$E_1^{i,j} = H^j(X, \Omega^i_{X/k}) \Rightarrow H^{i+j}_{DR}(X/k)$$

degenerates at E_1.

(PO) *The F-crystal $\left(H^*_{\text{cris}}(X), F\right)$ is polygonally ordinary.*

(PHW) *The F-crystal $\left(H^*_{\text{cris}}(X), F\right)$ is polygonally Hodge-Witt.*

Then we have

I. *X is ordinary if and only if X satisfies (HDR) and (PO).*

II. *X is Hodge-Witt if and only if X satisfies (HDR) and (PHW).*

To complete the reduction,we also need

Lemma 2. *Let X, Y be proper smooth non-empty k-schemes. Then $X \times Y$ satisfies (HDR) if and only if both X, Y satisfy (HDR).*

Proof. Clearly X satisfies (HDR) if and only if the universal inequality

$$\sum_{i,j} \dim_k H^j(X, \Omega^i_{X/k}) \geq \sum_l \dim_k H^l_{DR}(X/k)$$

is an *equality*. The result now follows from the Kunneth formula in both Hodge and DeRham cohomology, which gives

$$\sum_{i,j} h^{i,j}(X \times Y) = \Big(\sum_{i,j} h^{i,j}(X)\Big)\Big(\sum_{i,j} h^{i,j}(Y)\Big)$$

$$\sum_l h^l_{DR}(X \times Y) = \Big(\sum_i h^l_{DR}(X)\Big)\Big(\sum h^l_{DR}(Y)\Big),$$

and the observation that

$$\sum_{l} h^{l}_{DR}(X) \geq h^{0}_{DR}(X) \geq 1 > 0,$$

and similarly for Y.

Q.E.D.

Combining Lemma 2 with the key result of Illusie-Raynaud and Ekedahl, our theorem is a special case of the following theorem about F-crystals, applied to $(H^{*}_{\text{cris}}(X), F)$ and $(H^{*}_{\text{cris}}(Y), F)$.

Polygon Theorem. *Given two non-zero F-crystals (M_1, F_1) and (M_2, F_2), their tensor product $(M_1 \otimes M_2, F_1 \otimes F_2)$ is polygonally Hodge-Witt if and ony if one of the (M_i, F_i) is polygonally ordinary, and the other is polygonally Hodge-Witt.*

II. Further Reduction of the Polygon Theorem

In view of Lemma 2, we are immediately reduced to the case where both of the F-crystals (M_i, F_i) are *indecomposable* as F-crystals.

If (M_1, F_1) is polygonally ordinary as well, then (k being algebraically closed), there is an integer $j \geq 0$ such that

$$(M_1, F_1) \simeq (W, p^j \circ \sigma),$$

whence

$$(M_1 \otimes M_2, F_1 \otimes F_2) = (M_2, p^j F_2),$$

and the assertion of the theorem is obvious in this case.

It remains to show that when both of (M_i, F_i) are indecomposable, as F-crystals, and neither is polygonally ordinary, then their tensor product is *not* Hodge-Witt. Since neither of the (M_i, F_i) is polygonally ordinary, each of them has at least two distinct Hodge slopes. Therefore their tensor product has at least *three dintinct Hodge slopes* (for if (M_i, F_i) admits both a_i and $b_i > a_i$ as Hodge slopes, for $i = 1, 2$, then their tensor product admits $a_1 + a_2$, $b_2 + a_2$, $b_1 + b_2$ (as well as $a_1 + b_2$) as Hodge slopes, and by assumption

$$a_1 + a_2 < b_1 + a_2 < b_1 + b_2.$$

So it suffices to show that $(M_1 \otimes M_2, F_1 \otimes F_2)$ is *indecomposable* for the Newton-Hodge decomposition, i.e., that its Newton polygon lies strictly above its Hodge polygon, except at the endpoints. (For having at least three distinct Hodge slopes, it will not be Hodge-Witt!)

Each of the (M_i, F_i), being indecomposable and not polygonaly ordinary, has its Newton polygon strictly above its Hodge polygon, except at the endpoints. Let us say that a non-zero F-crystal satisfis $(N > H)$ if its Newton polygon is strictly above its Hodge polygon, except at the endpoints.

In view of the above discussion, the polygon theorem results from the more precise

$N>H$ **Theorem.** *The tensor product of two non-zero F-crystals which both satisfy $(N > H)$ again satisfies $(N > H)$.*

III. Interlude: Generalities on Polygons of Hodge and Newton Type

We fix as basic data a collection of $r \geq 1$ distinct non-negative real numbers, called "slopes,"

$$0 \leq \lambda_1 < \lambda_2 < \cdots < \lambda_r$$

together with strictly positive integers called "multiplicities"

$$n_1, \ldots, n_r.$$

The integer

$$N = \sum n_i$$

is called the "rank." Equivalent to such data is the corresponding convex polygon in the (X, Y) plane (illustrated below with $r = 3$ distinct slopes). For brevity, we will refer indifferently to either the data

$$(\lambda_1, \ldots, \lambda_r; n_1, \ldots, n_r)$$

or to the corresponding convex polygon as "a polygon."

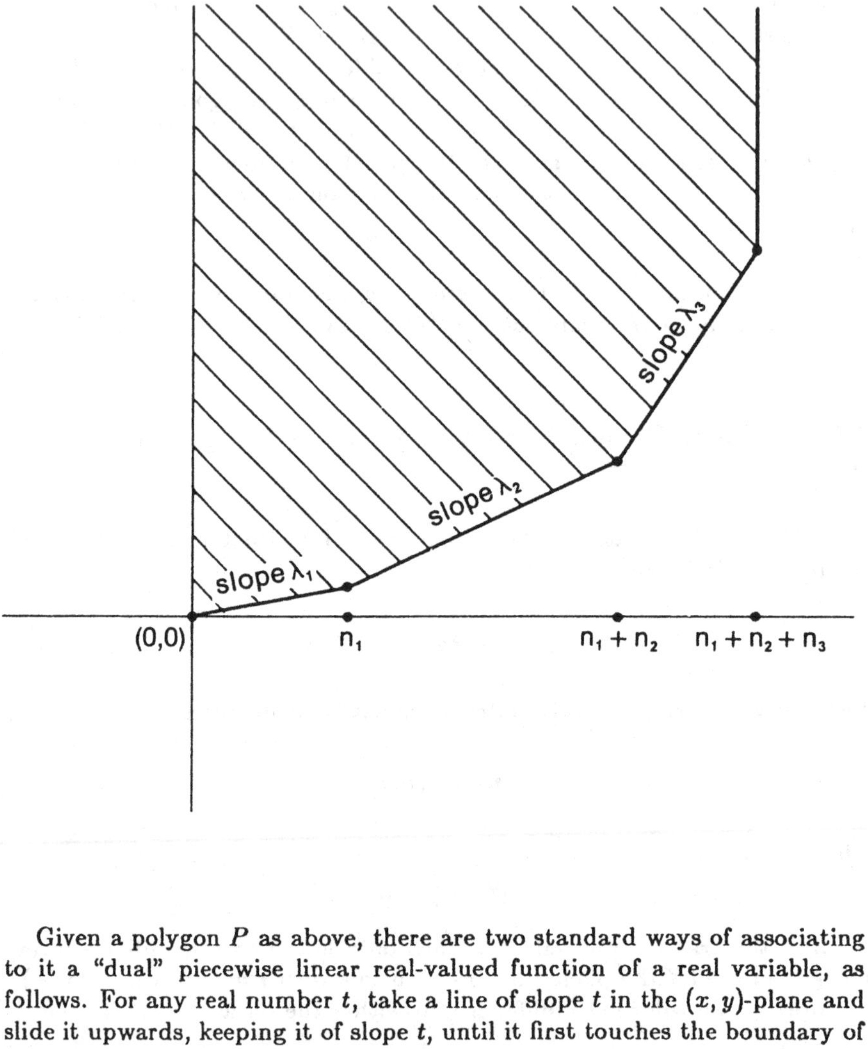

Given a polygon P as above, there are two standard ways of associating to it a "dual" piecewise linear real-valued function of a real variable, as follows. For any real number t, take a line of slope t in the (x, y)-plane and slide it upwards, keeping it of slope t, until it first touches the boundary of the convex polygon. When it first touches the boundary, record either

$D_P(t)$=the height of its intersection with the line "$x = N$".

$BO_P(t)$ = the negative of the height of its intersection with the y-axis.

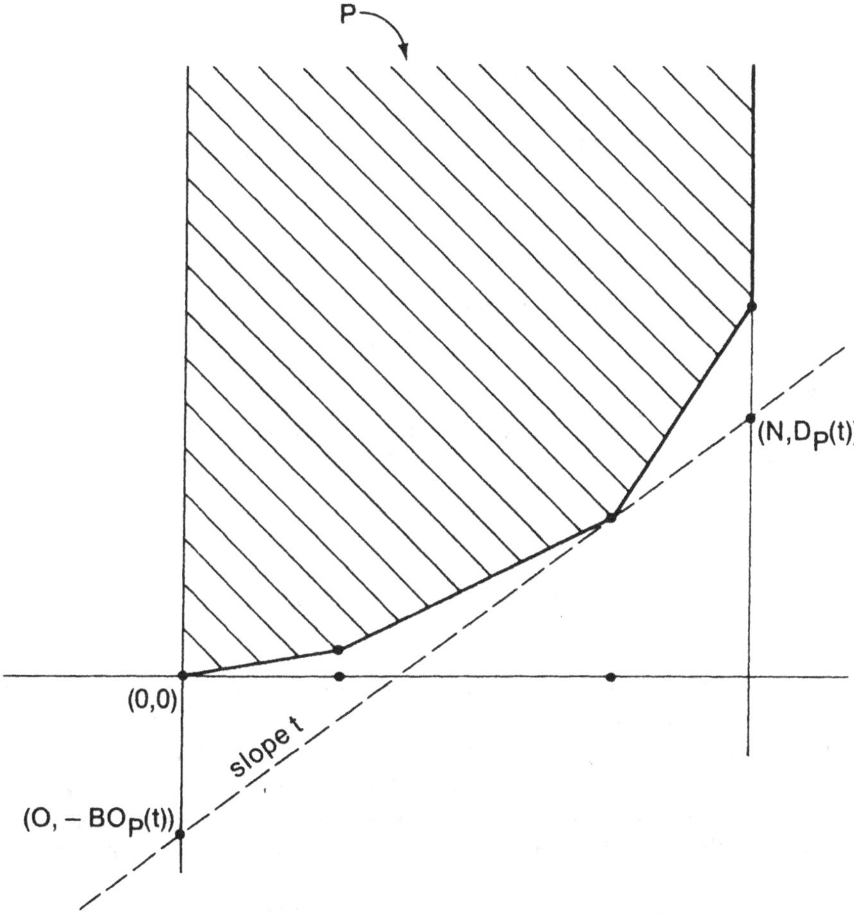

The function D_P is explicitly used in Demazure ([De]), while Berthelot-Ogus ([B-O]) implicitly use the function BO_P.)

In terms of that data $(\lambda_1, \ldots, \lambda_r; n_1, \ldots, n_r)$, we have the explicit formulas

$$D_P(t) = \sum_i n_i \min(t, \lambda_i)$$

$$BO_P(t) = \sum_i n_i \max(0, t - \lambda_i).$$

As is obvious from the picture, we have

$$D_P(t) + BO_P(t) = Nt$$

$$BO_P(t) \text{ is } \geq 0 \text{ and monotone increasing.}$$

As is also obvious from the picture, we have

Lemma 3. *Let P and Q be two polygons. Then:*

I. *Q lies above (i.e., is contained in) P if and only if*

$$BO_P(t) \geq BO_Q(t)$$

for all t.

II. *If P and Q have the same rank and the same terminal point, then Q lies strictly above P, except at the endpoints, if and only if we have*

$$BO_P(t) > BO_Q(t)$$

for all t in the interval

$$\min \text{slope}(P) < t < \max \text{slope}(P).$$

Remarks.

(1) The *second derivative* of $BO_P(t)$ is the sum over the slopes λ_i, counted with multiplicity, of Dirac delta functions supported at λ_i: $BO''_P = \sum n_i \delta_{\lambda_i}$ when P is given by data $(\lambda_1, \ldots, \lambda_r; n_1, \ldots, n_r)$.

(2) If the polygons in question have integer slopes, e.g., Hodge polygons, the dual functions BO_P and D_P are linear between successive integers, so one may verify inequalities among them just by checking at integer values of t.

(3) If P is the Hodge polygon of an F-crystal (M, F), one verifies easily that for any *integer* $k \geq 0$, we have

$$\begin{aligned} BO_P(k) &= \text{length}\left((FM + p^k M)/p^k M\right) \\ &= \text{length}\left(FM/FM \cap p^k M\right) \\ D_P(k) &= \text{length}\left(M/(FM + p^k M)\right). \end{aligned}$$

We summarize this in a diagram:

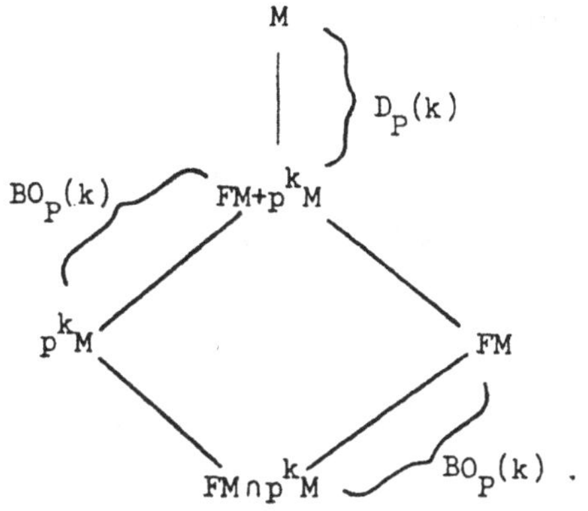

(4) If P corresponds to data $(\lambda_1, \ldots, \lambda_r; n_1, \ldots, n_r)$ with all $\lambda_i \in \mathbf{Q}$, consider the polynomials

$$\begin{cases} f(T) = \prod_i (1 - p^{\lambda_i} T)^{n_i} \\ F(T) = \prod_i (T - p^{\lambda_i})^{n_i}. \end{cases}$$

Then for any rational number $t \notin \{\lambda_1, \ldots, \lambda_r\}$, we have

$$D_P(t) = \mathrm{ord}_p(F(p^t))$$
$$BO_P(t) = -\mathrm{ord}_p(f(p^{-t})).$$

IV. Direct Sums and Tensor Products of Polygons

Let P and Q be polygons, corresponding to data

$$P: (\lambda_1, \ldots, \lambda_r; n_1, \ldots, n_r); \quad \text{rank } \sum n_i = N$$
$$Q: (\mu_1, \ldots, \mu_s; m_1, \ldots, m_s); \quad \text{rank } \sum m_j = M.$$

Then $P \oplus Q$ is the polygon whose rank is $N + M$ and whose slopes, counted with multiplicities, are the *union* of those of P and Q.

The tensor product $P \otimes Q$ is the polygon of rank NM whose slopes, counted with multiplicities, are the numbers $\lambda_i + \mu_j$, counted with multiplicity $n_i m_j$.

The dual functions are given by the formulas

$$BO_{P\oplus Q} = BO_P + BO_Q$$
$$BO_{P\otimes Q}(t) = ((BO_P)\otimes(BO_Q))(t) \overset{\text{def}}{=} (BO_P) * (BO_Q)'',$$

(where $*$ denotes convolution, or more usefully,)

$$\begin{aligned} BO_{P\otimes Q}(t) &= \sum_j m_j BO_P(t-\mu_j) \\ &= \sum_i n_i BO_Q(t-\lambda_i) \\ &= \sum_{i,j} n_i m_j \max(0, t-\lambda_i-\mu_j). \end{aligned}$$

Given non-zero F-crystals (M_1, F_1) and (M_2, F_2), let us denote by HP_1 and HP_2 their respective Hodge polygons, and by NP_1 and NP_2 their respective Newton polygons. Let us denote by $HP_{1\oplus 2}$ and $HP_{1\oplus 2}$ the Hodge and Newton polygons of $(M_1, F_1)\oplus(M_2, F_2)$, and let us denote by $HP_{1\otimes 2}$ and $NP_{1\otimes 2}$ the Hodge and Newton polygons of $(M_1\otimes M_2, F_1\otimes F_2)$. Then we have the expected compatibilities

$$\begin{cases} HP_{1\oplus 2} = HP_1 \oplus HP_2 \\ NP_{1\oplus 2} = NP_1 \oplus NP_2 \end{cases}$$

$$\begin{cases} HP_{1\otimes 2} = HP_1 \otimes HP_2 \\ NP_{1\otimes 2} = NP_1 \otimes NP_2. \end{cases}$$

V. Proof of the $N > H$ Theorem

Let (M_1, F_1) and (M_2, F_2) be two non-zero F-crystals, both of which satisfy $(N > H)$. Let $a_i < b_i$ denote the smallest and largest Hodge slopes of (M_i, F_i), for $i = 1, 2$. Then $a_1 + a_2 < b_1 + b_2$ are the smallest and largest Hodge slopes of $(M_1 \otimes M_2, F_1 \otimes F_2)$. To show that this tensor product satisfies $(N > H)$, it is equivalent (by Lemma 3, II) to show that

$$BO_{HP_{1\otimes 2}}(t) - BO_{NP_{1\otimes 2}}(t) > 0$$

for $a_1 + a_2 < t < b_1 + b_2$.

We rewrite this difference as *one-half* of

$$(BO_{HP_1} - BO_{NP_1}) \otimes (BO_{HP_2} + BO_{NP_2}) \\ + (BO_{HP_2} - BP_{NP_2}) \otimes (BO_{HP_1} + BO_{NP_1}).$$

Because the (M_i, F_i) both satisfy $(N > H)$, it follows from Lemma 3, II, that the functions (for $i = 1, 2$)

$$f_i = BO_{HP_i} - BO_{NP_i}$$

are *strictly positive* in the open interval $]a_i, b_i[$, and zero elsewhere.

For $i = 1, 2$, let α_i be the smallest Newton slope of (M_i, F_i). Then

$$a_i < \alpha_i < b_i,$$

because (M_i, F_i) satisfies $(N > H)$. Therefore each of the functions (for $i = 1, 2$)

$$g_i = BO_{HP_i} + BO_{NP_i}$$

has g_i'' a sum with *positive* integer multiplicities of Dirac delta functions supported at various points, *including* the points a_i, α_i, b_i. Because the functions f_i are non-negative, we have

$$(f_1 \otimes g_2)(t) \geq f_1(t - a_2) + f_1(t - \alpha_2) + f_1(t - b_2), \\ (f_2 \otimes g_1)(t) \geq f_2(t - a_1) + f_2(t - \alpha_1) + f_2(t - b_1);$$

where

$$(f_1 \otimes g_2 + f_2 \otimes g_1)(t) \geq [f_1(t - a_2) + f_2(t - b_1)] \\ + [f_2(t - a_1) + f_1(t - b_2)] + f_1(t - \alpha_2).$$

Because f_i is strictly positive in $]a_i, b_i[$, and zero outside, the first sum after the inequality sign,

$$(*) \qquad f_1(t - a_2) + f_2(t - b_1),$$

is strictly positive in $]a_1 + a_2, b_1 + b_2[$ except for a zero at $b_1 + a_2$. Similarly, the second sum

$$(**) \qquad f_2(t - a_1) + f_1(t - b_2)$$

is strictly positive in $]a_1 + a_2, b_1 + b_2[$ except for a zero at $b_2 + a_1$. So in the case

$$b_1 + a_2 \neq b_2 + a_1,$$

we have $f_1 \otimes g_2 + f_2 \otimes g_1$ strictly positive in $]a_1 + a_2, b_1 + b_2[$, as required. In the remaining case,

$$b_2 + a_1 = b_1 + a_2,$$

we have

$$b_2 - a_2 = b_1 - a_1.$$

The final summand

$$(***) \qquad f_1(t - \alpha_2)$$

is strictly positive in $]a_1 + \alpha_2, b_1 + \alpha_2[$. The picture of the regions of strict positivity of the three sums we have considered is

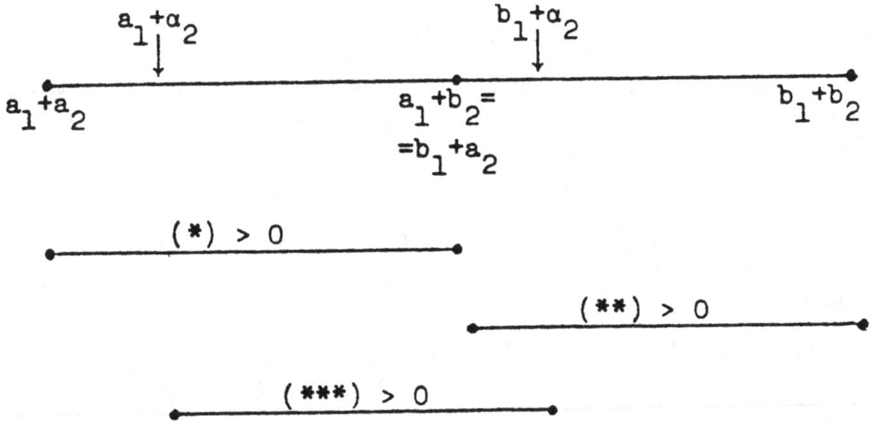

Therefore we find that $f_1 \otimes g_2 + f_2 \otimes g_1$ is strictly positive in $]a_1 + a_2, b_1 + b_2[$ in this case also.

Q.E.D.

References

[B-O] Berthelot, P., and Ogus, A., *Notes on Crystalline Cohomology*, Mathematical Notes 21, Princeton University Press, 1978

[Bl] Bloch, S., Algebraic K-theory and crystalline cohomology, Pub. Math. IHES, 47, 1978, 187–268.

[B-G-K] Bloch, S., Gabber, O., and Kato, manuscript in preparation on Hodge-Tate decompositions for ordinary varieities.

[Dem] Demazure, M., *Lectures on p-divisible groups*, Springer Lecture Notes in Mathematics, 1972.

[Ek] Ekedahl, T., manuscript in preparation.

[Ill 1] Illusie, L., Complexe de deRham-Witt, Journées de Géometrie Algébriques de Rennes, 1978, Astérisque, 63, S.M.F., 1979.

[Ill 2] Illusie, L., Complexe de deRham-Witt et cohomologie crystalline, Ann. Scient. Ec. Norm. Sup. 4e série, t. 12, 1979, 501–661.

[Ill-Ray] Illusie, L., and Raynaud, M., Les suites spectrales associées au complexe de deRham-Witt, to appear in Pub. Math. IHES.

[Ka] Katz, N., Slope filtrations of F-crystals, Journées de Géometrie Algébraique de Rennes, 1978, Astérisque, No. 63, S.M.F., 1979, 113–164.

[Se] Serre, J.-P. Quelque propriétes des variétes abeliennes en car. p., American Journal of Mathematics, 80, 1958, 715–739.

Received October 7, 1982

Professor Nicholas M. Katz
Department of Mathematics
Princeton Univesity
Princeton, New Jersey 08540

Conjectured Diophantine Estimates on Elliptic Curves

Serge Lang

To I.R. Shafaravich

Let A be an elliptic curve defined over the rational numbers $\mathbf{Q}$. Mordell's theorem asserts that the group of points $A(\mathbf{Q})$ is finitely generated. Say $\{P_1, \ldots, P_r\}$ is a basis of $A(\mathbf{Q})$ modulo torsion. Explicit upper bounds for the heights of elements in such a basis are not known. The purpose of this note is to conjecture such bounds for a suitable basis. Indeed, $\mathbf{R} \otimes A(\mathbf{Q})$ is a vector space over $\mathbf{R}$ with a positive definite quadratic form given by the Néron-Tate height: if A is defined by the equation

$$y^2 = x^3 + ax + b, \quad a, b \in \mathbf{Z},$$

and $P = (x, y)$ is a rational point with $x = c/d$ written as a fraction in lowest form, then one defines the x-height

$$h_x = \log \max(|c|, |d|).$$

There is a unique positive definite quadratic form h such that

$$h(P) = \frac{1}{2} h_x(P) + O_A(1) \quad \text{for } P \in A(\mathbf{Q}).$$

Then $A(\mathbf{Q}) \bmod A(\mathbf{Q})_{\mathrm{tor}}$ can be viewed as a lattice in the vector space $\mathbf{R} \otimes A(\mathbf{Q})$ endowed with the quadratic form h.

I assume that the reader is acquainted with the Birch-Swinnerton Dyer conjecture. For an elegant self contained presentation in the form which I shall use, see Tate [Ta 1]. Starting with the Birch-Swinnerton Dyer formula, and several conjectures from the analytic number theory associated with the L-function of the elliptic curve, I shall give arguments showing how one can get bounds for the heights of a suitably selected basis $\{P_1, \ldots, P_r\}$ of the lattice. A classical theorem of Hermite gives bounds for an almost-orthogonalized basis of a lattice in terms of the volume of the fundamental domain, so the problem is to estimate this volume. Hermite's theorem will be recalled for the convenience of the reader in the last section.

The conjectured estimates will also tie in naturally with a conjecture of Marshall Hall for integral points, which we discuss in §2. I shall give some numerical examples in §3.

Manin [Ma] gave a general discussion showing how the Birch-Swinnerton Dyer conjecture and the Taniyama-Weil conjecture that all elliptic curves over **Q** are modular give effective means to find a basis for the Mordell-Weil group $A(\mathbf{Q})$ and an estimate for the Shafarevich-Tate group. The Taniyama-Weil conjecture will play no role in the arguments of this paper, and I shall propose a much more precise way of estimating the regulator and the heights in a basis as in Hermite's theorem. Manin also attributes a "gloomy joke" to Shafarevich in the context of his Theorem 11.1. It is therefore a pleasure to dedicate the conjectures and this paper to Shafarevich.

1. Rational Points

We let $\langle P, Q\rangle$ be the bilinear symmetric form associated with the Néron-Tate height, namely

$$\langle P, Q\rangle = h(P+Q) - h(P) - h(Q).$$

Then

$$|P| = \sqrt{\langle P, P\rangle} \quad \text{and} \quad h(P) = \frac{1}{2}|P|^2.$$

If r is the rank of $A(\mathbf{Q})$, that is the dimension over $\mathbf{R}$ of $\mathbf{R} \otimes A(\mathbf{Q})$, then Birch-Swinnerton Dyer conjecture that the L-function $L_A(s) = L(s)$ has a zero of order r at $s = 1$, and that the coefficient of $(s-1)^r$ in the Taylor expansion is given by the formula

$$\frac{1}{r!}L^{(r)}(1) = |\text{Ш}|\frac{2^{-r}\ |\det\langle P_i, P_j\rangle|}{|A(\mathbf{Q})_{\text{tor}}|^2}\pi_\infty \prod_{p|\Delta} \pi_p.$$

On the right-hand side, $|\text{Ш}|$ is a positive integer, the order of the Shafarevich-Tate group.

The determinant is taken with respect to any basis $\{P_1, \ldots, P_r\}$. Its absolute value is called the **regulator**,

$$R_A = R = |\det\langle P_i, P_j\rangle|,$$

and is the square of the volume of a fundamental domain of the lattice $A(\mathbf{Q})/A(\mathbf{Q})_{\text{tor}}$ in $\mathbf{R} \otimes A(\mathbf{Q})$.

As usual, the discriminant is given by

$$\Delta = -16(4a^3 + 27b^2).$$

The terms π_p for $p|\Delta$ are given by the integral of $|\omega|$ over $A(\mathbf{Q}_p)$, where the differential form ω is the one associated with the global minimum model. As Tate pointed out, π_p is an integer for each p. This property is all we need to know about π_p for our purposes. Essentially.

$$\omega = dx/y,$$

except that the equation $y^2 = x^3 + ax + b$ is not the equation of a minimal model, and one has to introduce extra coefficients, cf. [Ta 1] where the matter is explained in detail. For simplicity of notation, I shall continue to write the equation in standard Weierstrass form as above.

The period π_∞ is given by the integral

$$\pi_\infty = \int_{A(\mathbf{R})} |\omega|.$$

Finally $|A(\mathbf{Q})_{\text{tor}}|$ is the order of the (finite) torsion group.

We are now interested in estimating all the terms to get a bound for $|\text{Ш}|R$, which will yield a bound for R since $|\text{Ш}|$ is an integer.

Since the π_p are intergers, they work for us in getting an upper bound for R.

Mazur [Ma] has shown that $|A(\mathbf{Q})_{\text{tor}}|$ is bounded by 16, and hence this torsion number has a limited effect on the desired estimates.

Concerning the r-th derivative of the L-function. I am indebted to H. Montgomery and David Rohrlich for instructive discussions on the analytic number theory of more classical zeta functions and L-series. In particular, for the Riemann zeta function $\zeta_{\mathbf{Q}}(s)$, the Riemann hypothesis implies that

$$\zeta_{\mathbf{Q}}\left(\frac{1}{2} + it\right) \ll t^{\epsilon(t)},$$

where $\epsilon(t)$ tends to 0 as $t \to \infty$. More precisely, Montgomery conjectures [Mo], formula (10) that one can take

$$\epsilon(t) = c(\log t \log\log t)^{-1/2}$$

with some constant c. Without the log log t, Titchmarsh already observed that one cannot improve the exponent $(\log t)^{-1/2}$, see [Ti], Theorem 8.12. The Riemann hypothesis implies by Theorem 14.14 that

$$\zeta_{\mathbf{Q}}\left(\frac{1}{2}+it\right) \ll t^{c/\log\log t},$$

but is not known to imply Montgomery's conjecture, or even the weaker formulation with the Titchmarsh exponent without the log log t.

For Dirichlet L-series over the rationals, with conductor q, Montgomery also conjectures that one has

$$\frac{1}{r!}L_q^{(r)}\left(\frac{1}{2}\right) \ll q^{\epsilon(q)}c_2^r(\log q)^r$$

with a universal constant c_2.

It is known in the case of elliptic curves with complex multiplication, that $L_A(s)$ is a Hecke L-series, which is so normalized that $L_A^{(r)}(1)$ corresponds to the r-th derivative of the Hecke L-series in its usual form at $s = 1/2$. So to get bounds for $L_A^{(r)}(1)$, on needs bounds for the Hecke L-series at $s = 1/2$. I once gave a general principle which allows one to prove (or conjecture) what happens for more general L-series, on the basis of what happens for the Riemann zeta function. To quote [L 2]: whenever you see a t in an estimate, with a logarithm, then replace it by $d_\chi t^n$, where $d_\chi = d_K \mathbf{N}\mathfrak{f}_\chi$, d_K is the absolute value of the discriminant of a number field K, and $n = [K : \mathbf{Q}]$. In line with this principle, I expect

$$\frac{1}{r!}L_A^{(r)}(1) \ll N^{\epsilon(N)}c_3^r(\log N)^r$$

where N is the conductor of the curve, cf. [Ta 1], end of §6; and $\epsilon(N)$ tends to 0 as $N \to \infty$. Furthermore, $\epsilon(N)$ may have the same shape as in Montgomery's conjecture, namely

$$\epsilon(N) = c(\log N \log\log N)^{-1/2}.$$

Note that the conductor is divisible by the same primes as the descriminant Δ, and $\operatorname{ord}_p N \leq \operatorname{ord}_p \Delta$ so $N \leq |\Delta|$. For the precise order of N at a prime p, see [Ta 1], end of §6, and [Ta 2] for a systematic way of computing N, which is an isogeny invariant. Thus the above inequalities for N in terms of Δ hold when Δ is replaced by a minimal discriminant $\Delta_{\min}$.

This leaves the real period π_∞ to be estimated from below. The Néron differential of a minimal model will not differ too much from dx/y. Let us compute roughly the integral

$$\int_{-\infty}^{\infty} \frac{dx}{\sqrt{x^3 + ax + b}}$$

where intervals of x such that $x^3 + ax + b$ is negative are to be disregarded in this integral. If $b^2 \geq |a|^3$ we change variables and let $x = |b|^{1/3}u$, $dx = |b|^{1/3}du$. If $b^2 < |a|^{1/2}u$.

For definiteness, say $b^2 \geq |a|^3$. Then the integral becomes

$$\frac{1}{|b|^{1/6}} \int_{-\infty}^{\infty} \frac{du}{\sqrt{u^3 + cu + 1}}$$

where $-1 \leq c \leq 1$. The integral for $|u| \geq 2$ is bounded from below by a universal constant. The integral

$$\int_{-2}^{2} \frac{du}{\sqrt{u^3 + cu + 1}}$$

is also bounded from below because $|u^3 + cu + 1|$ is bounded from above for c, u lying in bounded intervals. A similar anslysis if $b^2 \leq |a|^3$ yields the lower bound

$$|\pi_\infty| \gg 1/H^{1/12} \quad \text{where } H = \max(|a|^3, |b|^2).$$

Putting all this together, we arrive at the following conjecture.

Conjecture 1. *Let $H(A) = \max(|a|^3, |b|^2)$. For all elliptic curves $y^2 = x^3 + ax + b$ with $a, b \in \mathbf{Z}$ we have*

$$|\text{Ш}|R \ll H(A)^{1/12} N^{\epsilon(N)} c^r (\log N)^r$$

with some universal constant c, and $\epsilon(N) \to 0$ as $N \to \infty$. In fact, $\epsilon(N)$ may have the explicit form

$$\epsilon(N) = c'(\log N \log\log N)^{-1/2}.$$

Note that if either a or $b = 0$, then $H(A)^{1/12}$ can be replaced by $|\Delta|^{1/12}$. In light of the Hall conjecture below, it should always be the case that

$$H(A) \ll |\Delta|^{6+\epsilon}.$$

In any case, $H(A)$ is absolutely homogeneous of degree 12, in the sense that $H(cA) = |c|^{12}H(A)$, where cA denotes the elliptic curve obtained by multiplying all the quantities a, b, x, y with the appropriate power of c, corresponding to their weight.

The shape of the inequality in Conjecture 1 is analogous to the upper bound for the regulator of a number field K by easy estimates of the residue of the Dedekind zeta function at $s = 1$, apparently first noted by Landau, of the form

$$h_K R_K \leq c^{[K:\mathbf{Q}]} d_K^{1/2} (\log d_K)^{[K:\mathbf{Q}]-1}$$

where d_K is the absolute value of the discriminant. Siegel [Si] more carefully determined precise and very good constants in this estimate. In [L 1] I made the following conjecture:

Conjecture 2. *For all elliptic curves over the integers as above, one has the lower bound for the canonical height:*

$$h(P) \gg \log|\Delta_{\min}|$$

for any rational point P which is not a torsion point.

This conjecture was proved by Silverman [Si] when the j-invariant is an integer. Silverman [Sil 3] has also shown that the conjecture is best possible. Taking into account the Hermite theorem, we may now give an upper bound for an appropriate basis of the Mordell-Weil group. I leave out the integral factor $|Ш|$ on the left for simplicity.

Conjecture 3. *There exists a basis $\{P_1, \ldots, P_r\}$ for $A(\mathbf{Q})$ modulo torsion, ordered by ascending height, such that:*

$$h(P_1) \ll H(A)^{1/12r} N^{\epsilon(N)/r} \log N \left(\tfrac{2}{\sqrt{3}}\right)^{(r-1)/2}$$
$$h(P_r) \ll H(A)^{1/12} N^{\epsilon(N)} (\log N) c^{r(r-1)/2}$$

Indeed, we have $h(P) = \frac{1}{2}|P|^2$, so the upper bound for $h(P_1)$ is precisely the Hermite bound in Theorem 4.2. As to the highest point, we use

Theorem 4.2 (1) and the *lower bound* of Conjecture 2, which allows us to divide by $(\log|\Delta_{\min}|)^{r-1}$ on both sides so that we end up with only one factor of $\log N$ on the right hand side.

The above conjectures are phrased for the canonical height h. To apply them to the ordinary height h_x of the x-coordinate, we need an upper bound on $|h - \frac{1}{2}h_x|$ which will be discussed in the next section.

On the other hand, observe that the lower bound of Conjecture 2 allows us to estimate $|\text{Ш}|$ from above. Indeed, referring to Hermite's theorem in §4, we have

$$\begin{aligned} R = \det(L)^2 &= |u_1|^2 \cdots |u_r|^2 \\ &\geq \left(\frac{3}{4}\right)^{(r-1)} (2h(P_1))^r \\ &\gg c_0^{r^2} (\log|\Delta_{\min}|)^r \end{aligned}$$

by Conjecture 2. Hence the upper bound for $|\text{Ш}|R$ in Conjecture 1 yields an upper bound for $|\text{Ш}|$ of the same nature, except that we can cancel $(\log|\Delta_{\min}|)^r$.

The above arguments do not seem to give further insight in the question raised in [L 1], p. 92, concerning the ratios of successive minima, or equivalently the lengths of elements in an almost orthogonalized basis.

It is a good question to determine to what extent the factor $c^r(\log N)^r$ in Conjecture 1 must really be there. After extending $\mathbf{Q}$ to the field of 2-torsion points the standard proof of the Mordell-Weil theorem of course bounds r in terms of the number of prime factors of N but also in terms of the 2-rank of the class number or the p-rank of the class number of p-torsion points, for any prime p. The question is whether this second part of the bound is, in fact, significant over the rational numbers.

Supposing r is like the number of distinct prime factors of N, one knows that if N is a product of the first r primes, then r is asymptotic to $\log N / \log\log N$. On the other hand, the average value of r is $\log\log N$.

2. Integral Points

We begin by recalling an estimate for $|h - \frac{1}{2}h_x|$, see Demjanenko [De], Zimmer [Zi] or [L 1], p. 99, where I piece together the local estimates to give a global one as follows.

Theorem 2.1. *Let $y^2 = x^3 + ax + b$ be an elliptic curve with $a, b \in \mathbf{Z}$. Let h be the canonical height. Then*

$$\left|h - \frac{1}{2}h_x\right| \leq \frac{1}{6}\log|\Delta| + \frac{1}{6}\log\max(1, |j|) + O(1).$$

In particular,

$$\left|h - \frac{1}{2}h_x\right| \leq \frac{1}{6}\log H(A) + O(1).$$

Proof. Let h_v be the local canonical height, or in other words the Néron function at v, normalized as in Tate, cf. [L 1] Chapter I, §7 and §8, and Chapter III, §4. Let $h_{x,v}$ be the local height of the x-coordinate, that is

$$h_{x,v}(P) = \log\max\{1, |x(P)|_v\}.$$

We now distiguish the two cases when v is archimedean or not.

First let v be the archimedean absolute value on $\mathbf{Q}$. Then in [L 1], Theorem 8.4 of Chapter I, I proved

$$\left|h_v - \frac{1}{2}h_{x,v}\right| \leq \begin{cases} \frac{1}{12}\log|\Delta| + O(1) & \text{if } |j| \leq C_0 \\ \frac{1}{12}\log|\Delta| + \frac{1}{6}|v(j)| + O(1) & \text{if } |j| \geq C_0 \end{cases}$$

where C_0 is some appropriate constant. The two cases correspond to whether j is small or large at the archimedian absolute value. In the present case, Δ is an integer so $\log|\Delta| \geq 0$. Furthermore $j = -2^{12}3^3a^3/\Delta$, so we get $\log|j| = \log|a^3| - \log|\Delta| + O(1)$, (To get all bounds in terms of Δ, see Conjecture 4.)

On the other hand, if v is non-archimedean, Tate has shown that

$$\min\left\{0, \frac{1}{24}v(j)\right\} \leq h_v - \frac{1}{2}h_{x,v} \leq \frac{1}{12}v(\Delta).$$

See Theorem 4.5 of Chapter III. Recall that $v(z) = -\log|z|_p$ if $v = v_p$ is the p-adic absolute value, whence in this case

$$\left|h_v - \frac{1}{2}h_{x,v}\right| \leq \frac{1}{12}v_p(\Delta)$$

because $j = -2^{12}3^3a^3/\Delta$ and a, Δ are p-integral. Taking the sum we then obtain a bound

$$\left|h - \frac{1}{2}h_x\right| = \left|\sum\nolimits_v \left(h_v - \frac{1}{2}h_{x,v}\right)\right| \leq \sum\nolimits_v \left|h_v - \frac{1}{2}h_{x,v}\right|.$$

The $O(1)$ comes only from the archimedean v, there is no such constant for $v = v_p$ with a prime number p. The theorem follows at once. (*Note:* one can get something slightly better, see the Remark in [L 1], p. 32.)

Theorem 2.1 suffices amply to apply the bound of Conjecture 3 to the ordinary height h_x. However, I want the estimate for $|h - \frac{1}{2}h_x|$ to be entirely in terms of Δ. For this purpose, among many others, one can use a conjecture of Marshall Hall [Hall].

Conjecture 4. *If (x, y) is an integral point on the elliptic curve $y^2 = x^3 + b$ with $b \in \mathbf{Z}$, then $|x| \ll b^2$.*

Actually, Stark and Trotter after some theoretical probabilistic considerations have suggested that the conjecture should be with an exponent $2 + \epsilon$, or at least some appropriate powers of $\log|b|$ occurring as a factor of b^2. I shall refer to the Hall conjecture so modified. Hall also expressed his conjecture as an inequality

$$|t^2 - u^3| \gg \max\{|u|^{1/2}, |t|^{1/3}\}$$

whenever t, u are integers. Silverman has pointed out to me that [B-C-H-S] yields parametrizations of curves with integral points having the order of magnitude $|\Delta|^5$, by polynomials $g^2 - f^3$ with integer coefficients, having degree $1 + \frac{1}{2}\deg f$. Presumably $|\Delta|^6$ is the best possible power of Δ. Davenport [Da] has proved the Hall conjecture for polynomials in its original form, without the ϵ.

If we now use the Hall conjecture on the equation $4a^3 + 27b^2 = \Delta_0$ (see the remark below), we can apply Theorem 2.1 to get

$$\left|h - \frac{1}{2}h_x\right| \leq (1 + \epsilon)\log|\Delta| + O(1).$$

This is of special interest in that $\log|\Delta|$ occurs on the right hand side with multiplicity $1 + \epsilon$. As already mentioned, in some cases one might get an even better estimate.

Remark. In using the Hall conjecture, the "variables" a, b have fixed coefficients, namely 4 and 27. Trivial changes of coordinates (and of Δ_0 by a bounded small factor) reduce such equations to the case when the coefficients are equal to 1. For example we multiply the equation by $4^2 3^3$ and let $u = -12a$, $t = 2^2 c^3 b$ to get $t^2 - u^3 = 4^2 3^3 \Delta_0$.

It is not completely clear how the Hall conjecture should extend to an arbitrary equation

$$y^2 = x^3 + ax + b$$

with $a, b \in \mathbf{Z}$. It seems reasonable to expect:

Conjecture 5. *If (x, y) is an integral point on this elliptic curve, then*

$$|x| \ll \max(|a|^3, |b|^2)^k = H(A)^k$$

for some fixed positive number k independent of $a, b \in \mathbf{Z}$.

The analogous statement has been proved in the function field case by Schmidt [Sch]. It is not clear to me what the exponent k should be. Added in proofs: Stark has tentatively suggested $k = 5/3 + \epsilon$, but he has counterexamples showing that probabilistic arguments can be very misleading, and that one cannot make the conjecture until it is checked in the function field case. For instance, he gives the example

$$y^2 = x^3 + x + \frac{1}{4}t^2,$$

which has the solution $x = t^6 + 2t^2$ and $y = t^9 + 3t^5 + \frac{3}{2}t$.

3. Some Numerical Examples

There exist some tables in the literature giving bases for the Mordell-Weil group, mostly when the rank is 1. Selmer [Se 1], [Se 2] found some solutions which at first sight appear quite large. He works with the curve in Fermat form

$$X^3 + Y^3 = DZ^3, \quad 0 < D \leq 500.$$

I shall give here his two largest solutions. I am indebted to J. Brette for using the computer to put Selmer's solutions in the Weierstrass form

$$y^2 = x^3 + b, \quad \text{with } b = -2^4 3^3 D^2.$$

Since we are interested in the height, it is best to clear denominators and consider the homogeneous form

$$w^2 z = u^3 - 2^4 3^3 D^2 z^3 = u^3 + bz^3$$

to be solved in relatively prime integers (u, w, z). Then

$$x = u/z \quad \text{and} \quad y = w/z.$$

The transformation is given by the formulas

$$u' = 2^2 3DZ, \quad w' = 2^2 3^2 D(Y - Z), \quad z' = X + Y.$$

One still has to divide u', z' by their g.c.d. to get x in lowest form. The rank of $A(\mathbf{Q})$ is 1 in each of the following two cases, and the given values for (u, w, z) yield a generator of $A(\mathbf{Q})$ mod torsion. Note how the height h_x drops to about two-thirds of the height of $h_{X/Z}$, as it should for theoretical reasons since X/Z has degree 3 and x has degree 2. In this special case, one might apply Theorem 2.1 to h_x and an analogue to $h_{X/Z}$ to see this. The behavior is remarkably uniform for this particular algebraic family depending on D.

$D = 346$

$$u = 35,208,298,044,859,842,638,816,896,575,916$$
$$z = 94,295,149,257,506,211,891,484,409,025$$
$$(u', z') = 956,232,024,756,251,670$$

$$\Delta_0 = 3^3(2^4 3^3 D^2)^2 \sim 7.2 \times 10^{16}$$
$$\log \Delta_0 \sim 38.8$$

For the conductor, we use the formula in Stephens [St], p. 125, which for $N \not\equiv \pm 2 \bmod 9$ gives the conductor of the Hecke character as

$$\mathfrak{f}_X = 3 \prod_{p|D} p,$$

and therefore

$$N = d_{\mathbf{Q}(\sqrt{-3})}\mathrm{Nf}_X = 3\mathrm{Nf}_X.$$

Since $D = 2 \times 173 \equiv 4 \bmod 9$ we find:

$$N = 2^2 3^3 173^2 = 3,232,332$$
$$\log N \sim 15$$

Using $\epsilon(N) = (\log N \log\log N)^{-1/2}$, we also find:

$$H(A)^{1/12} N^{\epsilon(N)} \log N \sim 3.04 \times 10^3$$
$$h_x \sim 72.5$$

This puts the bound at about 40 times h_x.

$D = 382$

$$u = 96,793,912,150,542,047,971,667,215,388,941,033$$
$$z = 195,583,944,227,823,667,629,245,665,478,169$$
$$(u', z') = 384,647,097,245,468,469,552$$
$$\Delta_0 = 27(2^4 3^3 D^2)^2 \sim 1.07 \times 10^{17}$$
$$\log \Delta_0 \sim 38.9$$
$$N = 2^2 3^3 191^2 = 3,939,948$$
$$\log N \sim 15.2$$
$$H(A)^{1/12} N^{\epsilon(N)} \log N \sim 4.9 \times 10^3$$
$$h_x \sim 80.4$$

The estimate is again roughly the same as in the preceding example.

I owe the next example to Andrew Bremner, who considers the curve

$$y^2 = x^3 + 317x.$$

This is the "other" case with complex multiplication. Bremner parametrizes a generator of the Mordell-Weil group in the form

$$x = (a/b)^2, \quad y = ac/b^3$$

where

$$\begin{aligned} a^2 &= 317v^4 - 4u^4 \\ b &= 2uv \\ c &= 317v^4 + 4u^4 \\ u &= 48,869 \text{ and } v = 73,265. \end{aligned}$$

We have

$$\begin{aligned} \Delta_0 &\sim 1.3 \times 10^8 \\ \log \Delta_0 &\sim 18 \quad . \end{aligned}$$

Now we use the formula in Birch-Stephens [B-S], p. 297, giving the conductor of χ_D for $y^2 = x^3 - Dx$, namely

$$\mathfrak{f}_\chi = 4 \prod_{p|D} p$$

If $D \equiv 3 \bmod 4$. Here $D = -317 \equiv 3 \bmod 4$, so $N = d_{\mathbf{Q}(i)} \mathrm{N}\mathfrak{f}_\chi$, and we find:

$$\begin{aligned} N &= 2^6 317^2 = 6,431,296 \\ \log N &\sim 15.7 \\ H(A)^{1/12} N^{\epsilon(N)} \log N &\sim 718 \\ h_x &\sim 81 \end{aligned}$$

So h_x is about one tenth of its presumed upper bound.

It should be noted that most generators in Selmer's tables, for instance, are considerably smaller than the upper bound given by the conjecture. Statistically, it would seem that they follow quite a different distribution, namely

$$R \ll \log|\Delta|.$$

For the exponential height, $H_x = \exp h_x$, this would mean that on the average, one has

$$H_x \ll |\Delta|^k$$

for some constant k. Even the two largest Selmer points have heights compatible with such a better bound taking $k = 2$; while the Bremner point fits with $k = 3$ or 4. Most other points in Selmer's tables, and all the points in Cassels' early table [Ca] or Podsypanin [Po] fit with $k = 1$. Within the range of such tables, it seems impossible to distinguish a polynomial estimate as above from an exponential estimate (for the exponential height) as in Conjecture 3. It would be interesting to have a more precise statistical analysis of the possibility of a polynomial estimate.

We note that if $f(x,y)$ is an irreducible homogeneous polynomial of degree ≥ 3 over Z then there exists a constant k such that for any *integral* solution (x,y) of the equation

$$f(x,y) = m \in \mathbf{Z}, \quad m \neq 0$$

one has

$$\max(|x|, |y|) \ll |m|^k,$$

so the estimate is known to be polynomial in this case. This was proved by Feldman [Fe], following Baker's method. On the other hand, for the standard Weierstrass equation $y^2 = x^3 + b$, the best known result is exponential and due to Stark [St], using similar methods, namely

$$\max(|x|, |y|) \leq \exp(C_\epsilon |b|^{1+\epsilon}).$$

As to the constants which appear throughout in the estimates, I don't know any explicit bounds for them, although in number theory, one does not expect such constants to be "large" (whatever that means). However, Silverman analyzed the constant in Conjecture 2, and found that in the inequality

$$h(P) \geq c_1 \log|\Delta_{\min}|$$

one had $c_1 < 5 \times 10^{-4}$ in some examples [Si].

The estimate of $O(1)$ in Theorem 2.1, which comes from the archimedean solution value, could be given explicitly, but I have not done so. Zimmer [Zi] gives a good explicit form as $2\log 2$ in his version.

The constants which come up in Conjecture 1 are of a much more complicated nature, what with Riemann hypothesis estimates coming up. For Hecke L-series, if one is satisfied with the $\epsilon(N)$ which is implied by the Riemann hypothesis, then the constants could be estimated following the general approach of [L 2]. Similar analytic techniques might work more generally for the L-function of an elliptic curve, say assuming the functional equation and Riemann hypothesis. Cf. the remarks at the end of [L 2]. Of course, the Montgomery conjecture (or a weaker form along Titchmarsh) seems to lie outside the range of such techniques. These considerations anyhow lie at the periphery of this kind of analytic number theory.

4. The Hermite Theorem

This section is essentially an appendix, to state the Hermite theorem concerning lattices in euclidean space.

Theorem 4.1. *Let L be a lattice in a vector space V of dimension r over $\mathbf{R}$, with a positive definite quadratic form. Then there exists an orthogonal bases $\{u_1, \ldots, u_r\}$ of V, and a basis $\{e_1, \ldots, e_r\}$ of L having the following properties.*

(i) $e_1 = u_1$ *is a vector of minimal length in* L.

(ii) $e_i = b_{i,1}u_1 + \cdots + b_{i,i-1}u_{i-1} + u_i$
with $b_{i,j} \in \mathbf{R}$ *for* $i = 1, \ldots, r$ *and* $|b_{i,j}| \leq 1/2$ *for all* i, j.

(iii) $|u_i| \leq |e_i| \leq \left(\frac{2}{\sqrt{3}}\right)^{i+1} |u_i|$.

(iv) $|u_i| \leq \left(\frac{2}{\sqrt{3}}\right)|u_{i+1}|$.

The theorem is proved by induction, essentially by Gram-Schmidt orthogonalization, and taking minimal vectors successively.

A basis of the lattice as in Theorem 4.1 is called **almost orthogonalized**. One gets the following bounds for the lengths of such basis elements.

Theorem 4.2. *Let $\{e_1, \ldots, e_r\}$ be an almost orthogonalized basis of L. Then:*

(1) $|e_1| \cdots |e_r| \leq \left(\frac{2}{\sqrt{3}}\right)^{r(r-1)/2} \det(L)$

where $\det(L)$ *is the volume of a fundamental domain.*

(2) $|e_1| \leq \left(\frac{2}{\sqrt{3}}\right)^{(r-1)/2} \det(L)^{1/r}$.

Proof. The basis $\{e_1, \ldots, e_r\}$ of L is obtained from the orthogonal basis V by a triangular matrix all of whose diagonal elements are equal to 1. The volume of the rectangular box spanned by $u_1, \ldots, u_r$ is equal to the product of the lengths of the sides, so

$$\det(L) = |u_1| \cdots |u_r|.$$

The first assertion follows by using (ii) in Theorem 4.1. The second assertion follows immediately.

Note. The appearance of $r(r-1)/2$ (so essentially r^2) surprised me at first, but I saw no way of reducing it to the first power of r.

References

[B-C-H-S] B. BIRCH, S. CHOWLA, M. HALL, A. SCHINZEL, *On the difference* $x^3 - y^2$, Norske Vid. Selsk. Forrh. 38 (1965), pp. 65–69.

[B-S] B. BIRCH and N. STEPHENS, *The parity of the rank of the Mordell-Weil group*, Topology 5 (1966) pp. 295–299.

[Ca] J. W. CASSELS, *The rational solutions of the Diophantine equation* $Y^2 = X^3 - D$, Acta Math. 82 (1950) pp. 243–273. *Addenda and corrigenda to the above*, Acta Math. 84 (1951), p. 299.

[Da] H. DAVENPORT, *On* $f^3(t) - g^2(t)$, Kon. Norsk Vid. Selsk. For. Bd. 38 Nr. 20 (1965) pp. 86–87.

[De] V. A. DEMJANENKO, *Estimate of the remainder term in Tate's formula*, Mat. Zam. 3 (1968), pp. 271–278.

[Fe] N. I. FELDMAN, *An effective refinement of the exponent in Liouville's theorem*, Izv. Akad. Nauk 35 (1971) pp. 973–990, AMS Transl. (1971), pp. 985–1002.

[H] M. HALL, *The diophantine equation* $x^3 - y^2 = k$, *Computers in Number Theory*, Academic Press (1971), pp. 173–198.

[L 1] S. LANG, *Elliptic curves: diophantine analysis*, Springer Verlag, 1978.

[L 2] S. LANG, *On the zeta function of number fields*, Invent. Math. 12 (1971), pp. 337–345.

[M] J. MANIN, *Cyclotomic fields and modular curves*, Russian Mathematical Surveys Vol. 26 No. 6, Published by the London Mathematical Society, Macmillan Journals Ltd, 1971.

[Mo] H. MONTGOMERY, *Extreme values of the Riemann zeta function*, Comment. Math. Helvetici 52 (1977), pp. 511–518.

[Po] V. D. PODSYPANIN, *On the equation* $x^3 = y^2 + Az^6$, Math. Sbornik 24 (1949), pp. 391–403 (See also Cassels' corrections in [Ca]).

[Sch] W. SCHMIDT, *Thue's equation over function fields*, J. Australian Math. Soc. (A) 25 (1978) pp. 385–422.

[Se 1] E. SELMER, *The diophantine equation* $ax^3 + by^3 + cz^3$, Acta Math. (1951), pp. 203–362.

[Se 2] E. SELMER, *Ditto, Completion of the tables*, Acta. Math. (1954), pp. 191–197.

[Sie] C. L. SIEGEL, *Abschätzung von Einheiten*, Nachr. Wiss. Göttingen (1969), pp. 71–86.

[Sil 1] J. SILVERMAN, *Lower bound for the canonical height on elliptic curves*, Duke Math. J. Vol. 48 No. 3 (1981), pp. 633–648.

[Sil 2] J. SILVERMAN, *Integer points and the rank of Thue Elliptic curves*, Invent. Math. 66 (1962), pp. 395–404.

[Sil 3] J. SILVERMAN, *Heights and the Specialization map for families of abelian varieties*, to appear.

[St] H. STARK, *Effective estimates of solutions of some diophantine Euations*, Acta. Arith. 24 (1973), pp. 251–259.

[Ste] N. STEPHENS, *The diophantine equation* $X^3+Y^3 = DZ^3$ *and the conjectures of Birch-Swinnerton Dyer*, J. reine angew. Math. 231 (1968), pp. 121–162.

[Ta 1] J. TATE, *The arithmetic of elliptic curves*, Invent. Math. (1974), pp. 179–206.

[Ta 2] J. TATE, *Algorithm for determining the Type of a Singular Fiber in an Elliptic Pencil*, Modular Functions of One Variable IV, Springer Lecture Notes 476 (Antwerp Conference) (1972), pp. 33–52.

[Ti] E. C. TITCHMARSH, *The theory of the Riemann zeta function*, Oxford Clarendon Press, 1951.

[Zi] H. ZIMMER, *On the difference of the Weil height and the Néron-Tate height*, Math. Z. 147 (1976), pp. 35–51.

Received August 24, 1982

Professor Serge Lang
Department of Mathematics
Yale University
Box 2155 Yale Station
New Haven, Connecticut 06520

Zeta-Functions of Varieties Over Finite Fields at s=1

Stephen Lichtenbaum

To I.R. Shafarevich

Let k be a finite field of cardinality $q = p^f$. Let $\bar{k}$ be a fixed algebraic closure of k. Let X be a smooth projective algebraic variety of dimension d over k such that $\overline{X} = X \underset{k}{\times} \bar{k}$ is connected.

It is by now well-known that the zeta-function of X may be written in the form $\zeta(X, s) = Z(X, q^{-s})$, where

$$Z(X,T) = \frac{P_1(X,T)P_3(X,T)\cdots P_{2d-1}(X,T)}{P_0(X,T)P_2(X,T)\cdots P_{2d}(X,T)},$$

where $P_i(X,T) = \det(1 - \phi_{i,l}T)$ is the characteristic polynomial of the endomorphism $\phi_{i,l}$ of the étale cohomology groups $H^i(\overline{X}, \mathbf{Q}_l)$ induced by the Frobenius endomorphism ϕ of X. Deligne has shown [De 1], that these polynomials $P_i(X,T)$ have rational integral coefficients, are independent of l, and (the Riemann hypothesis) have complex "reciprocal roots" of absolute value $q^{i/2}$.

Motivated by the analogue of the Birch-Swinnerton-Dyer conjecture for Jacobians of curves over function-fields, Tate and Artin [T1] produced a conjecture when $d = 2$ relating the behavior of $P_2(X,T)$ as $T \to q^{-1}$ to various cohomological invariants of X. Moreover, they proved that if the l-part $\mathrm{Br}(X)(l)$ of the Brauer group $H^2(X, \mathbf{G}_m)$ of X was finite for one prime l different from p then the l-part of the Brauer group prime to p was finite, and their conjecture was true for X up to a power of p. Milne, in [M1], completed the work of Tate and Artin for surfaces by removing the restrictions concerning the characteristic.

In this paper we propose to begin to extend the work of Tate, Artin, and Milne to varieties of arbitrary dimension. There no longer seems to be any reason to expect a particularly elegant formula for $P_2(X,T)$, so we are led to turn our attention to the full zeta-function $Z(X,T)$. If we let $\rho_1(X)$ be the rank of the Neron-Severi group of X, we now obtain the conjectured formula:

$$(*)\ Z(X,T) \sim \pm c_X(1-qT)^{-\rho_1(X)} \text{ as } T \to q^{-1}, \text{ where } c_X = \frac{\chi(X,G_a)}{\chi(X,G_m)}.$$

This formula, although remarkably simple, needs some explanation. We mean by $\chi(X,G_a)$ the Euler characteristic of the sheaf G_a for the étale (or flat) topology on X, namely

$$\chi(X,G_a) = \frac{\#H^0(X,G_a)\#H^2(X,G_a)\ldots}{\#H^1(X,G_a)\#H^3(X,G_a)\ldots}$$

In view of the standard fact that $H^i(X,G_a)$ is isomorphic to $H^i(X,O_X)$, where the cohomology groups are now taken with respect to the Zariski topology, we see that $\chi(X,G_a) = q^{\chi(X,O_X)}$. We would like to define $\chi(X,G_m)$ in the analogous way, but run into the obvious problem that the $H^i(X,G_m)$ are not all finite.

However, at least in good cases, it is still possible to define an Euler characteristic. The following seems to be a natural generalization of the usual notion: Suppose that we have a sequence of cohomology groups $H^i, i = 0,1,2,\ldots$ with the following properties:

(a) The H^i are zero for i large.
(b) The H^i are finite for all i except $i = a, i = a+2$.
(c) H^a is a finitely generated abelian group.
(d) H^{a+2} is the dual of a finitely generated abelian group D^{a+2}.
(e) There is a natural pairing $\langle\ ,\ \rangle$ from $D^{a+2} \times H^a \to \mathbf{Q}$.

Then we may define the Euler characteristic in the usual way, except that $\#H^a$ is replaced by $\#(H^a_{\text{tor}})$, $\#H^{a+2}$ is replaced by $\#(H^{a+2}_{\text{cotor}}) = \#(D^{a+2}_{\text{tor}})$, and we add a regulator term $R(G_m) = \{\det\langle h_i, d_j\rangle\}^{(-1)^{a+1}}$, where $\{h_i\}$ and $\{d_j\}$ are bases of H^a modulo torsion and D^{a+2} modulo torsion, respectively. In our case, $a = 1$, and $H^3(X,G_m)$ ought to be, up to a finite group, the dual of the finitely generated abelian group consisting of 1-cycles modulo numerical equivalence.

In this paper, we show that,

(1) $(*)$ is true for X of dimension 0 or 1.

(2) (*) is equivalent to Tate's conjecture for X of dimension 2. In particular (*) is true if the l-part of $H^2(X, G_m)$ is finite for any single prime l, by the results of Tate and Milne.

(3) (*) is true up to a power of p for X of any dimension, provided the l-part of $H^2(X, G_m)$ is finite for any single prime $l \neq p$.

Along the way we show that $H^i(X, G_m)$ is finite for $i \neq 1, 2$, and 3, and that if in addition $H^2(X, G_m)(l)$ is finite for one prime $l \neq p$, we have

(i) $H^2(X, G_m)(\text{non-}p)$ is finite, and

(ii) $H^3(X, G_m)$ is the dual of a finitely-generated abelian group (up to p-torsion).

I would like to thank S. Chase for simplifying my original proof of Lemma 1.2.

1. Basic Notations and Necessary Results

We retain the notations and definitions of the introduction, and recommend Milne's book [M2] as a general reference for facts mentioned here without proof. If M is an abelian group, M_{tor} denotes the torsion subgroup and M_{cotor} the quotient group of M by the maximal divisible subgroup. We denote by M_n the set of elements in M killed by n, so $M(l) = \cup_m M_{l^m}$ is the l-torsion subgroup of M. $A^\times$ is the group of units of the ring A, and $\#S$ is the cardinality of the finite set S.

If U is a prescheme over X, then the sheaves G_a, G_m and μ_n associate with U the groups $\Gamma(U, O_U)$, $\Gamma(U, O_U)^\times$, and $\Gamma(U, O_U)^\times_n$ respectively. The obvious sequence of sheaves on X

$$(1.1) \qquad 0 \to \mu_n \to G_m \xrightarrow{n} G_m \to 0$$

is exact in the étale topology if n is prime to p and exact in the flat topology for any n.

Let G be the Galois group of $\bar{k}$ over k. Then G is isomorphic to $\hat{\mathbf{Z}}$ and has the Frobenius automorphism σ of $\bar{k}/k$ as a canonical topological generator. If M is any topological G-module, we denote by M^G (resp. M_G) the kernel (resp. cokernel) of the homomorphism $(\sigma - 1)$: $M \longrightarrow M$.

So $H^0(G,M) = M^G$ and if M is a torsion module, $H^1(G,M) = M_G$ and $H^i(G,M) = 0$ for $i > 1$. It follows from the Hochschild-Serre spectral sequence in flat cohomology that we have the exact sequence of finite groups

$$(1.2) \qquad 0 \to H_f^{i-1}(\overline{X},\mu_n)_G \to H_f^i(X,\mu_n) \to H_f^i(\overline{X},\mu_n)^G \to 0.$$

Here $H_f^i(X,F)$ denotes flat cohomology. Let $H^i(X,F)$ denote étale cohomology. Recall that $H^i(X,G_m) = H_f^i(X,G_m)$ and $H^i(X,\mu_n) = H_f^i(X,\mu_n)$ if $p \not| n$. Let l be any prime number, including p. Let $H_f^i(X,T_l(\mu))$ (resp. $H_f^i(\overline{X},T_l(\mu)) = \varprojlim H_{f}^i,(X,\mu_{l^n})$ (resp. $\varprojlim H_f^i(\overline{X},\mu_{l^n})$). Since inverse limit is an exact functor on the category of finite groups, we have

$$(1.3) \quad 0 \to H_f^{i-1}(\overline{X},T_l(\mu))_G \to H_f^i(X,T_l(\mu)) \to H_f^i(\overline{X},T_l(\mu))^G \to 0.$$

Starting with (1.1) and taking cohomology, we find

$$(1.4) \qquad 0 \to H_f^{i-1}(X,G_m)/l^n \to H_f^i(X,\mu_{l^n}) \to H_f^i(X,G_m)_{l^n} \to 0.$$

Taking projective limits, we obtain

$$(1.5) \qquad 0 \to \widehat{H_f^{i-1}(X},G_m) \to H_f^i(X,T_l(\mu)) \to T_l(H_f^i(X,G_m)) \to 0$$

where the notation is self-explanatory. Now assume $l \neq p$. As Tate explains in [T1] and [T2], $|P_i(X,q^{-1}T)|_l = |Q_i(X,T)|_l$, where $Q_i(X,T) = \det(1-\sigma_i T)$ and σ_i is the automorphism of $H^i(\overline{X},T_l(\mu))$ induced by Frobenius. So 1 is a root of Q_i iff q is a root of P_i, which by the Riemann hypothesis can only occur if $i = 2$. It follows that if $i \neq 2$, $1-\sigma_i$ is a quasi-isomorphism of the finitely generated $\mathbf{Z}_l$-module $H^i(\overline{X},T_l(\mu))$, i.e., $\ker(1-\sigma_i)$ and coker $(1-\sigma_i)$ are both finite. Letting $k_i = k_{i,l} = \#\ker(1-\sigma_i)$ and $c_i = c_{i,l} = \#\operatorname{coker}(1-\sigma_i)$, we have $|P_i(x,q^{-1})|_l = k_i c_i^{-1}$, where $|\ \ |_l$ is the l-adic absolute value normalized so that $|\,l\,|_l = l^{-1}$. If we let $b_{i,l} = \#H^i(X,T_l(\mu))$ it follows from (1.3) that $b_{i,l} = k_i c_{i-1}$ for i different from 2 or 3.

Lemma 1.1. *For all $i \neq 1,2$ we have:*

$$\#H^i(\overline{X},\mu(l))^G = \#H^{i+1}(\overline{X},T_l)^G \mid P_i(q^{-1}) \mid_l^{-1}.$$

Proof. Starting with the exact sequence of sheaves on $\overline{X}$:

$$0 \to \mu_{l^n} \to \mu_{l^{n+k}} \to \mu_{l^k} \to 0$$

and taking cohomology, we obtain

$$H^i(\mu_{l^n}) \to H^i(\mu_{l^{n+k}}) \to H^i(\mu_{l^k}) \to H^{i+l}(\mu_{l^n}) \to H^{i+l}(\mu_{l^{n+k}}).$$

Taking inverse limits with respect to n, we obtain:

$$H^i(\overline{X}, T_l) \xrightarrow{l^k} H^i(\overline{X}, T_l) \to H^i(\overline{X}, \mu_{l^k}) \to H^{i+1}(\overline{X}, T_l) \xrightarrow{l^k} H^{i+1}(\overline{X}, T_l)$$

which yields

$$0 \to H^i(\overline{X}, T_l) \otimes \mathbf{Z}/l\mathbf{Z} \to H^i(\overline{X}, \mu_{l^k}) \to H^{i+1}(\overline{X}, T_l)_{l^k} \to 0.$$

Taking direct limits with respect to k, we get:

$$0 \to H^i(\overline{X}, T_l) \otimes \mathbf{Q}_l/\mathbf{Z}_l \to H^i(\overline{X}, \mu(l)) \to H^{i+1}(\overline{X}, T_l)(l\text{-tor}) \to 0$$

which gives rise to:

$$\begin{aligned} 0 \to \left(H^i(\overline{X}, T_l) \otimes \mathbf{Q}_l/\mathbf{Z}_l\right)^G &\to H^i(\overline{X}, \mu(l))^G \\ &\to H^{i+1}(\overline{X}, T_l)(l\text{-tors})^G \to \left(H^i(\overline{X}, T_l) \otimes \mathbf{Q}_l/\mathbf{Z}_l\right)_G \end{aligned}$$

By the Riemann hypothesis multiplication by $(\sigma_i - 1)$ is a quasi-isomorphism on $H^i(\overline{X}, T_l) \otimes \mathbf{Q}_l/\mathbf{Z}_l$ for $i \neq 2$, and since $H^i(\overline{X}, T_l) \otimes \mathbf{Q}_l/\mathbf{Z}_l$ is divisible, $(H^i(\overline{X}, T_l) \otimes \mathbf{Q}_l/\mathbf{Z}_l)_G = 0$ for $i \neq 2$. If $i \neq 1$, $H^{i+1}(\overline{X}, T_l)^G$ is torsion, so equal to $H^{i+1}(\overline{X}, T_l)(l\text{-tors})^G$. By definition of P_i, we have $|P_i(q^{-1})|_l^{-1} = \#\left(H^i(\overline{X}, T_l) \otimes \mathbf{Q}_l/\mathbf{Z}_l\right)^G$, which completes the proof.

We denote by $NS(X)$ (resp. $NS(\overline{X})$) the group $\mathrm{Pic}(X)/\mathrm{Pic}_0(X)$ (resp. $\mathrm{Pic}(\overline{X})/\mathrm{Pic}_0(\overline{X})$) of divisors modulo algebraic equivalence. $NS(X)$ modulo torsion $(\widetilde{NS}(X))$ is well-known to be the group of divisors modulo numerical equivalence. Starting with the exact sequence

$$0 \to \mathrm{Pic}_0(\overline{X}) \to \mathrm{Pic}(\overline{X}) \to NS(\overline{X}) \to 0,$$

and taking Galois cohomology, we obtain:

$$0 \to \mathrm{Pic}_0(\overline{X})^G \to \mathrm{Pic}(\overline{X})^G \to NS(\overline{X})^G \to H^1(G, \mathrm{Pic}_0(\overline{X})) = 0$$

where the last equality is Lang's theorem. We have the commutative diagram

$$\begin{array}{ccccccccc} 0 & \to & \mathrm{Pic}_0(X) & \to & \mathrm{Pic}(X) & \to & NS(X) & \to & 0 \\ & & \downarrow & & \downarrow & & \downarrow & & \\ 0 & \to & \mathrm{Pic}_0(\overline{X})^G & \to & \mathrm{Pic}(\overline{X})^G & \to & NS(\overline{X})^G & \to & 0 \end{array} \tag{1}$$

and also, if K and $\overline{K}$ are the function fields of X and $\overline{X}$,

$$\begin{array}{ccccccccc} 0 & \to & k^* & \to & K^* & \to & P(X) & \to & 0 \\ & & \downarrow & & \downarrow & & \downarrow & & \\ 0 & \to & \overline{k}^{*G} & \to & \overline{K}^{*G} & \to & P(\overline{X})^G & \to & 0 \end{array} \tag{2}$$

where $P(X)$ denotes the group of principal divisors of X, and the bottom row is exact by Hilbert's Theorem 90. Also, we have

$$\begin{array}{ccccccccc} 0 & \to & P(X) & \to & \mathrm{Div}(X) & \to & \mathrm{Pic}(X) & \to & 0 \\ & & \downarrow & & \downarrow & & \downarrow & & \\ 0 & \to & P(\overline{X})^G & \to & \mathrm{Div}(\overline{X})^G & \to & \mathrm{Pic}(\overline{X})^G & \to & H^1(G, P(\overline{X})) = 0. \end{array} \tag{3}$$

$H^1(G, P(\overline{X})) = 0$ because of Hilbert's Theorem 90 and $H^2(G, \overline{k}^*)$ equalling 0. Diagram (2) then shows $P(X) \simeq P(X)^G$, so diagram (3) shows $\mathrm{Pic}(X) \simeq \mathrm{Pic}(\overline{X})^G$, so diagram (1) shows that $NS(X)$ may be identified with its image $NS(\overline{X})^G$ in $NS(\overline{X})$.

If we let $A(X)$ be the group of 1-cycles modulo numerical equivalence, the pairing $A(X) \times \widetilde{NS}(X) \to \mathbf{Z}$ is nondegenerate, i.e. we claim that if x is a 1-cycle such that $(x \cdot y) = 0$ for every divisor y on X, then if z is a divisor on $\overline{X}$, $(x, z) = 0$. Observe that z comes from a finite extension X' of X and $(x, z) = (x, z^\sigma)$ for all σ implies

$$0 = (x, N_{X'/X}(z)) = [X' : X](x, z) \quad \text{implies} \quad (x, z) = 0$$

We conclude this section with some algebraic results.

Lemma 1.2. *Let $P(X)$ be a monic polynomial of degree r in $\mathbf{Z}[X]$. Write $P(X) = X^k P^*(X)$, where $P^*(0) \neq 0$. Let l be a prime number and* *let T: $\mathbf{Z}_l^r \longrightarrow \mathbf{Z}_l^r$ be a $\mathbf{Z}_l$-linear map with characteristic polynomial $P(X)$. Assume that the minimal polynomial $M(X)$ of T contains X to at most the first power, i.e. that T is "semisimple at 0". Then the order of the torsion subgroup of the cokernel of T divides $P^*(0)$. In particular, for all but finitely many primes l, the cokernel of T is torsion-free.*

Proof. If $T \otimes \mathbf{Q}_l$ is invertible, the cokernel of T is torsion, and its order is equal to the determinant of T, up to an l-adic unit, and so divides the constant term of the characteristic polynomial of T. If $T \otimes \mathbf{Q}_l$ is not invertible, look at the diagram

$$\begin{array}{ccccccccc} 0 & \rightarrow & \mathrm{Ker}(T) & \rightarrow & \mathbf{Z}_l^r & \rightarrow & \mathrm{Im}(T) & \rightarrow & 0 \\ & & \downarrow 0 & & \downarrow T & & \downarrow T' & & \\ 0 & \rightarrow & \mathrm{Ker}(T) & \rightarrow & \mathbf{Z}_l^r & \rightarrow & \mathrm{Im}(T) & \rightarrow & 0. \end{array}$$

where T' is $T \mid Im(T)$. Since $T \otimes \mathbf{Q}_l$ is "semisimple at 0", T' is injective, hence $T' \otimes \mathbf{Q}_l$ is invertible. The snake lemma shows that

$$0 \rightarrow \ker(T) \rightarrow \mathrm{coker}(T) \rightarrow \mathrm{coker}(T') \rightarrow 0$$

is exact. Since $\ker(T)$ is torsion-free, $\mathrm{Coker}(T)_{tor} \subseteq \mathrm{Coker}(T')_{tor}$. But the order of $\mathrm{Coker}(T')_{tor}$ is the determinant of T', so divides the constant term $P^*(0)$ of the characteristic polynomial of T'.

1.3 We now need some remarks about regulators. Let M and N be two free finitely generated $\mathbf{Z}_l$-modules, and assume that $\langle\, ,\rangle$ is a non-degenerate $\mathbf{Z}_l$-pairing from $M \times N$ to $\mathbf{Q}_l$. This obviously extends to a non-degenerate pairing from $(M \otimes \mathbf{Q}_l) \times (N \otimes \mathbf{Q}_l)$ to $\mathbf{Q}_l$.

If M' and N' are two other free finitely-generated $\mathbf{Z}_l$-modules contained in $(M \otimes \mathbf{Q}_l)$ and $(N \otimes \mathbf{Q}_l)$ respectively, we may restrict to get a pairing of $M' \times N'$ into $\mathbf{Q}_l$. If we choose bases $\{a_i\}$, $\{a_i'\}$, $\{b_j\}$ and $\{b_i'\}$ of M, M', N and N', then we may define $\mathrm{Reg}(M,N) = \det\langle a_i, b_j\rangle$, $\mathrm{Reg}(M',N') = \det\langle a_i', b_j'\rangle$, where these regulators are defined up to a unit in $\mathbf{Z}_l$. We may define the *index* of M in M', $(M' : M)$ to be $(M' : M'')/(M : M'')$, where M'' is any submodule of finite index in both M' and M. This is clearly independent of the choice of M'', and we have

$$\mathrm{Reg}(M,N) = (M' : M)(N' : N)\,\mathrm{Reg}(M',N').$$

2. The Cohomology of G_m

In this section we investigate the relationship between the behavior of our zeta-function $Z(X,T)$ as $T \to q^{-1}$, and the étale cohomology of G_m. Let l be a variable prime different from p. We begin with the following proposition:

Proposition 2.1. a) *$H^i(X,G_m)$ is a torsion group for $i \neq 1$.*

b) *$H^i(X,G_m) = 0$ for $i > 2d+1$.*

c) *$H^i(X,G_m)$ is finite for $i \neq 1,2,3$.*

d) *Let the order of $H^i(X,G_m)(l)$ for $i \geq 4$ be denoted by $r_{i,l}$, and let the partial Euler characteristic $r_{4,l}r_{5,l}^{-1}r_{6,l}\dots$ be $\chi_{4,l}$. Then*

$$\chi_{4,l} = k_{4,l}\frac{|P_5(X,q^{-1})|_l \dots |P_{2d-1}(X,q^{-1})|_l}{|P_4(X,q^{-1})|_l \dots |P_{2d}(X,q^{-1})|_l}$$

e) *$H^1(X,G_m)$ is a finitely-generated abelian group whose rank we will call ρ_1.*

f) *If $i = 2$ or 3, $H^i(X,G_m)(l) = C_{i,l} \oplus (\mathbf{Q}_l/\mathbf{Z}_l)^{\rho_{i,l}}$, where $C_{i,l}$ is a finite group and $\rho_{i,l}$ is a non-negative integer. $C_{i,l} = 0$ for all but finitely many l.*

Proof. a) If x is a point of codimension 1 on X, let i_x be the inclusion map of x into X. Let j be the inclusion of the generic point Spec K of X into X. Then we have the exact sequence of sheaves in the étale topology:

$$0 \to G_m \to j_*G_m \to \coprod_x (i_x)_*\mathbf{Z} \to 0 \tag{2.1}$$

where the direct sum is over all points of codimension 1. We have the Leray spectral sequence for j_*: $H^p(X,R^qj_*G_m) \Rightarrow H^{p+q}(K,G_m)$. Since the sheaves $R^qj_*G_m$ are torsion for $q>0$ and since $H^q(K,G_m)$ is torsion for $q > 0$, we see that $H^p(X,j_*G_m)$ is torsion for $p > 0$. Similarly the Leray spectral sequence for $(i_x)_*\mathbf{Z}$ shows that $H^p(X,(i_x)_*\mathbf{Z})$ is torsion for $p > 0$. Then the long exact squence of cohomology coming from (2.1) shows that $H^p(X,G_m)$ is torsion for $p > 1$. $H^0(X,G_m) = k^*$ is of course torsion.

b) For l prime $\neq p$, taking cohomology of the Kummer sequence

$$0 \to \mu_l \to G_m \xrightarrow{l} G_m \to 0$$

shows that $H^i(X,\mu_l)$ maps onto $H^i(X,G_m)_l$. Since the cohomological dimension of X at l is $\leq 2d+1$ (see [M2], Chapter VI, Corollary 1.4), $H^i(X,G_m)_l$ and hence $H^i(X,G_m)(l)=0$ for $i>2d+1$. On the other hand we have (by [M2], Chapter VI, Remark 1.5 b) and c)) that $H^i_f(X,\mu_p)=0$ for $i>d+2$. Therefore the Kummer sequence for flat cohomology shows that $H^i(X,G_m)(p)=0$ for $i>d+2$, which is $\leq 2d+1$ if $d\geq 1$. (If $d=0$, $H^i(X,G_m)$ is well-known to be zero for $i>0$).

c) It follows from (1.5) that $H^i(X,T_l(\mu))$ maps onto $T_l(H^i(X,G_m))$, and $H^i_f(X,T_p(\mu))$ maps onto $T_p(H^i(X,G_m))$. If $i>3$, we have seen that $H^i(X,T_l(\mu))$ is finite and hence $T_l(H^i(X,G_m))$, being torsion-free, is zero. Since $H^i(X,G_m)_l$ is finite, $H^i(X,G_m)(l)$ must be finite. Similarly, $H^i(X,G_m)(p)$ is finite.

Looking now at (1.5) with i replaced by $i+1$, we conclude that

$$H^i(X,G_m)(l)\simeq H^{i+1}(X,T_l(\mu)),$$

so $r_{i,l}=b_{i+1,l}$. Since $b_{i+1,l}=c_{i+1}k_i$, it follows that $r_{i,l}=1$ unless either $l|P_i(X,q^{-1})$ or there are torsion elements in either of the two finitely-generated $\mathbf{Z}_l$-modules $H^i(\overline{X},\mathbf{Z}_l)$ or $H^{i+1}(\overline{X},\mathbf{Z}_l)$. The first restriction only rules out finitely many primes, and O. Gabber has shown ([G]) that so does the second. So $r_{i,l}$ is finite and equal to 1 for almost all l, which implies that $H^i(X,G_m)$ is finite.

d) Again we use (1.5). Since we have seen in c) that $H^i(X,G_m)(l)$ is finite for $i\geq 4$, we have $T_l(H^i(X,G_m))=0$. We conclude that

$$H^i(X,G_m)(l)\simeq \widehat{H^i(X,G_m)}\simeq H^{i+1}(X,T_l(\mu))$$

for $i\geq 4$, and hence $r_{i,l}=b_{i+1,l}$ for $i\geq 4$. It follows that

$$\begin{aligned}\chi_{4,l}&=r_{4,l}r_{5,l}^{-1}r_{6,l}\cdots=b_{5,l}b_{6,l}^{-1}b_{7,l}\cdots\\&=(k_5c_4)(k_6c_5)^{-1}(k_7c_6)\cdots=c_4(k_5c_5^{-1})(k_6c_6^{-1})^{-1}(k_7c_7^{-1})\cdots\\&=c_4\frac{|P_5(X,q^{-1})|_l\,|P_7(X,q^{-1})|_l\cdots}{|P_6(X,q^{-1})|_l\,|P_8(X,q^{-1})|_l\cdots}\end{aligned}$$

But $c_4=k_4|P_4(X,q^{-1})|_l^{-1}$, so we are done.

e) is well-known.

f) Let $i=2$ or 3, and look at $H^i(X,G_m)(l)$. This is a countable l-torsion abelian group with only finitely many elements of order l. By the structure theory of such groups ([K], Theorems 4 and 9), $H^i(X, G_m)(l) = (Q_l/Z_l)^{\rho(i,l)} \oplus C_l^i$, where $\rho(i,l)$ is a non-negative integer and C_l^i is a finite l-group. Now look at (1.3) with $i = 4$. This yields

$$0 \to \varprojlim_n H^3(X, G_m)/l^n \to H^4(X, T_l(\mu)) \to T_l(H^4(X, G_m)) \to 0.$$

Since $C_l^3 \simeq \varprojlim_n H^3(X, G_m)/l^n$, the proof of c) suffices to show that if $\overline{X}$ has finite torsion, C_l^3 is trivial for almost all l, and so $C_3 = \sum_l C_l^3$ is a finte group.

In order to proceed further, we must (and do) assume henceforth that for a fixed prime $l \neq p$, the l-component of the cohomological Brauer group $H^2(X, G_m)$ is finite. We prove in fact the following theorem:

Theorem 2.2. *The following statements are equivalent:*

(i) *$H^2(X, G_m)(l)$ is finite.*

(ii) *The map h: $NS(X) \otimes Z_l \longrightarrow H^2(\overline{X}, T_l(\mu))^G$ is bijective*

(iii) *The rank $\rho_1(X)$ of $NS(X) = rk_{Z_l}(H^2(\overline{X}, T_l(\mu))^G$*

(iv) *$\rho_1(X)$ is the multiplicity of q as reciprocal root of the polynomial $P_2(X,T)$.*

(v) $P_2(X,T) \sim a_X(1 - qT)^{\rho_1(X)}$ as $T \to q^{-1}$, with

$$|a_X|_l = \frac{\# NS(X)(l)_{\mathrm{tor}} |R_l(G_m)|_l \# H^3(X, G_m)(l)_{\mathrm{cotor}}}{\# H^2(X, G_m)(l) \# H^3(\overline{X}, \mu(l))^G}$$

($R_l(G_m)$ is the determinant of the natural pairing between $H^3(X, G_m)(l)^*$ and $(NS(X) \otimes Z_l)$, when both groups are taken modulo torsion.)

(vi) $Z(X,t) \sim C_X(1 - qT)^{-\rho_1(X)}$ as $T \to q^{-1}$, where

$$|C_X|_l = \chi_l(G_m) = \frac{|R_l(G_m)|_l \# H^0(X, G_m)(l) \# H^2(X, G_m)(l) \ldots \# H^{2d}(X, G_m)(l)}{\# H^1(X, G_m)(l) \# H^3(X, G_m)(l)_{\mathrm{cotor}} \# H^5(X, G_m)(l) \ldots \# H^{2d+1}(X, G_m)(l)}$$

(The reader will recognize this as a generalization of first part of Theorem 5.2 in [T1]. We will get to the second part later).

Proof. Tate's proof in [T1] of the equivalence of (i), (ii), and (iii) works perfectly well in this case. (See also [Z], p.218). It is also clear, as in

[T1], that (iv) implies (iii). So we must show that (i), (ii) and (iii) imply (iv). Let $C(X)$ (resp. $C(\overline{X})$) be the free abelian group generated by the one-dimensional integral subschemes (curves) on X (resp. $\overline{X}$). As Tate discussed in his Woods Hole talk [T2], we have a homomorhism ϕ from $C(\overline{X})$ to $H^{2d-2}(\overline{X}, T_l(\mu)^{\otimes(d-1)})$, which we will abbreviate by $H^{2d-2}(\overline{X})$. (Tate states only that his map lands in $H^{2d-2}(\overline{X}) \otimes \mathbf{Q}_l$, but it is easily seen that in fact the map ϕ factors through $H^{2d-2}(\overline{X})$.)

We see that ϕ induces a map ϕ_l from $C(X)\otimes \mathbf{Z}_l$ to $H^{2d-2}(\overline{X})^G$. Poincaré duality induces a map θ from $H^{2d-2}(\overline{X})^G$ to $\operatorname{Hom}(H^2(X, T_l(\mu))^G, \mathbf{Z}_l)$ and the natural map from $NS(X)$ to $H^2(\overline{X}, T_l)^G$ induces a map ψ from $\operatorname{Hom}(H^2(\overline{X}, T_l)^G, \mathbf{Z}_l)$ to $\operatorname{Hom}(NS(X) \underset{\mathbf{Z}}{\otimes} \mathbf{Z}_l, \mathbf{Z}_l)$. Diagrammatically, we have:

$$C(X)\otimes \mathbf{Z}_l \xrightarrow{\phi_l} H^{2d-2}(\overline{X})^G \xrightarrow{\theta} \operatorname{Hom}(H^2(\overline{X})^G, \mathbf{Z}_l) \xrightarrow{\psi} \operatorname{Hom}(NS(X)\otimes \mathbf{Z}_l, \mathbf{Z}_l).$$

Letting σ be the composite $\psi \circ \theta \circ \phi_l$, we know that σ is induced by the intersection pairing on $C(X) \times NS(X)$ by the compatibility of intersection product and cup product. Since $NS(X)$ is, up to torsion, divisors modulo numerical equivalence, we see that, modulo torsion, σ is onto and its kernel is $C_n(X) \otimes \mathbf{Z}_l$ where $C_n(X)$ is the subgroup of 1-cycles numerically equivalent to zero. Since ψ is bijective modulo torsion by assumption, we see that θ is onto modulo torsion. However, by Poincaré duality, and the compatibility of cup-product with the action of G, we have

$$rk_{\mathbf{Z}_l}(H^{2d-2}(\overline{X})^G) = rk_{\mathbf{Z}_l}(H^2(\overline{X}, T_l)_G),$$

which is obviously equal to $rk_{\mathbf{Z}_l}(H^2(\overline{X}, T_l)^G)$. So θ also must be bijective modulo torsion, and $C(X)\otimes \mathbf{Z}_l \xrightarrow{\phi_l} H^{2d-2}(\overline{X})^G$ is surjective modulo torsion, i.e., the Tate conjecture is true for curves on X.

Let $B(X)$ be the image of ϕ_l, let h be the inclusion of $B(X)$ in $H^{2d-2}(\overline{X})^G$, and let $A(X)$ be the group of curves modulo numerical equivalence. Then we have the two exact sequences

$$0 \to F_l \to B(X) \to A(X) \otimes \mathbf{Z}_l \to 0, \text{and}$$
$$0 \to B(X) \xrightarrow{h} H^{2d-2}(\overline{X})^G \to D_l \to 0,$$

where F_l and D_l are finite groups. We see by looking at the sequence (1.1) and taking direct limits that we have

$$0 \to H^2(X, G_m) \otimes \mathbf{Q}_l/\mathbf{Z}_l \to H^3(X, \mu(l)) \to H^2(X, G_m)(l) \to 0$$

and since $H^2(X, G_m)$ is torsion, $H^2(X, \mu(l)) \overset{\sim}{\to} H^3(X, G_m)(l)$. Now the Hochschild-Serre spectral sequence yields:

$$0 \to H^2(\overline{X}, \mu(l))_G \to H^3(X, \mu(l)) \to H^3(\overline{X}, \mu(l))^G \to 0.$$

By Poincaré duality, we may identify the dual of $H^3(\overline{X}, \mu(l))_G$ with $H^{2d-2}(\overline{X})^G$.

Again, by Poincaré duality, $H^3(\overline{X}, \mu(l))^G$ is dual to

$$H^{2d-3}(\overline{X}, T_l(\mu)^{\otimes(d-1)})_G,$$

which is finite by the Riemann hypothesis.

Summing up, we have the following diagram:

$$\begin{array}{ccccccc}
 & & & & 0 & & \\
 & & & & \downarrow & & \\
 & & 0 & & E_l & & \\
 & & \downarrow & & \downarrow & & \\
 & & F_l & & H^3(X, G_m)(l)^* & & \\
 & & \downarrow & & \downarrow & & \\
0 & \to & B(X) & \xrightarrow{h} & H^{2d-2}(\overline{X})^G & \to D_l \to 0 & \\
 & & \downarrow & & \downarrow & & \\
 & & A(X) \otimes \mathbf{Z}_l & & 0 & & \\
 & & \downarrow & & & & \\
 & & 0 & & & &
\end{array} \tag{2.2}$$

where * denotes the $\mathbf{Q}_l/\mathbf{Z}_l$-dual, the row and columns are exact, E_l, D_l and F_l are finite groups. Recall that E_l is in fact $\left(H^3(\overline{X}, \mu(l))^G\right)^*$.

We now use Tate's Bourbaki talk [T1] as a model. We also use his talk as a source of notations and lemmas.

We begin with the analogue of Tate's diagram (5.12):

$$\begin{array}{ccccc}
B(X) & \xrightarrow{e} & \mathrm{Hom}(NS(X), \mathbf{Z}_l) & \approx & \mathrm{Hom}(NS(X) \otimes \mathbf{Q}_l/\mathbf{Z}, \mathbf{Q}_l/\mathbf{Z}_l) \\
h\downarrow & & & & \uparrow g^* \\
H^{2d-2}(\overline{X})^G & \xrightarrow{f} & H^{2d-2}(\overline{X})_G & \approx & \mathrm{Hom}(H^2(\overline{X}, \mu(l))^G, \mathbf{Q}_l/\mathbf{Z}_l)
\end{array} \tag{2.3}$$

(If $d = 2$, this reduces to (5.12)). We have already defined h, Tate's definition of g makes sense in our context as well, and f is induced by the identity on $H^{2d-2}(\overline{X})^G$. The map e is induced by the intersection pairing

$$A(X) \times NS(X) \to \mathbf{Z},$$

which we have seen in §1 is non-degenerate. Let $\{C_i\}$ (resp.$\{D_j\}$) be a basis of $A(X)$ (resp $NS(X)$ modulo torsion). By Tate's lemma z.1 e is a quasi-isomophism, with

$$z(e) = |\det(C_i, D_j)|_l \quad \#F_l.$$

We continue as in [T1] and conclude that g^* is a quasi-isomorphism and

$$z(g^*) = \frac{\#H^2(X, G_m)(l)}{\#NS(X)_{\text{tor}}(l)}.$$

We have seen that h is a quasi-isomorphism and

$$z(h) = \frac{1}{\#D_l}.$$

As in [T1], the diagram is commutative, so f is a quasi-isomorphism and $z(f) = z(e)z(g^*)^{-1}z(h)^{-1}$.

It also follows by Poincaré duality that f is a quasi-isomorphism iff $\tilde{f}$ is, where $\tilde{f}$ is the map from $H^2(\overline{X}, T_l(\mu))^G$ to $H^2(\overline{X}, T_l(\mu))_G$ induced by the identify on $H^2(\overline{X}, T_l(\mu))$. Tate's lemma z.4 implies that $z(f) = z(\tilde{f})$, as in [T1] f being a quasi-isomorphism implies that $\rho_1(X)$ is the multiplicity of q as reciprocal root of the polynomial $P_2(X, T)$, and that the operator $\sigma_2 - 1$ on $H^2(\overline{X}, T_l(\mu))$ acts "semisimply at zero," i.e., that its minimal polynomial $P(X)$ contains X to at most the first power. This completes the proof of (iv).

Notice that (iv) is independent of l, so $H^2(X, G_m)(l)$ finite for one l implies $H^2(X, G_m)(l)$ finite for any $l \neq p$. Also, letting

$$R(T) = P_2(X, T)(1 - qT)^{-\rho(X)},$$

we have

$$|R(q^{-1})|_l = z(\tilde{f}) = z(f).$$

We conclude from all this that

$$(2.1) \qquad |R(q^{-1})|_l = \frac{\#NS(X)(l)_{\mathrm{tor}}\#D_l\#F_l|\det(C_i,D_j)|_l}{\#H^2(X,G_m)(l)}.$$

We now apply §1.2 to our situation. Let $M = A(X)\otimes \mathbf{Z}_l$, M' be the image of $\left(H^3(X,G_m)(l)\right)^*$ in $H^{2d-2}(\overline{X})^G_{\mathbf{Z}_l}\otimes Q_l$, and

$$N = N' = \left(NS(X)\otimes \mathbf{Z}_l\right)/\text{torsion}.$$

Define $R_l(G_m)$ to be $\mathrm{Reg}(M',N)$, and observe that

$$\det(C_i,D_j) = \mathrm{Reg}(M,N),$$

so $\det(C_i,D_j)\mathbf{Z}_l = (M':M)R_l(G_m)$. A computation from diagram (2.2) easily yields that

$$(M':M) = \frac{\#E_l\#D_l\#F_l}{\#H^3(X,G_m)(l)_{\mathrm{cotor}}}$$

after we remark that $\#H^3(X,G_m)(l)_{\mathrm{cotor}} = \#\left(H^3(X,G_m)(l)\right)^*_{\mathrm{tor}}$. We now see that

$$|R(q^{-1})|_l = \frac{\#NS(X)_{\mathrm{tor}}(l)|R_l(G_m)|_l\#H^3(X,G_m)(l)_{\mathrm{cotor}}}{\#H^2(X,G_m)(l)\#E_l}.$$

So we have completed the proof of (v). By Lemma 1.1, $\#E_l = k_4|P_3(q^{-1})|_l^{-1}$ so

$$|R(q^{-1})|_l\,|P_3(q^{-1})|_l^{-1} = \frac{\#NS(X)_{\mathrm{tor}}(l)|R_l(G_m)|_l\#H^3(X,G_m)(l)_{\mathrm{cotor}}}{\#H^2(X,G_m)(l)k_4}$$

Now $H^1(\overline{X},T_l)$,by 1.5 with $i=1$, is isomorphic to $T_l(\mathrm{Pic}_0(\overline{X}))$. In particular, it is torsion-free, so

$$\begin{aligned}|P_1(q^{-1})|_l^{-1} &= \#H^1(\overline{X},T_l)_G = \#\left(T_l(\mathrm{Pic}_0(\overline{X}))\right)_G = \#\left(\mathrm{Pic}_0(\overline{X})(l)\right)^G \\ &= \#\mathrm{Pic}_0(X)(l) = \mathrm{Pic}_0(X)_{\mathrm{tor}}(l)/\#NS(X)_{\mathrm{tor}}(l).\end{aligned}$$

Finally, $|P_0(q^{-1})|_l^{-1} = \#H^0(X,G_m)(l)$. Taking everything together including Proposition 2.1f), we see that indeed $Z(X,T)\sim c_X(1-qT)^{-\rho_1(X)}$ as

$T \to q^{-1}$, where

$$|c_X|_l = \chi_l(G_m).$$

Theorem 2.3. *Assume that X satisfies the equivalent hypotheses of Theorem 2.2. Then*

(i) *The integers $\rho_{3,l}$ of Proposition 2.1f) are all equal to $\rho_1(X)$. (Of course all the $\rho_{2,l} = 0$).*

(ii) *$H^2(X, G_m)(\text{non } p)$ is a finite group.*

Proof. (i) The exact sequence (1.5) with $i = 3$ becomes

$$0 \to C_l^2 \to H^3(X, T(\mu)) \to T_l(H^3(X, G_m)) \to 0.$$

So $\rho_{3,l}$ is equal to the rank of $H^3(X, T_l(\mu))$. Now looking at (1.3) with $i = 3$, we have (since $(\sigma_3 - 1)$ induces a quasi-isomorphism on $H^3(\overline{X}, T_l(\mu))$, $\rho_{3,l} = rk_{\mathbf{Z}_l}(H^2(\overline{X}, T_l(\mu)))_G = rk_{\mathbf{Z}_l}(H^2(\overline{X}, T_l(\mu)))^G$. But this is caught between $\rho_1(X)$ and the multiplicity of q as reciprocal root of $P_2(X, T)$, and since these are equal they are also equal to $\rho_{3,l}$.

(ii) Now use (1.5) with $i = 2$ to obtain:

$$0 \to H^2(X, G_m)(l) \to H^3(X, T_l(\mu)) \to T_l(H^3(X, G_m)) \to 0$$

which in turn implies that $H^2(X, G_m)(l) \simeq H^3(X, T_l(\mu))_{\text{tor}}$. (1.3) with $i = 3$ yields

$$0 \to H^2(\overline{X}, T_l(\mu))_G \to H^3(X, T_l(\mu)) \to H^3(\overline{X}, T_l(\mu))^G \to 0$$

By the result of Gabber previously mentioned, $H^3(\overline{X}, T_l(\mu))^G_{\text{tor}} = 0$ for all but finitely many l. Lemma 1.2 applies to $\sigma_2 - 1$ acting on $H^2(\overline{X}, T_l(\mu))$, since we know the characteristic polynomials are independent of l, and we have seen at the end of the proof of Theorem 2.2 that $(\sigma_2 - 1)$ acts "semi-simply at zero," $(H^2(\overline{X}, T_l(\mu))_G)_{\text{tor}}$ is zero for all but finitely many l, which implies our result.

We now move on to the statement and proof of our main formula.

Theorem 2.4. *Let X be as in Theorem 2.3. Then*

(i) $|R_l(G_m)|_l = 1$ *for all but finitely many* l.
(ii) *If we define* $R'(G_m)$ *to be* $\prod_{l \neq p} |R_l(G_m)|_l^{-1}$, *then*

$$Z(X,T) \sim C_X(1-qT)^{-\rho_1(X)} \text{ as } T \to q^{-1}, \tag{2.4}$$

where $C_X = \pm p^\nu \chi'(X,G_m)^{-1}$, ν *is an integer, and* $\chi'(X,G_m) =$

$$\frac{\#H^0(X,G_m)(\text{non}\,p)\#H^2(X,G_m)(\text{non}\,p)\cdots\#H^{2d}(X,G_m)(\text{non}\,p)}{R'(G_m)\#H^1(X,G_m)_{\text{tor}}(\text{non}\,p)\#H^3(X,G_m)_{\text{cotor}}(\text{non}\,p)\cdots\#H^{2d+1}(X,G_m)(\text{non}\,p)}.$$

Proof. We have seen in Proposition 2.1 that $H^1(X,G_m)(\text{non}\,p)$ is a finite group for $i > 3$ and $i = 0$, that $H^3(X,G_m)_{\text{cotor}}(\text{non}\,p)$ is finite, and that $H^1(X,G_m)(\text{tor})$ is finite. We have seen in Theorem 2.3 that $H^2(X,G_m)(\text{non}\,p)$ is finite. Since C_X is a non-zero rational number, and since formula (2.4) is true after taking l-adic absolute values for each prime $l \neq p$, we must have $|R_l(G_m)|_l = 1$ for all but finitely many l. The truth of (2.4) then follows formally.

We still would like to interpret $R'(G_m)$ as an actual regulator, i.e., as the determinant of a pairing of finitely-generated abelian groups. Since we are ignoring p in this section, this is too much to ask for. However, we can interpret $R'(G_m)$ as the determinant of a pairing of finitely-generated $\mathbf{Z}[1/p]$-modules into $\mathbf{Q}$.

Referring back to diagram (2.2) in the proof of Theorem 2.2, we see that

$$|R_l(G_m)|_l = \frac{|\det(C_i, D_j)|_l \#H^3(X,G_m)(l)_{\text{cotor}}}{\#F_l \#D_l \#E_l},$$

where $\det(C_j, D_j)$ was the determinant of the intersection pairing between curves and divisors on X. Since we know from Proposition 2.1 that $\#H^3(X,G_m)(l)_{\text{cotor}}$ is 1 for all but finitely many l, it follows that $\#E_l$ is 1 for all but finitely many l. Theorem 2.4 then implies that $\#D_l$ and $\#E_l$ are both equal to 1 for all but finitely many l. Interpreting $H^3(X,G_m)(\text{non}\,p)$ as the dual of a finitely-generated $\mathbf{Z}[1/p]$-module then follows from the next two lemmas.

Lemma 2.5. *Let* A *be a finitely-generated* $\mathbf{Z}[1/p]$*-module. Let* B *be a torsion abelian group such that* $B(p) = 0$ *and there exist injections* $\phi_l : A \otimes \mathbf{Z}_l \longrightarrow \text{Hom}(B, \mathbf{Q}_l/\mathbf{Z}_l)$ *with finite cokernels* C_l *such that* $C_l = 0$ *for all but finitely many* l. *Then there exists a finitely generated* $\mathbf{Z}[1/p]$*-module*

A' such that A injects into A' and there are isomorphisms

$$\psi_l\colon A' \otimes \mathbf{Z}_l \longrightarrow \mathrm{Hom}(B, \mathbf{Q}_l/\mathbf{Z}_l) \quad \textit{compatible with the } \phi_l\textit{'s.}$$

Lemma 2.6. *Let A be a finitely-generated $\mathbf{Z}[1/p]$-module. Let B be a torsion abelian group such that $B(p) = 0$ and there exist surjections $\phi_l\colon \mathrm{Hom}(B, \mathbf{Q}_l/\mathbf{Z}_l) \longrightarrow A \otimes \mathbf{Z}_l$ with finite kernels K_l such that $K_l = 0$ for all but finitely many l. Then there exists a finitely generated $\mathbf{Z}[1/p]$-module A' such that A' surjects onto A and there are isomorphisms $\phi_l\colon \mathrm{Hom}(B, \mathbf{Q}_l/\mathbf{Z}_l) \longrightarrow A' \otimes \mathbf{Z}_l$ compatible with the ϕ_l's.*

Proof. Lemma 2.6 is straightforward, so we prove Lemma 2.5. We immediately reduce to the case when A and $\mathrm{Hom}(B, \mathbf{Q}_l/\mathbf{Z}_l)$ are torsion-free. Let $\hat{A} = \prod_{l \neq p}(A \otimes \mathbf{Z}_l)$. The hypotheses of the lemma imply that we have an exact sequence

$$0 \to \hat{A} \xrightarrow{\phi} \mathrm{Hom}(B, \mathbf{Q}/\mathbf{Z}(\mathrm{non}\, p)) \xrightarrow{\pi} C \to 0,$$

with C finite and ϕ induced by the ϕ_l's. Let $B^* = \mathrm{Hom}(B, Q/\mathbf{Z}(\mathrm{non}\, p))$, and let $A' = \{x \in B^*\colon$ some non-zero multiple of x lies in $A\}$. It is immediate that A' is a $\mathbf{Z}[1/p]$-module and that we have an exact sequence $0 \to A \xrightarrow{\phi'} A' \xrightarrow{\pi'} C$, where ϕ' and π' are induced from ϕ and π. (Use the fact that A is pure in $\hat{A}$.) In particular, we see that A' is finitely-generated. We wish to show π' is surjective.

By construction $A/mA \xrightarrow{\sim} \hat{A}/m\hat{A}$ for any integer m. Since the isomorphism factors through A'/mA', A'/mA' maps onto $\hat{A}/m\hat{A}$, which maps onto C/mC. If we let $m = \#C$ then $C/mC \simeq C$ and we see that π' is surjective. It is then immediate that A' satisfies the condition of the lemma.

Remark 2.7. It is now clear that the regulator formed by using the finitely-generated $\mathbf{Z}[1/p]$-module dual to $H^3(X, G_m)(\mathrm{non}\, p)$ agrees with $R'(G_m)$.

Remark 2.8. Note also that if $\chi(X, G_m)$ makes sense, then $\chi'(X, G_m)$ differs from it by a power of p.

3. The Cohomology of G_a

In the previous paragraphs we have ignored the p-part of the zeta-function, and obtained a formula for the prime-to-p part under the assumption that $H^2(X,G_m)(l)$ was finite for one prime $l \neq p$. In this section we propose a conjecture on the p-part, and show that it is true for $d \leq 2$.

Conjecture 3.1. *Assume that $H^2(X,G_m)(l)$ is finite for some prime l. Then $H^2(X,G_m)(p)$ is finite, and*

$$C_X = \lim_{T \to q^{-1}} Z(X,T)(1-qT)^{\rho_1(X)} = \pm \frac{\chi(G_a)}{\chi(G_m)}.$$

Here

$$\chi(G_a) = \prod_i \#H^i(X,G_a)^{(-1)^i} = \prod_i \#H^i(X,O_x)^{(-1)^{-1}} = q^{\chi(X,O_X)}$$

$$\chi(G_m) = \frac{\#H^0(X,G_m)\#H^2(X,G_m)\cdots\#H^{2d}(X,G_m)}{\#H^1(X,G_m)_{\mathrm{tor}}\#H^3(X,G_m)_{\mathrm{cotor}}\#H^5(X,G_m)\cdots R(G_m)}.$$

Theorem 3.2. *Conjecture 3.1 is true if $d=0$ and 1, and true under the hypothesis that* $\mathrm{Br}(X)(l)$ *is finite for some l if $d=2$.*

Proof. We start with $d=0$. Then $X = \operatorname{Spec} k$, $Z(X,T) = \frac{1}{1-T}$ $\rho_1(X) = 0$, so

$$C_X = \lim_{T \to q^{-1}} Z(X,T)(1-qT)^{\rho_1(X)} == \frac{1}{1-q^{-1}} = \frac{q}{q-1}.$$

$H^i(X,G_a) = k$ if $i=0$, and is 0 otherwise, so $\chi(G_a) = q$. $H^i(X,G_m) = k^*$ if $i=0$, and is 0 otherwise, so $\chi(G_m) = q-1$.

Now let $d=1$. Then

$$Z(X,T) = \frac{P(T)}{(1-T)(1-qT)}, \quad \rho_1(X) = 1,$$

so,

$$C_X = \lim_{T \to q^{-1}} \frac{P(T)(1-qT)^{\rho_1(X)}}{(1-T)(1-qT)} = \frac{P(q^{-1})}{1-q^{-1}} = \frac{q}{q-1} P(q^{-1}).$$

We have $P(T) = \prod_{i=1}^{2g}(1-\alpha_i T)$, so $P(q^{-1}) = \prod_{i=1}^{2g}(1-\alpha_i q^{-1})$. By the functional equation, this equals

$$\prod_{j=1}^{2g}(1-\alpha_j^{-1}) = \prod_{j=1}^{2g} \alpha_j^{-1} \prod_{j=1}^{2g}(1-\alpha_j) = q^{-g}P(1),$$

and since $P(1) = h = \#\mathrm{Pic}_0(X), C_X = \frac{q^{1-g}P(1)}{q-1} = \frac{hq^{1-g}}{q-1}$. On the other hand, $\chi(X, O_X) = 1-g$, so $\chi(G_a) = q^{1-g}$. In this case, we have

$$\chi(G_m) = \frac{\#H^0(X,G_m)\#H^2(X,G_m)R(G_m)}{\#H^1(X,G_m)_{\mathrm{tor}}\#H^3(X,G_m)_{\mathrm{cotor}}}. \tag{3.1}$$

We have $\#H^0(X,G_m) = q-1, \#H^1(X,G_m)_{\mathrm{tor}} = h$, $H^3(X,G_m) = \mathbf{Q}/\mathbf{Z}$, so $\#H^3(X,G_m)_{\mathrm{cotor}} = 1$, $H^2(X,G_m) = 0$. The dual of $H^3(X,G_m)$ is $\mathbf{Z} = H^0(X,\mathbf{Z})$ and the regulator pairing takes (n,D) to $\deg(nD)$. Its determinant is 1, since there always exists a divisor of degree 1. So $\chi(G_m) = h/q-1$, and we are done.

Now let $d=2$. We begin with a useful lemma on duality:

Lemma 3.3. *Let X be an algebraic surface. Assume that $H^2(X,G_m)(l)$ is finite for any l, including the characteristic. Then there is a natural pairing from $H^i(X,G_m) \times H^{4-i}(X,G_m)$ into $\mathbf{Q}/\mathbf{Z}$ such that the induced map from $H^i(X,G_m)$ to $\mathrm{Hom}(H^{4-i}(X,G_m),\mathbf{Q}/\mathbf{Z})$ is an isomorphism for $i=0,2,3$, and 4.*

Proof. The case $i=2$ contained in [M1]. The case when $i=0$ will follow from the case when $i=4$. So we assume $i=3$ or 4. It follows from Proposition 2.1 that it is enough to prove that, for each l, $H^i(X,G_m)(l)$ is isomorphic to $\mathrm{Hom}(H^{4-i}(X,G_m),\mathbf{Q}_l/\mathbf{Z}_l)$. It again follows from the Kummer sequence for G_m that we have

$$0 \to \varprojlim_n H^i(X,G_m) \otimes \mathbf{Z}/l^n\mathbf{Z}$$
$$\to H^{i+1}(X,T_l(G_m)) \to T_l(H^{i+1}(X,(G_m)) \to 0.$$

It follows from the Hochschild-Serre spectral sequence for μ_{l^n} that we have

$$0 \to H^i(\overline{X}, T_l(G_m))_G \to H^{i+1}(X, T_l(G_m)) \to H^{i+1}(\overline{X}, T_l(G_m))^G \to 0.$$

We see from the description of the G-action on $H^i(\overline{X}, T_l(G_m))$ in terms of the zeta-function of X that $H^i(\overline{X}, T_l(G_m))_G$ and $H^i(\overline{X}, T_l(G_m))^G$ are finite for $i \neq 2$, so $H^{i+1}(X, T_l(G_m))$ is finite for $i \neq 2, 3$. In particular $T_l(H^4(X, G_m))$ is finite, so zero, so $H^4(X, G_m)(l)$ is finite. But

$$H^4(X, G_m)(l) \simeq H^4(X, \mu(l)) \simeq \varinjlim_r H^4(X, \mu_{l^n}),$$

which is dual to $\varprojlim_n H^1(X, \mu_{l^n})$, which is easily seen to be $H^0(X, G_m)(l)$. Let $i = 3$. $H^3(X, G_m)(l) \simeq H^3(X, \mu(l))$ is dual to

$$\varprojlim_n H^2(X, \mu_{l^n}) = H^2(X, T_l(G_m)).$$

Since we have assumed that $H^2(X, G_m)(l)$ is finite, $T_l(H^2(X, G_m)) = 0$, so $H^2(X, T_l(G_m) \simeq H^1(X, G_m) \otimes \mathbf{Z}_l$, and we are done.

We wish to show that our formula is equivalent to that stated by Tate in [T1] and proved under our hypotheses by Milne in [M1]. We start with

$$Z(X, T) = \frac{P_1(X, T) P_3(X, T)}{(1 - T) P_2(X, T)(1 - q^2 T)}$$

let $P_1(X, T)$ have $\deg B = B_1(X)$ and reciprocal roots $\alpha_1, \ldots, \alpha_B$. By the functional equation $q^2/\alpha_1, \ldots, q^2/\alpha_B$ are the reciprocal roots of $P_3(X, T)$. Also, $P_1(X, T)$ has integral coefficients, so if α_i is a root so is $\overline{\alpha}_i = q/\alpha_i$ by the Riemann hypothesis.

We have

$$\begin{aligned} P_1(X, q^{-1}) &= \prod(1 - q^{-1}\alpha_i) = \prod(1 - \alpha_j^{-1}) = \prod \alpha_j^{-1} \prod(\alpha_j - 1) \\ &= \pm q^{-B/2} P_1(1). \end{aligned}$$

Also, $P_3(X, q^{-1}) = \prod(1 - q\alpha_1^{-1}) = \prod(1 - \alpha_j) = P_1(1)$. So

$$C_X = \frac{\pm P_1(X, 1)^2 q^{-B/2} q}{(q-1)^2} \lim_{T \to q^{-1}} \frac{(1 - qT)^{\rho_1(X)}}{P_2(X, T)}.$$

On the other hand, by the main theorem of Milne's paper [M1] we have

$$\lim_{T\to q^{-1}} P_2(X,T)(1-qT)^{-\rho_1(X)} = \frac{\pm[H_2(X,G_m)]\det(D_i\cdot D_j)}{q^{\alpha(X)}[NS(X)_{\text{tor}}]^2},$$

where the D_i form a base for $NS(X)$ modulo torsion, and

$$\alpha(X) = \chi(X,O_X) - 1 + \dim(\text{Pic Var}(X)).$$

Since by Lemma 3.2, $H^1(X,G_m)$ is dual to $H^3(X,G_m)$, it follows that $\det(D_i\cdot D_j) = R(G_m)$. Also, $\dim(\text{Pic Var}(X)) = B/2$. Since $\text{Pic}(\text{Alb}\,X) = \text{Pic}(X)$, $P_1(X,T) = P_1(\text{Alb}\,X,T) = P_1(\text{Pic Var}X,T)$. It is then well known (see [Mu], pp.180 and 206) that for an abelian variety A, $P_1(A,1) = \#A(k)$. In our case this implies $P_1(X,1) = \#\text{Pic}_0(X)$. Taking into account that $\#\text{Pic}_0(X)\#NS(X)_{\text{tor}} = \#\text{Pic}(X)_{\text{tor}} =$ (by Lemma 3.3) $\#H^3(X,G_m)_{\text{cotor}}$, and that $P_4(q^{-1}) = (1-q) = -\#H^0(X,G_m) = -\#H^4(X,G_m)$ by Lemma 3.3, we see that

$$C_X = \pm\frac{\#H^1(X,G_m)_{\text{tor}}\#H^3(X,G_m)_{\text{cotor}}R(G_m)_q^{\chi(X,O_X)}}{\#H^0(X,G_m)\#H^2(X,G_m)\#H^4(X,G_m)} = \pm\frac{\chi(G_a)}{\chi(G_m)}.$$

References

[De 1] Deligne, P. *La conjecture de Weil*, I. Publ. Math. I.H.E.S. 43 (1974), 273-307.

[G] Gabber, O. *Sur la torsion dans la cohomologie l-adique d'une variété* (to appear).

[K] Kaplansky, I. *Infinite Abelian Groups*, University of Michigan Press, Ann Arbor (1954).

[M1] Milne, J.S. *On a conjecture of Artin and Tate*, Ann of Math. 102 (1975), 517-533.

[M2] Milne, J.S. *Etale Cohomology*, Princeton University Press, Princeton, (1980).

[Mu] Mumford, D. *Abelian Varieties*, Oxford University Press, London (1970).

[T1] Tate, J. *On a conjecture of Birch and Swinnerton-Dyer and a geometric analogue*. Seminaire Bourbaki no. 306, 1965-66, W.A. Benjamin Inc. (1966).

[T2] Tate, J. *Algebraic cycles and poles of zeta-functions*. Arithmetic Algebraic Geometry, Harper and Row, New York, (1965).

[Z] Zarchin, Yu. G. *The Brauer group of abelian varieties over finite fields*, (in Russian) Izv. Akad. Nauk. USSR 46 (1980), 211-243.

Received June 30, 1982
Partially supported by N.S.F. grants

Professor Stephen Lichtenbaum
Department of Mathematics
Cornell University
Ithaca, New York 14853

Canonical Height Pairings via Biextensions

B. Mazur and J. Tate

To I.R. Shafarevich

The object of this paper is to present the foundations of a theory of p-adic-valued height pairings

$$(*) \qquad A(K) \times A'(K) \to \mathbb{Q}_p,$$

where A is a abelian variety over a global field K, and A' is its dual. We say "pairings" in the plural because, in contrast to the classical theory of $\mathbb{R}$-valued) canonical height, there may be many canonical p-adic valued pairings: as we explain in § 4, up to nontrivial scalar multiple, they are in one-to-one correspondence with $\mathbb{Z}_p$-extensions L/K whose ramified primes are finite in number and are primes of ordinary reduction (1.1) for A.

When A also has good reduction at the primes of ramification for L/K, then a different method, introduced by Schneider (cf. [22] for the case of the cyclotomic $\mathbb{Z}_p$-extension) enables one to associate to L/K a p-adic valued pairing (*). We show this to be the same as our pairing.

Our method for the construction of the pairing is first to express the duality between A and A' via the "canonical biextension" of (A, A') by $\mathbb{G}_m$, and then to develop a theory of "canonical local splittings" of biextensions. Our pairings are then defined in a manner analogous to Bloch's definition of the classical $\mathbb{R}$-valued pairing. Whereas for Bloch it suffices to split certain local extensions, to obtain uniqueness we must ask for an especially coherent family of splittings of the local extensions, i.e., a splitting of the local biextension.

We treat simultaneously the $\mathbb{R}$-valued and p-adic valued theories, and express our results in a "uniform" manner in terms of the notion of a Y-valued canonical pairing for a general value group Y satisfying some axioms.

The connection between biextensions and heights is, to be sure, not surprising.

Firstly, Zarhin [24] pointed out that arbitrary (not necessarily canonical) splittings of the canonical biextension are equivalent to Néron type pairings between zero cycles and divisors.

Secondly, biextensions have been used to define theta (and sigma-) functions, as is explained of Breen [5] and in a manuscript in preparation by

Norman [19]. Both of these authors point out that, although the concept of biextensions is not explicitly mentioned in the theories of p-adic theta functions of Mumford, and of Barsotti [2] (see also Cristante [6]) it is directly related to these theories (via the theorem of the cube). One might also try to relate Néron's approach to p-adic theta functions [16], [17], [18] directly to biextensions.

Thirdly, the theory of p-adic heights for elliptic curves of complex multiplication (and p a prime of ordinary reduction) has been developed by Perrin-Riou [20] and Bernardi [4], using a p-adic version of the sigma-function. Here the p-adic sigma-function plays a role analogous to that of the classical sigma-function in Néron's theory [15] for archimedean primes.

Néron has also developed a theory of p-adic valued height pairings using his p-adic theta functions [18]. In the case of elliptic curves of complex multiplication, an explicit connection between Néron's definition and Bernardi's has not yet been made (to our knowledge). What is the relation (if any) between Néron's p-adic height and ours?

Since the explicit expression for the local terms of our canonical p-adic pairing involves the canonical p-adic theta functions (of Mumford and Barsotti), we would find it useful to have a practical algorithm for computing these functions. In a subsequent paper we will discuss this issue in the case of elliptic curves.

In this connection, one should also note that the beginnings of a (mod p)-valued theory of height for general elliptic curves (with ordinary reduction at p) can be found in [21].

Our construction of p-adic valued canonical heights requires ordinary reduction at the primes of ramification of the chosen $\mathbb{Z}_p$-extension. Can one find a generalization or replacement of our construction valid for all $\mathbb{Z}_p$-extensions? For elliptic curves with complex multiplication, B. Gross has some ideas on this; see J. Oesterlé, *Construction de hauteurs archimédiennes et p-adiques suivant la méthode de Bloch*, p. 175–192, in Séminaire de Theorie des Nombres (Séminaire Delange-Pisot-Poitou), Paris, 1980–81; Progress in Math., Vol. 22, Birkhäuser Boston, Basel, Stuttgart, 1982.

We are grateful to M. Artin, L. Breen, J. Coates, L. Moret-Bailly, P. Norman and B. Perrin-Riou for pleasant and informative conversations concerning algebraic spaces, biextensions, theta-functions, and p-adic heights.

We also thank O. Gabber for providing us with significant help in working out §5.

§ 1. Local Splittings

Let K be a field complete with respect to a place v which is either archimedean or discrete. If v is discrete, let $\mathfrak{o} = \mathfrak{o}_K$ denote the ring of v-integers in K, π a prime element in $\mathfrak{o}$ and $k = \mathfrak{o}/\pi\mathfrak{o}$ the residue field.

If $A_{/K}$ is an abelian variety over K, and v is discrete, we denote by A (or sometimes: $A_{/\mathfrak{o}}$) the Néron model of $A_{/K}$ over $\mathfrak{o}$. If v is archimedean, we let A denote $A_{/K}$.

In the non-archimedean case, A^o (or $A^o_{/\mathfrak{o}}$) denotes the connected component of zero in A, i.e., the open subgroupscheme of $A^o_{/\mathfrak{o}}$ whose closed fiber A^o_0 is connected. If $U_{/\mathfrak{o}}$ is any scheme over $\mathfrak{o}$, its closed fiber $U \times_{\mathfrak{o}} k$ is denoted U_0.

(1.1) Ordinary abelian varieties.

A and $A_{/K}$ are called *ordinary* if v is discrete, the characteristic of k is $\neq 0$, and the special fiber of A satisfies the following equivalent conditions.

(i) The formal completion A^f_0 of A_0 at the origin is of multiplicative type, i.e., is isomorphic to a product of copies of $\mathbb{G}^f_m$ over the algebraic closure $\bar{k}$ of k.

(ii) For $p = \operatorname{char} k$, the connected component of the kernel of the homomorphism $p\colon A^o_0 \longrightarrow A^o_0$ is the dual of an étale group scheme over k.

(iii) A^o_0 is an extension of an ordinary abelian variety by a torus T_A.

If L/K is a finite field extension and $A_{/K}$ is ordinary, so is $A_{/L}$ and formation of A^o commutes with the base change of rings $\mathfrak{o}_K \to \mathfrak{o}_L$.

If $A_{/K}$ is ordinary, then $A_{/K}$ has good reduction over k (equivalently: $A_{/\mathfrak{o}}$ is an abelian scheme) if an only if $T_A = 0$.

(1.2) Exponents.

By the *exponent* of a finite abelian group G we mean the smallest integer $m > 0$ such that $mG = 0$.

In this paragraph, suppose that v is discrete. Let $m_A = m_{A/K}$ denote the exponent of $A_0(k)/A^o_0(k)$. Now suppose that k is finite. Let T_A denote the "maximal torus" in A_0 which exists by, e.g., [SGA 3] exposé XIV, Thm. 1.1. Let $n_A = n_{A/K}$ denote the exponent of $A^o_0(k)/T_A(k)$.

We refer to the numbers m_A and n_A as the *exponents* of A.

The exponents are sensitive to isogenous change of A.

As for their dependence on the base field K, $m_{A/K'}$ admits a finite upper bound for all finite *unramified extensions* K'/K, while $n_{A/K'}$ is independent of K' provided A is ordinary and K'/K is a finite *totally ramified extension.* If we drop the assumption that A be ordinary, then $n_{A/K'}$ admits a finite upper bound for all finite totally ramified extensions K'/K.

(1.3) **Biextensions and paired abelian varieties.**

For an introduction to the concept of biextension, we suggest reading §2 and §3 of [12]. For a fuller treatment of this notion, see exposés VII and VIII of [SGA 7I]. A useful and pleasantly written introduction to this fuller treatment may be found in the first $5\frac{1}{2}$ pages of §1 of [5].

Let $A'_{/K}$ denote the dual of $A_{/K}$ and $E^A_{/K}$ the canonical biextension of $(A_{/K}, A'_{/K})$ by $\mathbb{G}_{m/K}$ expressing the duality ([SGA 7 I] Exposé VII, 2.9). If v is archimedean, let E^A denote the canonical biextension $E^A_{/K}$. If v is discrete, let E^A (or $E^A_{/\mathbb{O}}$) denote the canonical biextension of (A^o, A') by $\mathbb{G}_{m/\mathbb{O}}$, i.e., the unique such biextension whose general fiber is $E^A_{/K}$ (whose existence and uniqueness follow from [SGA 7 I] Exposé VII, 7.1b).

If $B_{/K}$ is any abelian variety, to give a biextension $E_{/K}$ of $(A_{/K}, B_{/K})$ by $\mathbb{G}_{m/K}$ is equivalent to giving a K-homomorphism $\lambda\colon B_{/K} \longrightarrow A'_{/K}$ (and $E_{/K}$ is the pullback of $E^A_{/K}$ by $(1,\lambda)$), or to giving a K-homomorphism $\lambda'\colon \longrightarrow A_{/K} B'_{/K}$ (the dual of λ).

Again, if v is archimedean, let E denote $E_{/K}$, while if v is discrete, E (or $E_{/\mathbb{O}}$) will denote the pullback of $E^A_{/\mathbb{O}}$, viewed as biextension of (A^o, B) by $\mathbb{G}_{m/\mathbb{O}}$.

The abelian varieties $A_{/K}, B_{/K}$ will be said to be *paired,* if a biextension $E_{/K}$ of $(A_{/K}, B_{/K})$ by $\mathbb{G}_{m/K}$ (equivalently: a K-homomorphism $\lambda\colon B_{/K} \longrightarrow A'_{/K}$) is fixed.

In what follows, we suppose $A_{/K}, B_{/K}$ are paired abelian varieties over K, with $E_{/K}$ the biextension expressing the pairing.

In all cases, archimedean or non-archimedean, the set $E(K)$ of points of E with coordinates in K is a set theoretical biextension of the groups $(A(K), B(K))$ by $\mathbb{G}_m(K) = K^*$. Our aim is to introduce canonical splittings of biextensions obtained from this one via various types of homomorphisms $\rho\colon K^* \longrightarrow Y$.

(1.4) ρ-**splittings**

Let U, V, W, and Y be abelian groups, X a biextention of (U,V) by W and $\rho\colon W \longrightarrow Y$ a homomorphism. A ρ-*splitting* of X is a map

$$\psi\colon X \longrightarrow Y$$

such that

(i) $\psi(w+x) = \rho(w) + \psi(x)$ for $w \in W$, $x \in X$.

(ii) For each $u \in U$ (resp. $v \in V$) the restriction of ψ to ${}_uX$ (resp. X_v) is a group homomorphism.

(Here ${}_uX$ (resp. X_v) denotes the part of X above $\{u\} \times V$ (resp. $U \times \{v\}$) and is a group extension of V (resp. U) by W.) Note that we are expressing the action of W on X additively. We will continue to do so even when $W = \mathbb{G}_m$.

(1.5) **Canonical** ρ-**splittings.**

Let Y be an abelian group and $\rho\colon K^* \longrightarrow Y$ a homomorphism.

Theorem. *There exists a canonical* ρ*-splitting*[1]

$$\psi_\rho\colon E(K) \longrightarrow Y,$$

in the following three cases:

(1.5.1) v *is archimedean and* $\rho(c) = 0$ *for* c *such that* $|c|_v = 1$.

(1.5.2) v *is discrete,* ρ *is unramified (i.e.,* $\rho(\mathfrak{o}^*) = 0$*), and* Y *is uniquely divisible by* m_A.

(1.5.3) v *is discrete,* k *is finite.* A *is ordinary, and* Y *is uniquely divisible by* $m_A m_B n_A n_B$.

If both (1.5.2) and (1.5.3) hold, they yield the same ψ_ρ.

The image of ψ_ρ satisfies the following inclusion relations:

$$\psi_\rho(E(K)) \subset \begin{cases} \rho(K^*) & \text{in case (1.5.1)} \\ \frac{1}{m_A}\rho(K^*), & \text{in case (1.5.2)} \\ \frac{1}{m_A}\rho(K^*) + \frac{1}{m_A m_B n_A n_B}\rho(\mathfrak{o}^*), & \text{in case (1.5.3).} \end{cases}$$

---

[1] The properties characterizing this ρ-splitting uniquely are explained in (1.9) below.

Before beginning the proof of the theorem, we give some lemmas on set-theoretical biextensions. Let U, V, W and X be as in (1.4), i.e., X a biextension of (U, V) by W. For integers m and n we define a map

$$(m, n): X \longrightarrow X$$

which for each $(u, v) \in U \times V$ takes the fiber ${}_uX_v$ of X over (u, v) to the fiber ${}_{mu}X_{nv}$. The map $(m, 1)$ is defined as multiplication by m in the group X_v for each $v \in V$, the map $(1, n)$ is defined as multiplication by n in the group ${}_uX$ for each $u \in U$, and finally,

$$(m, n) \overset{\text{defn}}{=} (m, 1) \circ (1, n) = (1, n) \circ (m, 1);$$

the commutativity of $(1, n)$ and $(m, 1)$ results from the compatibility axiom for the two laws of composition in a biextension. We have the rules

$$(m_1, n_1)(m, n) = (m_1 m, n_1 n)$$

and

$$(m, n)(x + w) = (m, n)x + mn\, w \qquad \text{for } w \in W.$$

If ρ: $W \longrightarrow Y$ is a homomorphism and ψ is a ρ-splitting, then

$$\psi((m, n)x) = mn\, \psi(x).$$

In particular, if Y is uniquely divisible by m and n, then we have

$$\psi(x) = \frac{1}{mn} \psi((m, n)x).$$

This leads to

(1.6) **Lemma.** *Suppose U^o and V^o are subgroups of U and V, and that m and n are integers > 0 such that $mU \subset U^o$ and $nV \subset V^o$. Let X^o be the part of X lying over $U^o \times V^o$ and let ρ: $W \longrightarrow Y$ be a homomorphism of W into a group Y which is uniquely divisible by mn. Then a ρ-splitting ψ_o: $X^o \longrightarrow Y$ extends uniquely to a ρ-splitting ψ of X.*

Indeed $(m, n)X \subset X^o$. Hence, if ψ extends ψ_o we must have

$$\psi(x) = \frac{1}{mn} \psi_o((m, n)x).$$

On the other hand, it is easy to check that this formula defines a ρ-splitting ψ on all of X, if ψ_o is a ρ-splitting of X^o.

Another case of unique extendibility of ρ-splittings is given by

(1.7) **Lemma.** *Suppose W' is a subgroup of W and X' a subset of X such that X' is a biextension of (U,V) by W'. Let $\rho\colon W \longrightarrow Y$ be a homomorphism and let $\rho' = \rho \mid W'$. A ρ'-splitting ψ' of W' extends uniquely to a ρ-splitting ψ of W.*

Let $W = \bigcup_i (w_i + W')$ be the expression of W as disjoint union of cosets of W'. Then $X = \bigcup(w_i + X')$ is a disjoint union because this is true on each fiber over $U \times V$. If $x = w_i + x' \in w_i + X'$ and ψ extends ψ', we must have

$$\psi(x) = \rho(w_i) + \psi'(x').$$

On the other hand, it is easy to check that this formula defines a ρ-splitting ψ of X, if ψ' is a ρ'-splitting of X'.

(1.8) **Proposition.** *Suppose U and V are compact topological abelian groups and X a topological biextension of (U,V) by $\mathbb{R}$, such that the projection $pr\colon X \longrightarrow U \times V$ has local sections. Then X has a unique continuous splitting $\psi\colon X \longrightarrow \mathbb{R}$.*

Since the space of points of a projective variety over a locally compact field is compact, an immediate consequence of 1.8 is:

(1.8.1) **Corollary.** *Define $v\colon K^* \longrightarrow \mathbb{R}$ by $v(x) = -\log|x|$. If K is locally compact, then $E(K)$ has a unique v-splitting which is continuous from the v-topology in $E(K)$ to the usual topology in $\mathbb{R}$.*

Proof of 1.8. Since the projection $X \to U \times V$ has continuous local sections and its fiber is real affine 1-space, we can use a partition of unity of $U \times V$ to obtain a continuous global section $s\colon U \times V \longrightarrow X$.

Unicity: If ψ_1 and ψ_2 are two splittings, then $(\psi_1 - \psi_2)\circ s$ is a continuous biadditive map of $U \times V$ into $\mathbb{R}$. The image of such a map is bounded and closed under multiplication by 2, so is 0.

Existence: Define $f_0: X \longrightarrow \mathbb{R}$ by

$$f_0(x) = x - s(pr(x)).$$

It is easy to check that

$$\psi(x) \stackrel{\text{defn}}{=} \lim_{n \to \infty} \frac{1}{2^n}\big(f_0(2^n, 1)x\big) \stackrel{\text{thm.}}{=} \lim_{n \to \infty} \frac{1}{2^n} f_0\big((1, 2^n)x\big)$$

is the desired splitting.

(1.9) Reduction of the theorem of (1.5) to §5.
We treat the three cases separately.

Case (1.5.1). (v is archimedean and $\rho(c) = 0$ if $|c|_v = 1$.) Then $K = \mathbb{R}$ or $\mathbb{C}$, and the homomorphism $v: K^* \longrightarrow \mathbb{R}$ defined by $v(c) = -\log|c|$ is surjective. Since $\rho(c)$ depends only on $v(c)$, it follows that there is a (unique) homomorphism $\rho_1: \mathbb{R} \longrightarrow Y$ such that $\rho(c) = \rho_1(v(c))$ for $c \in K^*$. We put then

$$\psi_\rho(x) = \rho_1(\psi_v(x)) \qquad \text{for } x \in E(K),$$

where

$$\psi_v: E(K) \longrightarrow \mathbb{R}$$

is the *unique v-splitting of $E(K)$ which is continuous* (cf. 1.8.1). Clearly, $\psi_\rho(E(K)) \subset \rho(K^*)$.

Case (1.5.2). (v is discrete, ρ is unramified (i.e., $\rho(\mathfrak{o}^*) = 0$), and Y is uniquely divisible by m_A). Since $A^\circ(\mathfrak{o})$ is the subgroup of $A(\mathfrak{o}) = A(K)$ consisting of those points $P \in A(\mathfrak{o})$ whose reduction mod π is contained in $A^\circ(k)$, we have $m_A \cdot A(K) \subset A^\circ(\mathfrak{o})$. Let $E^\circ(K)$ denote that part of $E(K)$ which lies over $A^\circ(\mathfrak{o}) \times B(K)$. We have then a tower of three biextensions as follows:

$$\begin{array}{ccccc}
\mathfrak{o}^* & \subset & K^* & = & K^* \\
\cap & & \cap & & \cap \\
E(\mathfrak{o}) & \subset & E^\circ(K) & \subset & E(K) \\
\downarrow & & \downarrow & & \downarrow \\
A^\circ(\mathfrak{o}) \times B(\mathfrak{o}) & = & A^\circ(\mathfrak{o}) \times B(K) & \subset & A(K) \times B(K)
\end{array}$$

Since $\rho(\mathfrak{o}^*) = 0$, the constant function 0 (i.e., neutral element in Y) is a $(\rho \mid \mathfrak{o}^*)$-splitting of $E(\mathfrak{o})$. Applying (1.7) with $W' = \mathfrak{o}^*$ and $W = K^*$, then

(1.6) with $U^o = A^o(\mathfrak{o})$, $U = A(K)$, $V = V^o = B(K)$, we see that this $(\rho \mid \mathfrak{o}^*)$-splitting of $E(\mathfrak{o})$ extends uniquely to $E(K)$. We can therefore, and do, define in this case ψ_ρ to be the *unique ρ-splitting of $E(K)$ such that* $\rho(E(\mathfrak{o})) = 0$.

Looking at the explicit constructions in the proofs of (1.7) and (1.6) we find that for $x \in E(K)$ there is a unique integer ν such that $(m,1)x \in \pi^\nu + E(\mathfrak{o})$, where $m = m_A$, and for this ν we have

$$\psi(x) = \frac{1}{m}\psi((m,1)x) = \frac{\nu}{m}\rho(\pi) \in \frac{1}{m}\rho(K^*).$$

Case (1.5.3). (v is discrete, k is finite, A is ordinary, and Y is uniquely divisible by $m_A n_A m_B n_B$.) Let T_A and T_B denote the maximal tori in the special fibers of A and B respectively. Let A^t (resp. B^t) denote the formal completion of A (resp. B) along the torus T_A (resp. T_B), and let E^t denote the formal completion of E along the inverse image of $T_A \times T_B$ in E. Then E^t is a biextension (in the category of formal group schemes over $\hat{\mathfrak{o}}$) of (A^t, B^t) by $\hat{\mathbb{G}}_m$. (Here $\hat{\mathfrak{o}}$ denotes $\mathfrak{o}$ viewed as adic-ring. See the technical description of adic-rings in §5. The formal spectrum of $\hat{\mathfrak{o}}$ is denoted by $\hat{S}$. By $\hat{\mathbb{G}}_m$ we mean the formal completion of $\mathbb{G}_{m/\mathfrak{o}}$ along its special fiber $\mathbb{G}_{m/k}$.) Since A is ordinary, it follows from 5.11.1 or 5.12 below that the formal group scheme biextension E^t has a unique splitting, $\psi\colon E^t \longrightarrow \hat{\mathbb{G}}_m$. Taking points with coordinates in $\hat{\mathfrak{o}}$ we obtain a canonical splitting $\psi_{\mathfrak{o}}\colon E^t(\hat{\mathfrak{o}}) \longrightarrow \hat{\mathbb{G}}_m(\hat{\mathfrak{o}}) = \mathfrak{o}^*$ of the set-theoretic biextension $E^t(\hat{\mathfrak{o}})$ of $(A^t(\hat{\mathfrak{o}}), B^t(\hat{\mathfrak{o}}))$ by $\mathfrak{o}^*$, and we define in this case ψ_ρ to be the *unique ρ-splitting of $E(K)$ such that $\psi_\rho \mid E^t(\mathfrak{o}) = \rho \circ \psi_{\mathfrak{o}}$*. Again, the existence and uniqueness of such a ρ-splitting follows from (1.6) and (1.7), once we note (cf. (5.1.1)) that $A^t(\hat{\mathfrak{o}})$ (resp. $B^t(\hat{\mathfrak{o}})$) is the subgroup of points $P \in A(\mathfrak{o})$ (resp. $P \in B(\mathfrak{o})$) whose reduction mod π is contained in $T_A(k)$ (resp. $T_B(k)$), and that $E^t(\hat{\mathfrak{o}})$ is simply the part of $E(\mathfrak{o})$ lying over $A^t(\hat{\mathfrak{o}}) \times B^t(\hat{\mathfrak{o}})$. Thus we can add a still smaller biextension to the left of the diagram under case 1.5.2:

$$\begin{array}{ccc}
\mathfrak{o}^* & = & \mathfrak{o}^* \\
\cap & & \cap \\
E^t(\hat{\mathfrak{o}}) & \subset & E(\mathfrak{o}) \\
\downarrow & & \downarrow \\
A^t(\hat{\mathfrak{o}}) \times B^t(\hat{\mathfrak{o}}) & \subset & A^o(\mathfrak{o}) \times B(K)
\end{array}$$

By (1.7), (1.6) and (1.7) in succession we extend the $\rho \mid \mathfrak{o}^*$-splitting $\rho \circ \psi_{\mathfrak{o}}$ of $E^t(\hat{\mathfrak{o}})$ to $E(\mathfrak{o})$, then to $E^o(K)$ and then to $E(K)$ obtaining finally our canonical ρ-splitting ψ_ρ. For $x \in E(K)$ there is an integer ν such that $(m_A, 1)x \in \pi^\nu + E(\mathfrak{o})$; writing $(m_A, 1)x = \pi^\nu + y$, we have

$$(n_A, m_B n_B)y \in E^t(\hat{\mathfrak{o}})$$

and then $\psi_\rho(x)$ is given explicitly by the formula

$$\psi_\rho(x) = \frac{1}{m_A}\Big(\nu\rho(\pi) + \frac{1}{m_B n_A n_B}\rho\Big(\psi_0((m_A, m_B n_B)y)\Big)\Big).$$

Hence

$$\psi_\rho(x) \in \frac{1}{m_A}\rho(K^*) + \frac{1}{m_A m_B n_A n_B}\rho(\mathfrak{o}^*).$$

If we are simultaneously in case (1.5.2) and (1.5.3), then $\rho \circ \psi_0 = 0$, hence the ψ_ρ we have just defined in case (1.5.3) does indeed coincide with that defined for (1.5.2). This concludes the proof of 1.5, or rather its reduction to our result (5.11, 5.12) on the existence of unique splittings of certain formal group scheme biextensions.

Remark. At the cost of increasing n_A and n_B a bit, one can avoid some of the technical complications of 5.11.1 by replacing the "toric completions" A^t, B^t, E^t by the formal completions A^f, B^f, E^f of A, B and E at the zero points of their special fibers. Then E^f is a biextension of (A^f, B^f) by $\mathbb{G}_m^f$ in the category of formal groups over $\hat{\mathfrak{o}}$, and the existence of a unique splitting for it is given by a proposition of Mumford; see (5.11.5). But then instead of n_A, n_B one must take integers N_A, N_B such that

$$N_A \circ A^\circ(k) = 0 = N_B \circ B^\circ(k).$$

(1.10) Functorial properties of ψ_ρ.

In this paragraph, without stating it explicitly, we suppose that, in each situation considered, the canonical ρ-splittings which occur in formulae are defined, i.e., that we are in one of the three cases (1.5.1-3). If no indication of a proof is given, it is because the stated formula follows immediately from the unique characterization of canonical ρ-splittings given in (1.9).

(1.10.1) *Change of value group, and linearity in ρ.*

Let $\rho\colon K^* \longrightarrow Y$ and $c\colon Y \longrightarrow Y'$ be homomorphisms. Then

$$\psi_{c\rho} = c\psi_\rho.$$

Let $\rho, \rho' : K^* \to Y$ be homomorphisms and $c, c' \in \operatorname{End} Y$. Then

$$\psi_{c\rho+c'\rho'} = c \cdot \psi_\rho + c'\psi_{\rho'}.$$

(1.10.2) *Change of field.*
Suppose $\sigma\colon K \longrightarrow L$ is a continuous homomorphisms of local fields. Let

$$\rho\colon L^* \longrightarrow Y$$

be a homomorphism. Then

$$\psi_{\rho\circ\sigma} = \psi_\rho \circ \sigma,$$

i.e., the diagram

$$\begin{array}{ccc} E(K) & \xrightarrow{\sigma} & E(L) \\ {\scriptstyle \psi_{\rho\circ\sigma}}\downarrow & & \downarrow{\scriptstyle \psi_\sigma} \\ Y & = & Y \end{array}$$

is commutative. Indeed, the right side has the characterizing properties of the left, in each of the three cases.

(1.10.3) *Change of abelian variety.*
First, let A_1, B_1 be abelian varieties over K and $f\colon A_1 \longrightarrow A$, $g\colon B_1 \longrightarrow B$ K-homomorphisms. Let E_1 be the biextension of (A_1, B_1) by $\mathbb{G}_{m/K}$ obtained from E by pullback via (f, g), so that we have a commutative diagram

$$\begin{array}{ccc} E_1 & \underset{\varphi}{\rightarrow} & E \\ \downarrow & & \downarrow \\ A_1 \times B_1 & \xrightarrow{f\times g} & A \times B \end{array}$$

Let $\rho\colon K^* \longrightarrow Y$ be a homomorphism and ψ_ρ (resp. $\psi_{1\rho}$) the canonical ρ-splitting of E (resp. of E_1). Then

$$\psi_{1\rho} = \psi_\rho \circ \varphi.$$

(1.10.4) *Symmetry.*
If E is the biextension pairing A and B, then its "mirror-image" ${}^sE$ (cf. [SGA 7 I], exp. VII, 2.7) is a biextension, pairing B and A. Moreover, there

is a canonical identification of sets ${}^{s}E(K) = E(K)$. Any ρ-splitting of E is a ρ-splitting of ${}^{s}E$.

(1.10.5) *Trace.*

Suppose $L \supset K$ is a finite extension, which we assume Galois for simplicity. Let $G = \mathrm{Gal}(L/K)$ and suppose $\rho\colon L^* \longrightarrow Y$ is a homomorphism. Let $E(L, K)$ be the part of E which lies over $A(L) \times B(K)$, and define a map Tr: $E(L, K) \longrightarrow E(K)$ as the map which is the *trace* $E_b(L) \to E_b(K)$ on the fibers of $E \to B$ for $b \in B(K)$. Then for $x \in E(L, K)$,

$$\psi_{(\rho|K^*)}(\mathrm{Tr}\, x) = \psi_{\sum_{\tau\in G} \rho\circ\tau}(x).$$

Indeed, if $x \in E_b$, $b \in B(K)$,

$$\psi_{(\rho|K^*)}(\mathrm{Tr}\, x) = \psi_{(\rho|K^*)}\Big(\sum_{\tau\in G} \tau x\Big) \qquad \text{(sum in } E_b(L)).$$

By (1.10.2) and (1.10.1) this equals

$$\begin{aligned}\psi_\rho\Big(\sum_\tau \tau x\Big) &= \sum_\tau \psi_\rho(\tau x) = \sum_\tau \psi_{\rho\tau}(x)\\ &= \psi_{\sum_\tau \rho\tau}(x).\end{aligned}$$

In particular, if $\rho\tau = \rho$ for all $\tau \in G$, then

$$\psi_{(\rho|K^*)}(\mathrm{Tr}\, x) = [L:K]\psi_\rho(x). \tag{1.10.5.2}$$

And if $\rho = \theta \circ \mathrm{N}_{L/K}$, where $\theta\colon K^* \longrightarrow Y$, then this becomes

$$[L:K]\psi_\theta(\mathrm{Tr}\, x) = [L:K]\psi_{\theta\circ \mathrm{N}_{L/K}}(x),$$

whence, if Y has no $[L:K]$-torsion,

$$\psi_\theta(\mathrm{Tr}\, x) = \psi_{\theta\circ \mathrm{N}_{L/K}}(x). \tag{1.10.5.3}$$

(1.11) Local universal norms.

In [22] Schneider defined a p-adic height pairing using an approach modelled on Bloch's definition of the archimedean height pairing. To define his pairing, Schneider makes use of "local splittings" obtained by consideration of local universal norms. In this paragraph we investigate the connection between the canonical biextension splittings of (1.5) and Schneider's splittings.

(1.11.1) *Local $\mathbb{Z}_p$-extensions.* Let K be a complete local field with finite residue field. Let $\rho: K^* \longrightarrow \mathbb{Q}_p$ be a non-trivial continuous homomorphism. Such a homomorphism extends uniquely to the profinite completion of K^*

$$\hat{\rho}: \hat{K}^* \longrightarrow \mathbb{Q}_p$$

and by local class field theory, $\hat{\rho}$ determines a $\mathbb{Z}_p$-extension L/K whose $\nu^{\underline{th}}$ layer K_ν/K is the unique cyclic extension of degree p^ν such that

$$\rho(N_{K_\nu/K}K_\nu^*) = p^\nu \cdot \rho(K^*) \subseteq \mathbb{Q}_p.$$

(1.11.2) *Universal ρ-norms.*

Fix a ρ as in (1.11.1). If $G_{/K}$ is any commutative group scheme, let the subgroup of *universal ρ-norms*, $\tilde{G}(K) \subset G(K)$, be defined as the intersection of the images

$$\mathrm{Tr}_{K_\nu/K}: G(K_\nu) \longrightarrow G(K)$$

for all ν.

Examples. If $G = \mathbb{G}_m$, then $\tilde{\mathbb{G}}_m(K)$ is the kernel of the homomorphism $\rho: \mathbb{G}_m(K) = K^* \longrightarrow \mathbb{Q}_p$. The group of universal ρ-norms $\tilde{A}(K)$ for our abelian variety A is of finite index in $A(K)$ if either $A_{/\mathbb{O}}$ is ordinary of good reduction or L/K is unramified, ([11] 4.39; and if L/K is unramified, 4.2 and 4.3).

(1.11.3) *Biextensions of universal ρ-norms.*

Recall that $E(K_\nu, K) \subset E(K_\nu)$ is the set of points which project to $A(K_\nu) \times B(K)$. Define the subset $\tilde{E}(K) \subset E(K)$ to be the intersection of the images of

$$\mathrm{Tr}_{K_\nu/K}: E(K_\nu, K) \longrightarrow E(K)$$

for all ν, (cf. 1.10.5).

(1.11.4) **Lemma.** *If $\tilde{A}(K)$ is of finite index in $A(K)$, then $\tilde{E}(K)$ inherits the structure of a biextension of $(\tilde{A}(K), B(K))$ by $\tilde{\mathbb{G}}_m(K)$ via the natural inclusion*

$$\begin{array}{ccc} \tilde{\mathbb{G}}_m(K) & \subset & \mathbb{G}_m(K) \\ \downarrow & & \downarrow \\ \tilde{E}(K) & \subset & E(K) \\ \downarrow & & \downarrow \\ \tilde{A}(K) \times B(K) & \subset & A(K) \times B(K). \end{array}$$

Proof. For each $b \in B(K)$ we must show that

$$0 \to \tilde{\mathbb{G}}_m(K) \to \tilde{E}_b(K) \to \tilde{A}(K) \to 0$$

is exact. This is (essentially) lemma 3 of §2 of [22].

Schneider's method for constructing his p-adic analytic height may be phrased in terms of biextensions as follows:

(1.11.5) **Lemma.** *Let $\tilde{A}(K)$ be of finite index in $A(K)$. Then there is a unique ρ-splitting of the biextensions $E(K)$ which takes $\tilde{E}(K)$ to zero.*

Proof. An easy application of (1.6) and (1.7). We refer to the above ρ-splitting (which exists when $\tilde{A}(K)$ is of finite index in $A(K)$) as *Schneider's ρ-splitting.*

(1.11.6) **Proposition.** *Suppose that either*

(a) *L/K is unramified,*

or

(b) *$A_{/\mathcal{O}}$ and $B_{/\mathcal{O}}$ have good reduction, and are ordinary.*

Then $\tilde{A}(K)$ is of finite index in $A(K)$ and the canonical ρ-splitting of $E(K)$ is equal to Schneider's ρ-splitting.

Proof. By (1.11.2), $\tilde{A}(K)$ is indeed of finite index in $A(K)$.
For each $\nu \geq 1$, let

$$\rho_\nu = \rho \circ \mathbb{N}_{K_\nu/K} \colon K_\nu^* \longrightarrow \mathbb{Q}_p$$

and $\psi_\nu = \psi_{\rho_\nu}$, the canonical ρ_ν-splitting of $E(K_\nu)$. We have by (1.10.5.3) a commutative diagram:

(1.11.6.1)
$$\begin{array}{ccc} E(K_\nu, K) & \xrightarrow{\psi_\nu} & \mathbb{Q}_p \\ \downarrow \mathrm{Tr}_{K_\nu/K} & & \downarrow id. \\ E(K) & \xrightarrow{\psi} & \mathbb{Q}_p \end{array}$$

where $\psi = \psi_1$ is the canonical ρ-splitting of $E(K)$.

(1.11.7) *Claim.*
There is an integer $c \in \mathbb{Z}$, such that, for all ν,

$$\psi_\nu E(K_\nu) \subseteq p^{-c}\rho_\nu(K_\nu^*).$$

Our proposition follows from (1.11.5) because

$$\rho_\nu(K_\nu^*) = p^\nu \rho(K^*) \subseteq \mathbb{Q}_p$$

and consequently by (1.11.6.1), $\psi\tilde{E}(K)$ is contained in $p^{\nu-c} \cdot \rho(K^*)$ for all ν, hence is zero.

To prove the claim, let $\mathfrak{o}_\nu = \mathfrak{o}(K_\nu)$ be the ring of integers in K_ν. Note that the Néron model of $A_{/K_\nu}$ over $\mathfrak{o}_\nu$ is, under assumption (a) or (b), the base-change to $\mathfrak{o}_\nu$ of A. Let $m_\nu = m_{A/K_\nu}$, $n_\nu = n_{A/K_\nu}$ and $m'_\nu = m_{B/K_\nu}$, $n'_\nu = n_{B/K_\nu}$ be the exponents of $A_{/K_\nu}$ and $B_{/K_\nu}$.

By the theorem in (1.5),

$$\psi_\nu E(K_\nu) \subset \frac{1}{m_\nu}\rho_\nu(K_\nu^*)$$

if L/K is unramified, and

$$\psi_\nu E(K_\nu) \subset \rho_\nu(K_\nu^*) + \frac{1}{n_\nu n'_\nu}\rho_\nu(\mathfrak{o}_\nu^*)$$

if $A_{/K}$ and $B_{/K}$ have good, ordinary reduction.

Note that, in case (a), m_ν is bounded independent of ν (by the number of components of $A_{/\bar{k}}$). If we are not in case (a), then $m_\nu = m'_\nu = 1$ and n_ν, (resp. n'_ν) is bounded independent of ν (by the number of points of A (resp. B) in the residue field of L). Our claim follows.

§ 2. Interpretation in Terms of Zero-Cycles and Divisors; Relation with Néron's Canonical Quasi-Functions

(2.1) The symbol $[\mathfrak{a}, D, c]$.

In the next two paragraphs, K can be any field; $A_{/K}$ is an abelian variety over K, $A'_{/K}$ its dual, and $E = E^A$, cf. (1.3).

Consider the set T of triples $(\mathfrak{a}, D, c)$ consisting of:

(i) A zero cycle of degree zero, $\mathfrak{a} = \sum_i n_i(a_i)$, $a_i \in A(K)$, on A_K, all of whose components are points a_i of A rational over K.

(ii) A divisor D algebraically equivalent to zero on $A_{/K}$, whose support is disjoint from $\mathfrak{a}$.

(iii) An element $c \in K^*$.

A triple $(\mathfrak{a}, D, c) \in T$ determines a point $[\mathfrak{a}, D, c] \in E(K)$ in a well-known manner, and every point of $E(K)$ is of this form. The symbol $[\mathfrak{a}, D, c]$ obeys the following rules.

(2.1.1) $[\mathfrak{a}, D, c]$ lies over the point $(a, b) \in A(K) \times B(K)$, where

$$a = S(\mathfrak{a}) \overset{\text{defn}}{=} \sum_i n_i a_i$$

and

$$b = Cl(D) \overset{\text{defn}}{=} \text{The point in } A'(K) \text{ representing the class of the divisor } D.$$

(2.1.2)
$$[\mathfrak{a}, D, c] = c + [\mathfrak{a}, D, 1] \quad \text{for } c \in K^*.$$

(2.1.3) Addition in ${}_aE(K)$ for $a = S(\mathfrak{a})$ is given by

$$[\mathfrak{a}, D_1, 1] + [\mathfrak{a}, D_2, 1] = [\mathfrak{a}, D_1 + D_2, 1].$$

(2.1.4) Addition in $E_b(K)$ for $b = Cl\ D$ is given by

$$[\mathfrak{a}_1, D, 1] + [\mathfrak{a}_2, D, 1] = [\mathfrak{a}_1 + \mathfrak{a}_2, D, 1].$$

(2.1.5) If f is a rational function on A_K whose support is disjoint from $\mathfrak{a}$ we have

$$[\mathfrak{a}, (f), 1] = [\mathfrak{a}, 0, f(\mathfrak{a})],$$

where

$$f(\mathfrak{a}) = \prod_i f(a_i)^{n_i}, \quad \text{if} \quad \mathfrak{a} = \sum n_i(a_i).$$

(2.1.6) For each D and each $a_0 \in (A - \operatorname{Supp} D)(K)$ there is a K-morphism

$$g = g_{a_0,D}\colon (A_{/K} - \operatorname{Supp} D) \longrightarrow E$$

such that

$$g(a) = [(a) - (a_0), D, 1].$$

The properties (2.1.1)-(2.1.6), for K and its algebraic extensions, *characterize* the symbol $[\mathfrak{a}, D, c]$. Indeed, suppose $[\ ]_1$ and $[\ ]_2$ are two such symbols with those properties. Let their difference be

$$\delta(\mathfrak{a}, D, c) = [\mathfrak{a}, D, c]_1 - [\mathfrak{a}, D, c]_2 \in K^*.$$

This makes sense by (2.1.1). By (2.1.2), $\delta(\mathfrak{a}, D, c)$ is independent of c, and by (2.1.5), it depends only on the class of D. Choosing divisors D_j in this class such that none of them passes through zero, and whose supports have an empty intersection, we find using (2.1.6) that there are morphisms

$$h_i\colon (A - \operatorname{Supp} D_i) \longrightarrow \mathbb{G}_m$$

such that

$$h_i(a) = \delta\big((a) - (0), D_i, 1\big).$$

These fit together to make a morphism

$$h\colon A \longrightarrow \mathbb{G}_m.$$

This h is constant since A is complete, and $h = 0$ because $h(0) = 0$. Since the elements $(a) - (0)$ generate the group of zero cycles of degree 0, this shows that $\delta(\mathfrak{a}, D, c) = 0$ for all $(\mathfrak{a}, D, c)$ and proves unicity.

A further property of the symbol is invariance under translation

$$[\mathfrak{a}_a, D_a, c] = [\mathfrak{a}, D, c], \tag{2.1.7}$$

if $\mathfrak{a}_a$ and D_a denote the images of $\mathfrak{a}$ and D under translation by a. Indeed, the left side has the properties characterizing the right.

(2.2) Interpretations of splittings as pairings between disjoint zero cycles and divisors.

In this paragraph we recall the connection between ρ-splittings and Néron type pairings; cf. Zarhin [24], where also conditions are given for ρ to admit some splitting.

Let P be the set of all pairs $(\mathfrak{a}, D)$ such that $(\mathfrak{a}, D, 1) \in T$. Suppose now $\rho: K^* \longrightarrow Y$ is a homomorphism, and $\psi: E(K) \longrightarrow Y$ is a ρ-splitting of $E(K)$. Put

$$[\mathfrak{a}, D]_\psi \overset{\text{defn}}{=} \psi([\mathfrak{a}, D, 1]).$$

Then it is easy to check that the symbol $[\mathfrak{a}, D]_\psi$, which is defined for $(\mathfrak{a}, d) \in P$, and takes values in Y, satisfies the rules

(2.2.1) $$[\mathfrak{a}, D] \text{ is biadditive.}$$

(2.2.2) $$[\mathfrak{a}, (f)] = \rho(f(\mathfrak{a})).$$

(2.2.3) $$[\mathfrak{a}_a, D_a] = [\mathfrak{a}, D].$$

Conversely, given a symbol satisfying (2.2.1), (2.2.2), and (2.2.3), we obtain a ρ-splitting ψ of $E(K)$ by putting

(2.2.4) $$\psi([\mathfrak{a}, D, c]) \doteq \rho(c) + [\mathfrak{a}, D].$$

This is clear, once one proves that $\psi(x)$ is well defined, i.e., that (2.2.4) is independent of the representation of x as a triple $x = [\mathfrak{a}, D, c]$. We leave the details of this to the reader. *Thus, a ρ-splitting of $E(K)$ is the same as a pairing $[\mathfrak{a}, D]$ satisfying the above three properties.*

(2.3) Canonical pairings.

Now suppose we are in the situation of the theorem of (1.5). In particular, we suppose that K is local as in §1. Define the *canonical ρ-pairing* (on P, with values in Y) by

$$[\mathfrak{a}, D]_\rho = \psi_\rho([\mathfrak{a}, D, 1])$$

where ψ_ρ is the *canonical ρ-splitting* of (1.5).

(2.3.1) **Proposition.** *Suppose $v(x) = -\log|x|$. Then the canonical v-pairing $[\mathfrak{a}, D]_v$ coincides with Néron's symbol $(D, \mathfrak{a})_v$, defined in §9 of [15].*

Indeed, our conditions (2.2.1). (2.2.2), (2.2.3) are Néron's conditions (i), (ii) and (iii′), and according to Néron's Theorem 3 and his remark (d) after its proof, his symbol is characterized by those three conditions together with his condition (iv), which states that the function $a \mapsto [(a)-(a_o),D]_\rho$ is bounded on bounded subsets of $(A - \mathrm{Supp}\ D)(K)$. We have (cf. 2.16)

$$[(a)-(a_o),D]_\rho = \psi_v\big((a)-(a_o),D,1\big) = g_{a_o,D}(a).$$

A morphism like $g_{a_o,D}$ takes bounded sets to bounded sets. It suffices therefore to show that our canonical splitting ψ_v is bounded on bounded subsets of $E(K)$. In the archimedean case (1.5.1) this is true because bounded sets are compact and ψ_v is continuous. In case (1.5.2), it is a consequence of the following lemma, whose proof we leave to the reader.

(2.3.2) **Lemma.** *A bounded subset T of $E(K)$ which lies in $E^0(K) = \bigcup_{\nu\in\mathbb{Z}}(\pi^\nu + E(R))$ is contained in a finite union of the sets $\pi^\nu + E(R)$.*

In cases (1.5.1) and (1.5.2) the canonical ρ-splitting ψ_ρ is determined by ψ_v, because $\rho(x)$ depends only on $v(x)$. Hence, those two cases of the theorem of 1.5 are simply the expression in terms of biextensions of Bloch's interpretation in terms of extensions of Néron's theory of canonical quasi-functions and his pairing $(X,\mathfrak{a})_v$. Thus these two cases are not really new; nor are biextensions essential for them, since the characterizing properties of ψ_v can be expressed in terms of single extensions.

§ 3. Global Fields

In this section K denotes a "global" field, by which we mean at first only this, that there is given a set $\mathcal{V}$ of places of K such that each $v \in \mathcal{V}$ is either archimedean or discrete, and such that, for each $c \in K^*$, we have $|c|_v = 1$ for all but a finite number of $v \in \mathcal{V}$. In particular, the set $\mathcal{V}_\infty$ of archimedean places in $\mathcal{V}$ is finite.

For each $v \in \mathcal{V}$, let K_v denote the completion of K at v, and for $v \notin \mathcal{V}_\infty$, let $\mathfrak{o}_v$ be the ring of integers in K_v, π_v a uniformizer and $k(v) = \mathfrak{o}_v/\pi_v\mathfrak{o}_v$.

Let $A_{/K}$ and $B_{/K}$ be abelian varieties and fix $E_{/K}$ a biextension of $(A_{/K}, B_{/K})$ by $\mathbb{G}_{m/K}$, i.e., $A_{/K}$ and $B_{/K}$ are paired abelian varieties. For each v of $\mathcal{V}$, we consider the local theory for $A_{/K_v}$ discussed in §1, and let the symbols A_v, E_v, for all $v \in \mathcal{V}$; A_v^o, m_{A_v}, for $v \notin \mathcal{V}_\infty; T_{A,v}, n_{A_v}$ for v

such that $k(v)$ is finite have the meanings explained in the beginning of §1; and similarly for the abelian variety $B_{/K}$.

Let Y be a commutative group, and suppose that we are given a family $\rho = (\rho_v)_{v\in\mathcal{V}}$ of homomorphisms $\rho_v \colon K_v^* \longrightarrow Y$, such that $\rho_v(O_v^*) = 0$ for all but a finite number of v's, and such that the "sum formula" $\sum_{v\in\mathcal{V}} \rho_v(c) = 0$ holds for all $c \in K^*$.

Define the topological ring $\mathbb{A}_K$ as the restricted product $\prod'_{v\in\mathcal{V}} K_v$ where a vector $x = (x_v)_{v\in\mathcal{V}}$ is in $\mathbb{A}_K$ if $x_v \in \mathfrak{o}_v$ for all but a finite number of $v \in \mathcal{V} - \mathcal{V}_\infty$. We have a canonical homomorphism $K \to \mathbb{A}_K$. It is convenient to view a family $\rho = (\rho_v)$ as above, as a homomorphism

$$\rho \colon \mathbb{A}_K^* \longrightarrow Y$$

which annihilates the image of $\mathfrak{o}_v^*$ for all but a finite number of $v \in \mathcal{V} - \mathcal{V}_\infty$ as well as the image of K^*.

Suppose that we are given for each $v \in \mathcal{V}$ a ρ_v-splitting, ψ_v, of $E(K_v)$, and that $\psi_v(E(\mathfrak{o}_v)) = 0$ for all but a finite number of v's:

(3.1) Lemma. *Let $\rho = (\rho_v)$ and let (ψ_v) be as just described. There is a unique pairing*

$$(\ ,\ ) \colon A(K) \times B(K) \longrightarrow Y$$

such that if $x \in E(K)$ lies above $(a,b) \in A(K) \times B(K)$, then

$$(a,b) = \sum_{v\in\mathcal{V}} \psi_v(x_v), \tag{3.1.1}$$

where $x_v \in E(K_v)$ is the image of x under the inclusion $K \subset K_v$.

Proof. We first note that (3.1.1) is a finite sum. Indeed, our abelian varieties A, B and the biextension E come from abelian schemes $A_{/R}$, $B_{/R}$ and a biextension E of $(A_{/R}, B_{/R})$ by $\mathbb{G}_{m/R}$ for some finitely generated subring $R \subset K$.

This R is contained in $\mathfrak{o}_v$ for almost all v by the fact that $|c|_v = 1$ for almost all v, for $c \in K^*$. We have $A(K) = \bigcup A(R)$, the union over all such R's, and for $x \in E(R)$ we have $x_v \in E(\mathfrak{o}_v)$ for all but a finite number of v.

We can therefore define a map $\psi \colon E(K) \longrightarrow Y$ by

$$\psi(x) = \sum_{v\in\mathcal{V}} \psi_v(x_v).$$

For each $c \in K^*$ we have then

$$\psi(x+c) = \psi(x)$$

because $\psi_v(x_v + c) = \psi_v(x_v) + \rho_v(c)$ and $\sum_v \rho_v(c) = 0$. Thus the right side of (3.1.1) is independent of the choice of $x \in E(K)$ above (a, b). Moreover, the map $E(K) \to A(K) \times B(K)$ is surjective, because there are local sections (E is a line bundle on $A \times B$, minus its 0-section). Hence (3.1.1) defines a map $A(K) \times B(K) \to Y$. The map so defined is biadditive, because the ψ_v are splittings.

(3.2) *Definition.* If, in the situation of (3.1), the ρ_v-splitting ψ_v is the canonical ρ_v-splitting of Theorem 1.3, for each v, then the pairing (3.1.1) is called the *canonical ρ-pairing*, and is denoted by $(\ ,\ )_\rho$.

(3.3) Thus, the conditions for the canonical ρ-pairing

$$A(K) \times B(K) \to Y$$

to be defined for a family $\rho = (\rho_v)$ of homomorphisms $\rho_v \colon K_v^* \longrightarrow Y$ are

(3.3.1) For each $v \in \mathcal{V}_\infty$, we have $\rho_v(c) = 0$ if $|c|_v = 1$.

(3.3.2) There is a finite subset $S \subset \mathcal{V} - \mathcal{V}_\infty$ (possibly empty) such that $\rho_v(\mathfrak{o}_v^*) = 0$ for $v \notin S \cup \mathcal{V}_\infty$ and such that A_v is ordinary and $k(v)$ finite for $v \in S$.

(3.3.3) *Sum formula* $\sum_{v \in \mathcal{V}} \rho_v(x) = 0$ for $x \in K^*$.

(3.3.4) Y is uniquely divisible by MN, where

$$M = \prod_{v \notin \mathcal{V}_\infty} m_{A_v} \quad \text{and}$$

$$N = \prod_{v \in S} m_{B_v} \cdot n_{A_v} \cdot n_{B_v}.$$

A homomorphism $\rho \colon \mathbb{A}_K^* \longrightarrow Y$ defined by such a family (ρ_v) where (3.3.1-4) are satisfied is called *admissible*. This notion depends upon the "global" field K with its $\mathcal{V}$, $A_{/K}$, $B_{/K}$ and Y. If Y is a uniquely divisible group, then the notion does not depend on $B_{/K}$.

The values of the canonical ρ-pairing are in the following subgroup of Y:

$$\sum_{v \in \mathcal{V}_\infty} \rho_v(K_v^*) + \sum_{v \in S} \frac{1}{m_{A_v}} \left[\rho_v(K_v^*) + \frac{1}{m_{B_v} n_{A_v} n_{B_v}} \rho_v(\mathfrak{o}_v^*) \right] + \sum_{v \notin S, \mathcal{V}_\infty} \frac{1}{m_{A_v}} \rho_v(K_v^*).$$

(3.4) Functorial Properties.

In this paragraph, we let A, B be paired abelian varieties over a "global" field K, and without stating it explicitly we suppose that, in each situation considered, the canonical ρ-pairings which occur in formulae are defined, i.e., that the ρ's which occur are all admissible. Each subparagraph follows directly from its local counterpart in (1.10).

(3.4.1) *Change of value group, and linearity in* ρ.

Let $\rho\colon \mathbb{A}_K^* \longrightarrow Y$ and $c\colon Y \longrightarrow Y'$ be homomorphisms. Then

$$c \cdot (a,b)_\rho = (a,b)_{c\rho}$$

for $a \in A(K)$, $b \in B(K)$.

The group of homomorphisms

$$\rho\colon \mathbb{A}_K^* \longrightarrow Y$$

which are admissible for A, B form an End(Y)-module, and we have:

$$(a,b)_{c\rho + c'\rho'} = c \cdot (a,b)_\rho + c' \cdot (a,b)_{\rho'}$$

for $c, c' \in \operatorname{End}(Y)$, $a \in A(K)$, $b \in B(K)$.

(3.4.2) *Change of field*

Let $\sigma\colon K \longrightarrow L$ be a homomorphism of "global" fields such that L is of finite degree over K, and the chosen set of places $\mathcal{V}_L$ for L is the full "inverse image" of the chosen set of places $\mathcal{V}_K$ for K.

Denote by the same letter σ the induced mapping $\mathbb{A}_K^* \to \mathbb{A}_L^*$ and also the mappings induced on groups of rational points.

Then for an admissible $\rho\colon \mathbb{A}_L^* \longrightarrow Y$ we have

$$(a,b)_{\rho\sigma} = (\sigma a, \sigma b)_\rho$$

for $a \in A(K)$, $b \in B(K)$.

(3.4.3) *Change of abelian variety.*

First, let A_1, B_1 be abelian varieties over K and $f\colon A_1 \longrightarrow A$, $g\colon B_1 \longrightarrow B$ K-homomorphisms. Let E_1 be the biextension of $(A_{1/K}, B_{1/K})$ by $\mathbb{G}_{m/K}$ obtained from E via pullback via (f,g). Then

$$(fa_1, gb_1)_\rho = (a_1, b_1)_\rho$$

for $a_1 \in A_1(K)$, $b_1 \in B_1(K)$.

An important corollary of this rule is the following. Suppose A and B are abelian varieties over K, A' and B' their duals, and E^A (resp. E^B) the biextension of (A, A') (resp. (B, B')) by $\mathbb{G}_m$ expressing the duality. Suppose $f: A \longrightarrow B$ is a homomorphism and $f': B' \longrightarrow A'$ its dual. This means that the pullbacks $(f \times 1_{B'})^* E^B$ and $(1_A \times f')^* E^A$ are canonically isomorphic biextensions of (A, B') by $\mathbb{G}_m$. Hence for $a \in A(K)$ and $b' \in B'(K)$ we have

$$(fa, b')_\rho = (a, f'b')_\rho,$$

where the canonical ρ-pairing on the left is relative to E^B and that on the right is relative to E^A.

(3.4.4) *Symmetry.*

Let ${}^sE$ be the "mirror-image" of the biextension E (cf. [SGA 7 I] exp. VII 2.7). Thus ${}^sE$ is a biextension of $(B_{/K}, A_{/K})$ by $\mathbb{G}_{m/K}$. Then

$$(a, b)_\rho = (b, a)_\rho$$

for $a \in A(K)$, $b \in A(K)$, where the left-hand side refers to the canonical ρ-pairing $A(K) \times B(K) \to Y$ (coming from E) and the right-hand side refers to the canonical ρ-pairing $B(K) \times A(K) \to Y$ (coming from ${}^sE$).

If A is *symmetrically paired* with itself (by a biextension E of $(A_{/K}, A_{/K})$ by $\mathbb{G}_{m/K}$ such that there is an isomorphism of biextension $E \cong {}^sE$) then the canonical ρ-pairing

$$A(K) \times A(K) \to Y$$
$$(a_1, a_2) \mapsto (a_1, a_2)_\rho$$

is a symmetric bilinear form.

Remark. This is notably the case when A is the jacobian of a smooth projective curve $X_{/K}$ and the biextension E is the one determined by the canonical θ-divisor (cf. [13], §2, §3).

(3.4.5) *Trace.*

Let $K \subset L$ be a finite Galois extension of "global" fields. Assume again that $\mathcal{V}_L$ is the full inverse image of $\mathcal{V}_K$. We have a natural norm homomorphism $N_{L/K}: \mathbb{A}_L^* \longrightarrow \mathbb{A}_K^*$ compatible with local norms $N_{L_w/K_v}: L_w^* \longrightarrow K_v^*$

for $v \in \mathcal{V}_K$ and $w \in \mathcal{V}_L$ "lying above" v, and with the global norm $N_{L/K}\colon L^* \longrightarrow K^*$. Assume that Y has no $[L : K]$-torsion. Then for $a \in A(L)$ and $b \in B(K)$

$$(\mathrm{Tr}_{L/K} a, b)_\rho = (a, b)_{\rho \circ N_{L/K}}.$$

(3.5) **Examples of canonical ρ-pairings.**

(3.5.1) *The Néron pairing*: Suppose our places v have absolute values $c \mapsto |c|_v$ satisfying the product formula

$$\prod_{v \in \mathcal{V}} |c|_v = 1.$$

Taking $Y = \mathbb{R}$ and $\rho_v(c) = -\log|c|_v$ for each v we obtain a canonical pairing

$$A(K) \times B(K) \to \mathbb{R}.$$

This is Néron's canonical pairing, as follows immediately from 2.3.1.

(3.5.2) A slight refinement of (3.5.1) which uses essentially the same local splittngs and therefore is in some sense an old story, but which does not seem to have been considered much and whose value we are unable to estimate, is obtained as follows. *Suppose that each* $m_v = 1$. Let

$$W = \bigoplus_{v \in \mathcal{V}} \mathbb{R} e_v$$

be a real vector space with basis elements e_v in one-one correspondence with the places $v \in \mathcal{V}$. Let

$$W_0 = \sum_{v \in \mathcal{V}_\infty} \mathbb{R} e_v + \sum_{v \notin \mathcal{V}_\infty} \mathbb{Z} \log|\pi_v| e_v$$

and let Z denote the subgroup of elements of W of the form

$$\sum_{v \in \mathcal{V}} \log|c|_v e_v, \qquad c \in K^*.$$

Put

$$Y = (W_0/Z)$$

and for each $v \in \mathcal{V}$, put

$$\rho_v(c) = -\log|c|_v + Z \in Y.$$

Then the canonical ρ-pairing

$$A(K) \times B(K) \to Y$$

is defined. Note that if K is a number field, $\mathcal{V}$ the set of all places, and $|c|_v$ the normed absolute value at $v \in \mathcal{V}$, then "dividing" the exact sequence

$$0 \to \sum_{v \in \mathcal{V}_\infty} \mathbb{R}e_v \to W_0 \to (\text{ideal group of } K) \to 0$$

by the sequence

$$0 \to \frac{\text{units of } K}{\text{roots of 1}} \to \frac{K^*}{\text{roots of 1}} \to \begin{pmatrix}\text{principal ideals} \\ \text{of } K\end{pmatrix} \to 0$$

we obtain an exact sequence

$$0 \to \frac{\bigoplus_{v \in \mathcal{V}_\infty} \mathbb{R}}{\text{image of units of } K} \to Y \to (\text{ideal class group of } K) \to 0.$$

Thus Y is (non-canonically) a product of a finite group, a real line, and (card $\mathcal{V}_\infty - 1$) circles.

(3.5.3) *Remarks.* 1) Y maps canonically to the ideal class group, so we get a canonical pairing

$$A(K) \times B(K) \to (\text{ideal class group of } K).$$

This pairing is not new; see [10], [23] for a function field version, and for something in number fields, see Duncan Buell's paper *Elliptic curves and class groups of quadratic fields*, J. London Math. Soc. (2), 15, (1977), 19–25.

2) Let us specialize (3.5.2) to the case of K a real quadratic field of class number one, and $A_{/\mathbb{Q}}$, $B_{/\mathbb{Q}}$ paired abelian varieties over $\mathbb{Q}$.

Then $\mathrm{Gal}(K/\mathbb{Q})$ operates on $A(K)$, $B(K)$ and Y. Let the superscript $^+$ or $^-$ refer to the maximal subgroup on which the nontrivial element of $\mathrm{Gal}(K/\mathbb{Q})$ acts as multiplication by $+1$ or -1. Thus $B(K)^+ = B(\mathbb{Q})$, and the canonical ρ-pairing induces a pairing

$$A(K)^- \times B(\mathbb{Q}) \to Y^- \cong \mathbb{R}/\mathbb{Z}.$$

What is the meaning of this pairing?

§ 4. p-adic Height Pairings

Fix paired abelian varieties $A_{/K}, B_{/K}$ and suppose that K is a global field in the strict sense, i.e., is either a finite extension of $\mathbb{Q}$ or of $F(T)$, F a finite field. Let $\mathcal{V}$ be the set of all places of K.

(4.1) Let $\rho\colon \mathbb{A}_K^* \longrightarrow \mathbb{Q}_p$ be a continuous admissible homomorphism, (for A, K) so that the canonical pairing

$$A(K) \times B(K) \to \mathbb{Q}_p$$
$$(a, b) \mapsto (a, b)_\rho$$

is defined. The space of such homomorphisms forms a $\mathbb{Q}_p$-vector space $V_p = V_p(A, K)$. If $p \neq \operatorname{char} K$, then V_p is finite-dimensional.

The canonical pairing $(\ ,\ )_\rho$ induces a trilinear functional

$$A(K) \otimes \mathbb{Q}_p \times B(K) \otimes \mathbb{Q}_p \times V_p \to \mathbb{Q}_p$$
$$(\alpha, \beta, \rho) \mapsto (\alpha, \beta)_\rho.$$

For $a \in A(K)$, $b \in B(K)$ we have:

(4.1.1) $$(a, b)_\rho \in p^{-\nu} \cdot \rho(\mathbb{A}_K^*) \subset \mathbb{Q}_p$$

where ν is the maximum of the two numbers

(4.1.2) $$\operatorname*{Max}_{v\,finite}(ord_p m_{A_v}) \quad \text{and} \quad \operatorname*{Max}_{v \in S}(ord_p m_{A_v} m_{B_v} n_{A_v} n_{B_v})$$

where S is as in (3.3.2).

(4.2) *Admissible $\mathbb{Z}_p$-extensions.*

Let

$$\rho_{\mathrm{Gal}}\colon G_K \longrightarrow \mathbb{Q}_p$$

be a continuous homomorphisms of the Galois group G_K of an algebraic closure $\overline{K}$ of K into the additive group of p-adic numbers. Then by the reciprocity law, ρ_{Gal} induces a continuous homomorphism

$$\rho\colon \mathbb{A}_K^* \longrightarrow \mathbb{Q}_p.$$

We say that ρ_{Gal} is *admissible* (for A, K) if ρ is.

Since $\mathbb{Q}_p$ is totally disconnected and 2-torsion free, $\rho_v = 0$ for archimedean v. For nonarchimedean v, ρ_v is unramified unless $p = \text{char}\,(k(v))$, because otherwise $\mathfrak{o}_v^*$ has no infinite pro-p-group as quotient.

The homomorphism ρ (and ρ_{Gal}) is admissible for A, if and only if A is ordinary at v if ρ_v is ramified (and if, in the function field case, such places are finite in number).

If ρ_{Gal} is nontrivial, then ρ_{Gal} cuts out a $\mathbb{Z}_p$-extension L/K defined by the condition that ρ_{Gal} factors:

$$\begin{array}{ccc} G_K & \xrightarrow{\rho_{\text{Gal}}} & \mathbb{Q}_p \\ & \searrow \quad \nearrow & \\ & \text{Gal}(L/K) & \end{array}$$

Such a $\mathbb{Z}_p$-extension is called *admissible for* A if ρ_{Gal} is admissible. Thus a $\mathbb{Z}_p$-extension L/K is admissible for A if and only if it is ramified at only a finite set S of places of K, and A_v is ordinary for each $v \in S$. The admissible $\mathbb{Z}_p$-extensions are in $(1:1)$-correspondence with one-dimensional subspaces of $V_p(A, K)$.

(4.3) The determinant form.

For this paragraph and the next, let $B = A'$, and $E = E^A$.

By the Mordell-Weil theorem, the groups $A(K)$ and $A'(K)$ are finitely generated. They are of the same rank since A and A' are isogenous.

Let the common rank be r and let $(P_i)_{1 \leq i \leq r}$, (resp. $(Q_j)_{1 \leq j \leq r}$) be a basis for $A(K)$ (resp. $A'(K)$) mod torsion.

Then for admissible ρ the determinant

$$\delta(A, \rho) \overset{\text{defn}}{=} det_{1 \leq i,j \leq r}(P_i, Q_j)_\rho$$

is defined, and is, up to sign, an invariant of A and ρ. Since $(\ ,\ )_\rho$ is linear in ρ, the determinant $\delta(A)$ is a homogeneous form of degree r on the $\mathbb{Q}_p$-vector space V_p. A line in this space, and also the corresponding $\mathbb{Z}_p$-extension is called *singular (for A)* if $\delta(A, \rho) = 0$ for ρ in the line.

(4.4) Schneider's p-adic analytic height.

Suppose, in this paragraph, that K is a number field and that A has good, ordinary reduction at all places v of K, of residual characteristic p.

Then Schneider [22] has defined a height pairing

$$A(K) \times A'(K) \to \mathbb{Q}_p$$
$$(a, b) \mapsto \langle a, b \rangle_p$$

using "local splittings" defined by universal norms in the cyclotomic $\mathbb{Z}_p$-extension. Define the homomorphism $\rho_c: G_K \longrightarrow \mathbb{Q}_p$ cutting out the cyclotomic $\mathbb{Z}_p$-extensions by $\rho_c(\sigma) = \log_p(u)$ where $u = u(\sigma) \in \mathbb{Z}_p^*$ is defined by $\varsigma^\sigma = \varsigma^u$ for all p-power roots of unity $\varsigma \in \overline{K}$. Here $\log_p$ is the p-adic logarithm.

From (1.11.6) and the definition of Schneider's height pairing $\langle\ ,\ \rangle_p$ we immediately have:

Proposition. *Schneider's height pairing is equal to our canonical ρ_c-pairing.*

Schneider conjectures that this height pairing is nondegenerate.

Remark. Schneider's construction of a p-adic height pairing generalizes immediately to the following kind of ρ_{Gal}: $G_K \longrightarrow \mathbb{Q}_p$; for K any global field.

(4.4.1) *The homomorphism ρ_{Gal} is ramified only at a finite set of valuations v of k, and for each such v, A has good ordinary reduction.*

Then (1.11.6) again insures that Schneider's ρ-pairing coincides with the canonical ρ-pairing.

(4.5) **Global universal norms.**

Let Γ be a topological group isomorphic to $\mathbb{Z}_p$, and written multiplicatively. Let $\Gamma_n = \Gamma^{p^n}$. Let W be a $\mathbb{Z}_p$-module admitting an action of Γ such that if $W_n = W^{\Gamma_n}$ (the fixed submodule under the action of Γ_n) then W_n is a $\mathbb{Z}_p$-module of finite type for each $n \geq 0$, and $W = \bigcup_n W_n$.

Let

$$N_{m,n} = \sum_{\sigma \in \Gamma_n/\Gamma_m} \sigma \in \mathbb{Z}_p[\Gamma_n/\Gamma_m]$$

for integers $m \geq n \geq 0$. Set $N_m = N_{m,0}$

Form the projective limit $\varprojlim_m W_m$, compiled via the mappings

$$N_{m,n}: W_m \longrightarrow W_n \qquad (m \geq n \geq 0).$$

(4.5.8) **Proposition:** *Let $w \in W_0 = W^{\Gamma}$. These conditions are equivalent.*

(1) *The element w is in the image of the natural projection*

$$\lim_{\substack{\leftarrow \\ m}} W_m \to W_0.$$

(1′) *There is a family of elements $w_m \in W_m$ such that $N_{m,n} w_m = w_n$ for $m \geq n \geq 0$, and $w_0 = w$.*

(2) *For each $m \geq 0$ there is an element $w_m \in W_m$ such that $N_m w_m = w$.*

Proof. (1) and (1′) are clearly equivalent, and (1′) $\Rightarrow$ (2) trivially. A standard compactness argument gives that (2) $\Rightarrow$ (1′).

Call an element $w \in W_0$ satisfying the above conditions a *universal norm.*

Let K be a global field.

If $\rho_{\mathrm{Gal}}: G_K \longrightarrow \mathbb{Q}_p$ is a nontrivial continuous homomorphism cutting out the $\mathbb{Z}_p$-extension L/K set $\Gamma = \mathrm{Gal}(L/K)$ and let K_n/K denote the $n^{\underline{th}}$ layer, so that $\Gamma_n = \mathrm{Gal}(L/K_n)$.

Suppose that $A_{/K}$ is an abelian variety; set

$$W = A(L) \otimes \mathbb{Z}_p; \quad W_n = A(K_n) \otimes \mathbb{Z}_p.$$

The universal norms (for the Γ-module W) in $W_0 = A(K) \otimes \mathbb{Z}_p$ form a $\mathbb{Z}_p$-submodule:

$$U_\rho(A_{/K}) \subseteq A(K) \otimes \mathbb{Z}_p$$

which we refer to as the module of *universal ρ-norms.* Fix $B_{/K}$ an abelian variety and $E_{/K}$ a biextension of (A, B) by $\mathbb{G}_m$.

(4.5.2) **Proposition.** *Let ρ satisfy (4.4.1). The universal ρ-norms are degenerate for the canonical ρ-pairing:*

$$(u, \beta)_\rho = 0$$

for all $u \in U_\rho(A_{/K})$ and $\beta \in B(K) \otimes \mathbb{Z}_p$.

Proof. By the functorial property (3.4.3) we may assume that B is the dual of A and $E = E^A$. For each $n \geq 0$, let $a_n \in A(K_n) \otimes \mathbb{Z}_p$ be such that

$Tr_{K_n/K}(a_n) = u$. By (3.4.5)

$$(u,\beta)_\rho = (a_n,\beta)_{\rho_n}$$

where $\rho_n = \rho \circ N_{K_n/K}$.

By (4.1.1), $(a_n,\beta)_{\rho_n} \in \frac{1}{p^{\nu_n}}\rho_n(\mathbb{A}^*_{K_n})$, where ν_n is given in terms of the exponents for $A_{/K_n}$, $B_{/K_n}$ as in (4.1.2). By the same reasoning as in (1.11.7) ν_n admits an upper bound (say N) of n, giving:

$$(u,\beta)_\rho \in p^{n-N}\rho(\mathbb{A}^*_K)$$

for all n.

Remarks. 1. Let $\mathcal{K}$ be a quadratic imaginary field, and p a prime number. The *anti-cyclotomic* $\mathbb{Z}_p$*-extension* of $\mathcal{K}$ is the unique $\mathbb{Z}_p$-extension $\mathcal{L}/\mathcal{K}$ such that $\mathcal{L}/\mathbb{Q}$ is Galois with non-abelian (necessarily "dihedral") Galois group. By considering Birch-Heegner points on factors of jacobians of modular curves one may produce (for any quadratic imaginary field $\mathcal{K}$) examples of prime p, abelian varieties $A_{/\mathbb{Q}}$ and finite field extensions $K/\mathcal{K}$ such that if $\rho\colon G_K \longrightarrow \mathbb{Q}_p$ cuts out the anti-cyclotomic $\mathbb{Z}_p$-extension L/K (where $L = \mathcal{L}\cdot K$) the $\mathbb{Z}_p$-module $U_\rho(A_{/K})$ of universal ρ-norms is of positive rank. ([7], [9]).

We have no examples where $rank_{\mathbb{Z}_p} U_\rho(A_{/K})$ is positive, and ρ cuts out a $\mathbb{Z}_p$-extension different from (a base-change of) an anti-cyclotomic $\mathbb{Z}_p$-extension.

2. Let $D_\rho(A_{/K})$ denote the left-kernel of the canonical ρ-pairing, where ρ is admissible. By (4.5.2)

$$rank_{\mathbb{Z}_p} U_\rho(A_{/K}) \le rank_{\mathbb{Z}_p} D_\rho(A_{/K}).$$

Although at present, we have no examples where $B = A'$, $E = E^A$ and where this is a strict inequality, we expect that such examples exist.

3. An interesting special case to investigate is that in which A is an elliptic curve over $\mathbb{Q}$, with rank $A(\mathbb{Q}) = 1$. Let $P \in A(\mathbb{Q})$ be a generator mod torsion. Let p be a prime of ordinary reduction for A.

Let K be a quadratic imaginary field and $\rho_{a,K}\colon G_K \longrightarrow \mathbb{Q}_p$ a continuous homomorphism which cuts out the anti-cyclotomic $\mathbb{Z}_p$-extension of K.

Note that $V_p(A,K)$ is 2-dimensional, generated by ρ_c (cf. 4.4) and $\rho_{a,K}$.

Using the behavior of our canonical pairing under the action of Gal$(K/\mathbb{Q})$ (cf. (3.4.2)) one sees that

$$(P,P)_{\rho_{a,K}} = 0.$$

Consider these two special cases.

CASE 1. *rank* $A(K) = 1$.
Then the anti-cyclotomic canonical pairing $(\ ,\ )_{\rho_{a,K}}$ is totally degenerate. What are the universal norms? If $(P,P)_{\rho_c} \neq 0$ it follows that the anti-cyclotomic $\mathbb{Z}_p$-extension is the *only* $\mathbb{Z}_p$-extension of K, singular for A.

CASE 2. *rank* $A(K) = 2$.
Let $Q \in A(K)$ be such that P, Q generate $A(K)$ mod torsion, and $\sigma Q = -Q$ for $\sigma \neq 1 \in \mathrm{Gal}(K/\mathbb{Q})$.
Writing $\rho = u\rho_c + v\rho_{a,K} \in V_p$ for $(u,v) \in \mathbb{Q}_p \times \mathbb{Q}_p$, the determinant, $\delta(A,\rho)$, defined in (4.3), is easily seen to be this quadratic form:

$$(P,P)_{\rho_c} \cdot (Q,Q)_{\rho_c} \cdot u^2 - (P,Q)^2_{\rho_{a,K}} \cdot v^2,$$

which represents zero if $(P,Q)_{\rho_{a,K}} = 0$ or if $(P,P)_{\rho_c} \cdot (Q,Q)_{\rho_c}$ is a square in $\mathbb{Q}_p$. It would be interesting to have some cases where $(P,Q)_{\rho_{a,K}} \neq 0$ and $(P,P)_{\rho_c} \cdot (Q,Q)_{\rho_c}$ is a nonzero square in $\mathbb{Q}_p$.[2] In such a case one would have precisely two $\mathbb{Z}_p$-extensions of K which are singular for A. It would then be especially interesting to investigate the arithmetic of A over the various finite layers of these singular $\mathbb{Z}_p$-extensions.

§ 5. Biextensions of Formal Group Schemes

(5.1) **Review of formal schemes.**
Our references for this paragraph are [EGA I] §10 and Knutson's [8], Ch. V, §1. All our rings are assumed to be *noetherian* (and our schemes, formal schemes, etc., will be locally noetherian).
A (noetherian) topological ring A is called *adic* if A has an ideal I (called an *ideal of definition*) such that the topology on A is the I-adic topology, and A is separated and complete for its topology. Since A is noetherian,

---

[2] In this regard, see forthcoming publications of Gudrun Brattström.

A has a largest ideal of definition I (which is the radical of any ideal of definition).

If A is adic, the topological space $\operatorname{Spec}(A/I)$ (for I any ideal of definition) is independent of I, and is called the *formal spectrum* of A.

The formal spectrum of A may be endowed with a canonical local adic-ringed space structure $(\mathcal{X}, \mathfrak{o}_{\mathcal{X}})$, denoted $\operatorname{Spf}(A)$ such that there is a canonical isomorphism $\Gamma(\mathcal{X}, \mathfrak{o}_{\mathcal{X}}) \cong A$ and such that the functor $A \mapsto \operatorname{Spf}(A)$ is a fully faithful functor from the category of adic rings to the category of local adic-ringed spaces. By definition, an *affine formal scheme* is a local adic-ringed space isomorphic to $\operatorname{Spf}(A)$ for some adic A.

The topological ring A is called the *affine coordinate ring* of the affine formal scheme $\operatorname{Spf}(A)$.

The category of *formal schemes* is the (fully faithful) subcategory of local (adic) ringed spaces "modelled on" affine formal schemes. (Cf. [8], [EGA I] §10.)

Recall that $\mathfrak{o}$ is a complete discrete valuation ring with uniformizer π. Let $S = \operatorname{Spec} \mathfrak{o}$, and $S_n = \operatorname{Spec} \mathfrak{o}/\pi^{n+1}\mathfrak{o}$ for $n \geq 0$. In particular, $S_0 = \operatorname{Spec}(\mathfrak{o}/\pi\mathfrak{o}) = \operatorname{Spec} k$ where k is the residue field of $\mathfrak{o}$.

We let $\hat{\mathfrak{o}}$ denote the adic ring $\mathfrak{o}$ with $\pi\mathfrak{o}$ as ideal of definition, and $\hat{S} = \operatorname{Spf}(\hat{\mathfrak{o}})$.

Let $X_{/S}$ be an S-scheme and set $X_n = X \times_S S_n$, viewed as S_n-scheme.

Let $Y \subset X_0$ be a closed subscheme.

If X^Y denotes the completion of X along Y, then X^Y is a formal $\hat{S}$-scheme. Let R be a local $\mathfrak{o}$-algebra separated and complete for the π-adic topology.

Let $\hat{R}$ be R viewed as adic local $\hat{\mathfrak{o}}$-algebra with ideal of definition $\pi\hat{R}$.

Then $\operatorname{Spf} \hat{R}$ is an affine formal scheme and it makes sense to consider $X^Y(\hat{R})$, the $\hat{R}$-valued points of the formal $\hat{S}$-scheme X^Y. This set may be viewed in a natural way as a subset of $X(R)$, the set of R-valued points of the S-scheme X. Which subset? Since R is local, one can reduce this question to the case where X is affine, where it is easily resolved. Specifically, if $x \in X(R)$, denote by $x \times_S S_0 \in X_0(R_0)$ its specialization to the closed fiber. Then

$$X^Y(\hat{R}) = \{x \in X(R) \mid x \underset{S}{\times} S_0 \in Y(R_0)\}.$$

(5.2) Formal group schemes and formal groups.

If S is a formal scheme, a *formal group scheme over* S is a group-object in the category of formal schemes over S. From now on, *formal group scheme* means commutative formal group scheme.

If $\mathcal{G}_{/S}$ is a (formally) *smooth* formal group scheme over the formal scheme S such that the structural morphisms $\mathcal{G} \to S$ induces an isomorphism on underlying topological spaces, we say that $\mathcal{G}$ is a *formal group over* S.

(5.3) *Examples.*

Most of our examples of formal schemes can be obtained by completing a scheme X along a closed subscheme Y.

(5.3.1) *Completion along the closed fiber.*

Let $X_{/S}$ be an S-scheme locally of finite type and set $X_n = X \times_S S_n$. Let $Y = X_{0/S_0}$ be the closed fiber.

Denote by $\hat{X}$ the completion of X along Y. Then $\hat{X}$ is a formal $\hat{S}$-scheme. In the notation of ([EGA I] §10) it is given as an *adic* (S_n)*-system of schemes* $(X_{n/S_n})_n$.

If $X'_{/S}$ is another S-scheme locally of finite type, we have a bijection

$$\mathrm{Hom}_{\hat{S}}(\hat{X}, \hat{X}') \to \varprojlim_{n} \mathrm{Hom}_{S_n}(X_n, X'_n).$$

(5.3.2) *Formal completion at zero.*

Let $X_{/S}$ be a smooth commutative group scheme of dimension d. Let e_0 denote the identity element of the closed fiber X_{0/S_0}. Form X^f, the completion of X at the point e_0. Then X^f is a formal group over the formal scheme $\hat{S}$. Moreover, X^f is an affine formal scheme whose affine coordinate ring is isomorphic to a power series ring in d variables over $\hat{\mathfrak{o}}$ with the maximal ideal as ideal of definition.

(5.3.3) *Formal completion along subtori.*

Let $X_{/S}$ be a smooth (commutative) group scheme, and $T_0 \subset X_0$ a torus over k, contained in the closed fiber. Let X^t denote the formal group scheme over $\hat{S}$ obtained by completing X along T_0. Since T_0 is an affine k-group scheme, a straightforward application of Serre's criterion (also compare [8] V thm. 2.5) shows that X^t is an affine formal group scheme over $\hat{S}$.

There are natural homomorphisms of formal group schemes

$$X^f \to X^t \to \hat{X}.$$

(5.4) *Tori.*

By definition, a *torus* (of dimension d) over any base scheme S is a group scheme over S which, locally for the étale topology, is isomorphic to a product of d copies of $\mathbb{G}_m$.

Let k^s be a separable algebraic closure of k, and $G = \mathrm{Gal}(k^s/k)$. Let R be a local 'artinian ring' with k as residue field. If T is a torus over R, let $T_{0/k}$ denote its closed fiber. The category of tori over R is equivalent to the category of tori over k, the equivalence being given by passage to closed fiber $T \mapsto T_0$; the latter category is anti-equivalent to the category of free abelian groups of finite rank endowed with continuous G-module structures, the anti-equivalence being given by $\mathbb{G}_m$-duality (cf. [SGA 3], exp. VIII).

Consequently, given $T_{0/k}$, for each $n \geq 0$, there is a torus T_{n/S_n} given up to canonical isomorphism, such that $T_n \times_{S_n} S_0 = T_0$.

The system of tori $(T_{n/S_n})_{n\geq 0}$ has the property that $T_n \times_{S_n} S_m \cong T_m$ for $n \geq m$, and determines a formal group scheme $\hat{T}_{/\hat{S}}$ [3] which we refer to as the *formal torus over* $\hat{S}$ determined by $T_{0/k}$.

(5.5) **The geometry of toric completions.**

Let $X_{/S}$ be a commutative smooth group scheme over S, $T_{0/k}$ a torus, and

$$\varphi_0\colon T_0 \longrightarrow X_0$$

a closed immersion of k-group schemes. Let $\hat{T}_{/\hat{S}} = (T_{n/S_n})_{n\geq 0}$ be the formal torus over $\hat{S}$ determined by $T_{0/k}$ as discussed in (5.4).

For each $n \geq 0$, there is a unique closed immersion

$$\varphi_n\colon T_n \longrightarrow X_n$$

of S_n-group schemes extending φ_0 (SGA 3 exp. VIII thm. 5.1).

Invoking [SGA 3] exp VI, we may form the quotient group scheme X_n/T_n. The sequence

$$(5.5.1) \qquad 0 \to T_n \overset{\varphi_n}{\to} X_n \overset{\rho_n}{\to} X_n/T_n \to 0$$

is exact in the sense that ρ_n is faithfully flat (it is, in fact, smooth) with kernel T_n.

The sequence (5.5.1) induces a sequence:

$$(5.5.2) \qquad 0 \to T_n \to (X_n)^t \to Y_n^f \to 0$$

[3] The notation is not misleading. This $\hat{T}$ is, in fact, the completion along the closed fiber of a torus T over S (determined uniquely up to canonical isomorphism), but this is irrelevant to us.

where $(X_n)^t$ is the formal completion of X_n along T_0 and Y_n^f is the formal completion at zero in X_n/T_n.

Since $(X_n)^t$ is canonically $(X^t)_n$, passage to the limit gives a sequence of affine formal group schemes over $\hat{S}$:

$$(5.5.3) \qquad 0 \to \hat{T} \to X^t \to Y^f \to 0.$$

The sequence is exact in the sense that the morphism $X^t \to Y^f$ is formally smooth, with kernel $\hat{T}$. The affine coordinate ring of the $\hat{S}$-formal group Y^f[4] is a power series ring over $\mathfrak{o}$ in ν variables where ν is the codimension of T_0 in X_0.

(5.5.4) *Split tori.*

If T_0 is a split torus over k, then the affine coordinate ring of X^t is isomorphic to an $\mathfrak{o}$-algebra of the form:

$$\varprojlim_n \ \mathfrak{o}/(\pi^n)[[y_1, \ldots, y_\nu]][\tau_1, \tau_1^{-1}, \ldots, \tau_\mu, \tau_\mu^{-1}]$$

where μ is the dimension of T_0. Since T_0 is split $\mathrm{Pic}(T_0) = 0$, and hence $\mathrm{Pic}(X^t) = 0$ ([SGA 2] exp. XI prop. 1.1, plus the fact that T_0 is affine).

(5.6) **Biextensions of formal group schemes over $\hat{S}$.**

Let $\mathcal{A}, \mathcal{B}, \mathcal{C}$ be formal group schemes over $\hat{S}$ and E a biextension of $(\mathcal{A}, \mathcal{B})$ by $\mathcal{C}$ over $\hat{S}$. Thus E is a formal $\hat{S}$-scheme which is a $\mathcal{C}$-torsor over $\mathcal{A} \times_{\hat{S}} \mathcal{B}$, locally trivial for the Zariski topology and, moreover, E is endowed with *two* group-extensions structures

$$(5.6.1) \qquad \begin{aligned} (\epsilon_B)&: 0 \to \mathcal{C}_{\mathcal{B}} \to E_{(\mathcal{B})} \to \mathcal{A}_{\mathcal{B}} \to 0 \\ (\epsilon_A)&: 0 \to \mathcal{C}_{\mathcal{A}} \to E_{(\mathcal{A})} \to \mathcal{B}_{\mathcal{A}} \to 0 \end{aligned}$$

where $\mathcal{A}_{\mathcal{B}} = \mathcal{A} \times_{\hat{S}} \mathcal{B}$ viewed as formal group scheme over $\mathcal{B}$, and $E_{(\mathcal{B})}$ denotes E as $\mathcal{B}$-formal scheme via its natural projection to $\mathcal{B}$, etc.

The group-extensions $(\epsilon_{\mathcal{A}})$, $(\epsilon_{\mathcal{B}})$ are required to be compatible with the $\mathcal{C}$-torsor structure of E, and with each other (cf. [SGA 7 I] exp. VII).

---

[4]The notation is not *totally* misleading. If X_0/T_0 is an abelian variety, there is, in fact, an abelian variety over S whose formal completion at the zero part of the closed fiber is Y^f. See ([SGA 7 I] exp. IX §7). However, we make no use of this fact.

In the specific case that the C-torsor E admits a global section $\sigma : \mathcal{A} \times_{\hat{S}} \mathcal{B} \to E$ (in the category of formal $\hat{S}$-schemes) the biextension structure on E determines, and is determined by two 2-cocycles

$$\text{(5.6.2)} \qquad \begin{aligned} \varphi_\sigma &: (\mathcal{A} \times \mathcal{A})_{\mathcal{B}} \longrightarrow C_{\mathcal{B}} \\ \psi_\sigma &: (\mathcal{B} \times \mathcal{B})_{\mathcal{A}} \longrightarrow C_{\mathcal{A}} \end{aligned}$$

where φ_σ is a morphism in the category of formal $\mathcal{B}$-schemes (a 2-cocycle) determining the group-law in the usual way on the $C_{\mathcal{B}}$-torsor

$$\begin{aligned} \mathcal{A}_{\mathcal{B}} \times_{\mathcal{B}} C_{\mathcal{B}} &\xrightarrow{\sim} E_{(\mathcal{B})} \\ (\alpha, \gamma) &\mapsto (\gamma \cdot \sigma\alpha) \end{aligned}$$

and ψ_σ is similar. For more details, see §2 of [12]. The 2-cocycles $\varphi_\sigma, \psi_\sigma$ satisfy the relations (a), (b), (c) of §2 of [12].

(5.7) **The category of biextensions.**

Let $\mathrm{BIEXT}_{\hat{S}}(\mathcal{A}, \mathcal{B}; C)$ denote the category of biextensions (of formal S-schemes) of $(\mathcal{A}, \mathcal{B})$ by C. See ([5], [SGA 7 I] exp. VII) for a discussion of this category.

Let $\mathrm{Biext}^1_{\hat{S}}(\mathcal{A}, \mathcal{B}; C)$ denote the group of isomorphisms classes of biextensions of $(\mathcal{A}, \mathcal{B})$ by C, and let $\mathrm{Biext}^0_{\hat{S}}(\mathcal{A}, \mathcal{B}; C)$ be the group of automorphisms of the trivial biextension.

(5.7.1) In general, we have that $\mathrm{Biext}^0_{\hat{S}}(\mathcal{A}, \mathcal{B}; C)$ is the group of $\hat{S}$-morphisms h: $\mathcal{A} \times \mathcal{B} \longrightarrow C$ which are bilinear in the sense that the induced mappings

$$h_{\mathcal{B}} : \mathcal{A}_{\mathcal{B}} \longrightarrow C_{\mathcal{B}}; \quad h_{\mathcal{A}} : \mathcal{B}_{\mathcal{A}} \longrightarrow C_{\mathcal{A}}$$

are homomorphisms (of formal $\mathcal{B}$-groups, and $\mathcal{A}$-groups, respectively).

(5.8) **Canonical trivializations.**

These statements are equivalent:

(5.8.1) $\mathrm{BIEXT}_{\hat{S}}(\mathcal{A}, \mathcal{B}; C)$ is equivalent to the punctual category.

(5.8.2) $\mathrm{Biext}^i_{\hat{S}}(\mathcal{A}, \mathcal{B}; C) = 0$, for $i = 0, 1$.

(5.8.3) Every biextensions E of $(\mathcal{A}, \mathcal{B})$ by C in the category of $\hat{S}$-schemes admits a unique trivialization (i.e., a 1_C-splitting ψ: $E \longrightarrow C$; cf. (1.4).

(5.9) Biextensions by $\hat{\mathbb{G}}_{m/\hat{S}}$.

Recall that $\hat{\mathbb{G}}_{m/\hat{S}}$ is the formal completion of $\mathbb{G}_{m/S}$ along its closed fiber $\mathbb{G}_{m/k}$. It is determined by the system $(\mathbb{G}_{m/S_n})_{n\geq 0}$.

If E is a biextension of $(\mathcal{A}, \mathcal{B})$ by $\hat{\mathbb{G}}_m$, the $\hat{\mathbb{G}}_m$-torsor E may be taken to be the complement of the zero-section in a line bundle $\mathcal{L}$ over $\mathcal{A} \times_{\hat{S}} \mathcal{B}$.

In particular, if $\mathrm{Pic}(\mathcal{A} \times \mathcal{B}) = 0$, as is the case if $\mathcal{A}$ and $\mathcal{B}$ are toric completions A^t, B^t along *split* tori, any such E admits a section $\sigma\colon \mathcal{A} \times \mathcal{B} \longrightarrow E$, and hence can be described by the 2-cocyles $\varphi_\sigma, \psi_\sigma$ as in (5.6).

(5.10) Canonical Reductions.

Let U, V be two smooth group schemes over S *with connected fibers.* Let $T(U)_0 \subset U_0$, $T(V)_0 \subset V_0$ denote k-tori contained in the special fibers. Let U^t, V^t be the completions of U, V along $T(U)_0$, and $T(V)_0$ respectively.

Let

$$0 \to \hat{T}(U) \to U^t \to Y^f \to 0 \tag{5.10.1}$$

$$0 \to \hat{T}(V) \to V^t \to Z^f \to 0 \tag{5.10.2}$$

denote the exact sequences of formal group schemes over $\hat{S}$ as provided by (5.5.3).

(5.10.3) **Propositions.** *The functor "inverse image of biextensions" induces an equivalence of categories*

$$\mathrm{BIEXT}_{\hat{S}}(Y^f, Z^f; \hat{\mathbb{G}}_m) \xrightarrow{\sim} \mathrm{BIEXT}_{\hat{S}}(U^t, V^t; \hat{\mathbb{G}}_m).$$

Note that this proposition is very close to the statement of Corollary 3.5 of ([SGA 7 I] exp. VIII). Indeed, it is identical except that we are dealing with formal group schemes rather than group schemes. We shall adapt the proof of that corollary to our situation.

Here we rely on some of the results and notions explained in [8], (especially chapter V). In particular, we shall consider the *global etale topology* of formal $\hat{S}$-schemes (as in loc. cit. V §1) and its induced topos, which we denote $\mathcal{T}$. We also shall consider *formal algebraic spaces* over $\hat{S}$ (loc. cit. II §2) and group objects in that category (*formal group algebraic spaces* over $\hat{S}$).

We may also consider biextensions for the topos $\mathcal{T}$ ([SGA 7 I] exp. VII §2). For example, $\mathrm{BIEXT}_{\mathcal{T}}(U^t, V^t; \hat{\mathbb{G}}_m)$ means the category of biextensions of (U^t, V^t) by $\hat{\mathbb{G}}_m$ in the topos $\mathcal{T}$, i.e., a biextensions E in this category is a *sheaf in* $\mathcal{T}$ (and consequently a formal algebraic space over $\hat{S}$) endowed with the structure of biextension.

A key step in the proof of (5.10.3) is:

(5.10.4) **Proposition.** $\mathrm{BIEXT}_{\mathcal{T}}(\hat{T}(U), V^t; \hat{\mathbb{G}}_m)$ *is equivalent to the punctual category.*

We prepare for its proof.

(5.10.5) **Proposition.** *Any formal group algebraic space over* $\hat{S}$ *is a formal group scheme over* $\hat{S}$.

Proof. Let $\mathcal{B}$ be a formal group algebraic space over $\hat{S}$, and, keeping to the notation of loc. cit., let $\mathcal{B}_1$ denote its *first truncation* (loc. cit. V Defn. 2.4). Then $\mathcal{B}_1$ is an algebraic space over S_0 and it inherits an algebraic space-group structure from that of $\mathcal{B}$. By Murre's theorem ([1], Theorem 4.1, [14]), $\mathcal{B}_1$ is a group scheme over S_0. By (loc. cit. V Theorem 2.5) $\mathcal{B}$ is a formal group scheme.

(5.10.6) **Corollary.** *Let* G_1, G_2 *be formal group schemes over* $\hat{S}$ *(commutative, as always). Then any element e in* $\mathrm{Ext}^1_{\mathcal{T}}(G_1, G_2)$ *is representable by an extension* $0 \to G_2 \to \mathcal{E} \to G_1 \to 0$ *in the category of formal group schemes over* $\hat{S}$.

Now form

$$\hat{M} = \underline{\mathrm{Hom}}_{\mathcal{T}}(\hat{T}(U), \hat{\mathbb{G}}_m)$$
$$\hat{N} = \underline{\mathrm{Ext}}^1_{\mathcal{T}}(\hat{T}(U), \hat{\mathbb{G}}_m),$$

where the *underline* means as sheaves for the étale topology over $\hat{S}$. One easily sees that $\hat{M}$ (the "Cartier dual" of $\hat{T}(U)$) is representable by an adic (S_n)-system of locally constant group schemes (torsion-free of finite rank). But $\hat{N} = 0$. To see this, we may suppose that k is separably algebraically closed, and (using (5.10.6)) we are reduced to proving that if

$$(e) \qquad 0 \to \hat{\mathbb{G}}_m \to \mathcal{E} \to \hat{T}(U) \to 0$$

is an exact sequence of formal $\hat{S}$-group schemes, then (e) splits. But, for each $n \geq 0$, $(e_n) = (e) \times_{\hat{S}} S_n$ may be seen to be an exact sequence of group schemes over S_n, and splits by ([SGA 7 I]exp. VIII, Prop. 3.3.1). Moreover, a splitting of (e_o) determines uniquely a compatible splitting of (e_n) for all n. Thus $\hat{N} = 0$.

It then follows from the general fact ([SGA 7 I] exp. VIII 1.5.2) that

$$\mathrm{Biext}^1_{\mathcal{T}}(\hat{T}(U), V^t; \hat{\mathbb{G}}_m) \cong \mathrm{Ext}^1_{\mathcal{T}}(V^t, \hat{M}).$$

To show that $\mathrm{Biext}^1_{\mathcal{T}}(\hat{T}(U), V^t; \hat{\mathbb{G}}_m)$ vanishes, we note that (5.10.2) represents an exact sequence of sheaves in the topos $\mathcal{T}$ and hence our problem is reduced to the following two vanishing statements:

(5.10.7) $$\mathrm{Ext}^1_{\mathcal{T}}(\hat{T}(V), \hat{M}) = 0.$$

(5.10.8) $$\mathrm{Ext}^1_{\mathcal{T}}(Z^f, \hat{M}) = 0.$$

But (5.10.7) follows immediately from (5.10.6) and the argument of proposition 3.4 of. [SGA 7 I], ext. VIII. To see (5.10.8), consider an exact sequence of formal group schemes over $\hat{S}$.

(5.10.9) $$0 \to \hat{M} \to \mathcal{E} \to Z^f \to 0.$$

Since $Z^f(k) = 0$, and $\mathcal{E} \to Z^f$ is formally étale, there is a unique lifting $\gamma\colon Z^f \longrightarrow \mathcal{E}$ sending the k-valued point of Z^f to zero in $\mathcal{E}(k)$. It is immediate that γ provides a splitting of (5.10.9).

Since the "Cartier dual," $\hat{M} = \underline{\mathrm{Hom}}_{\mathcal{T}}(\hat{T}(U), \hat{\mathbb{G}}_m)$ is étale, one easily sees that $\mathrm{Biext}^0_{\hat{S}}(\hat{T}(U), V^t; \hat{\mathbb{G}}_m) = 0$. The proposition then follows from (5.8).

(5.10.10) *Conclusion of the proof of the proposition* (5.10.3).

By ([SGA 7 I] VII 3.7.6) and the exact sequence (5.10.1), one sees that the category $\mathrm{BIEXT}_{\mathcal{T}}(Y^f, V^t; \hat{\mathbb{G}}_m)$ may be identified up to equivalence with the category whose objects are $\hat{S}$-biextensions E of (U^t, V^t) by $\hat{\mathbb{G}}_m$ supplied with trivializations of the induced biextensions $\mathcal{E}'$ of $(\hat{T}(U), V^t)$ by $\hat{\mathbb{G}}_m$. Morphisms are defined evidently. By (5.10.4) we then have an equivalence of categories

$$\mathrm{BIEXT}_{\mathcal{T}}(Y^f, V^t; \hat{\mathbb{G}}_m) \xrightarrow{\sim} \mathrm{BIEXT}_{\mathcal{T}}(U^t, V^t; \hat{\mathbb{G}}_m)$$

and a symmetrical reduction of V^t to Z^f establishes our proposition.

(5.11) Canonical trivializations.

Let $U, V, T(U)_0, T(V)_0$ be as in (5.10). Thus $U_{/S}, V_{/S}$ are smooth group schemes with connected fibers.

(5.11.1) **Proposition.** *Suppose that U^f (equivalently: Y^f) is of multiplicative type. Then* $\mathrm{BIEXT}_T(U^t, V^t; \hat{\mathbb{G}}_m)$ *is equivalent to the punctual category.*

Proof. Using (5.10.3) it suffices to show

(5.11.2) **Lemma.** *If Y^f is ordinary, then* $\mathrm{BIEXT}_T(Y^f, Z^f; \hat{\mathbb{G}}_m)$ *is punctual.*

This is a simple exercise which can be seen in a variety of ways. For example:

(5.11.3) *Proof of* (5.11.2): In analogy with the proof of (5.10.3), set

$$\begin{aligned} \hat{\hat{M}} &= \underline{\mathrm{Hom}}_T(Y^f, \hat{\mathbb{G}}_m) \\ \hat{\hat{N}} &= \underline{\mathrm{Ext}}^1_T(Y^f, \hat{\mathbb{G}}_m), \end{aligned}$$

Then $\hat{\hat{M}}$ is easily seen to be a projective limit of locally constant finite (formal) group schemes over $\hat{S}$.

Moreover, $\hat{\hat{N}}$ is seen to vanish by the proof of (SGA 7 I exp. VIII, Prop. 3.3.1). Thus $\mathrm{Biext}^1(Y, Z^f; \hat{\mathbb{G}}_m) \rightrightarrows \mathrm{Ext}^1(Z^f, \hat{\hat{M}})$ and the latter group is seen to be zero by the argument demonstrating (5.10.8).

Finally, since $\mathrm{Hom}_S(Z^f, \hat{\hat{M}}) = 0$, we see that

$$\mathrm{Biext}^0_S(Y^f, Z^f; \hat{\mathbb{G}}_m) = 0$$

as well and (5.11.1) follows from (5.8).

(5.11.4) *Paraphrase of part of the above proof, using cocycles.*

From the fact that $\hat{\hat{M}}$ is a projective limit of locally constant finite (formal) group schemes over $\hat{S}$, and that $\hat{\hat{N}} = 0$, the reader may verify that if E is a biextension of (Y^f, Z^f) by $\hat{\mathbb{G}}_m$, then there is a section $\sigma\colon Y^f \times Z^f \longrightarrow E$ such that the 2-cocycle (5.6.2)

$$\psi_\sigma\colon (Y^f \times Y^f)_{Z^f} \longrightarrow \hat{\mathbb{G}}_{m_{Z^f}}$$

is trivial. From the relations satisfied by $\psi_\sigma, \varphi_\sigma$ (cf. [12], §2 (a), (b), (c)) it follows that φ_σ may be viewed as an $\hat{S}$-morphisms from $Z^f \times Z^f$ to

$\hat{M} = \underline{\mathrm{Hom}}_{\hat{S}}(\hat{Y}^f, \hat{\mathbb{G}}_m)$ which takes the unique k-valued point of $Z^f \times Z^f$ to zero. Since $\hat{M}$ is a projective limit of locally constant group schemes, φ_σ is trivial as well and consequently E is the trivial biextension.

(5.11.5) *A mild weakening.*
If one is willing to assume that the formal group Z^f is of "finite height" which is indeed all that occurs in any serious application that we envision, then (5.11.2) also follows immediately from ([12], §5, Prop. 4).

References

[1] M. Artin, *Algebraization of formal moduli*: I, *Global Analysis*. Papers in honor of K. Kodaira. (1970) 21–71. Princeton Univ. Press.

[2] I. Barsotti, *Considerations on theta-functions*, Symposia Mathematica (1970) p. 247.

[3] S. Bloch, *A note on height pairings, Tamagawa numbers and the Birch and Swinnerton-Dyer conjecture*, Inv. Math. *58* (1980), 65–76.

[4] D. Bernardi, *Hauteur p-adique sur les courbes elliptiques*. Séminaire de Théorie des Nombres, Paris 1979-80, Séminaire Delange-Pisot-Poitou. Progress in Math. series vol. 12. Birkhäuser, Boston-Basel-Stuttgart. (1981) 1–14.

[5] L. Breen, *Fonctions thêta et théorème du cube*, preprint of the Laboratoire associé au C.N.R.S. N° 305, Université de Rennes I, 1982.

[6] V. Cristante, *Theta functions and Barsotti-Tate groups*, Annali della Scuola Normale Superiore di Pisa, Serie IV, 7 (1980) 181–215.

[7] M. Harris, *Systematic growth of Mordell-Weil groups of abelian varieties in towers of number fields*, Inv. Math. *51* (1979) 123–141.

[8] D. Knutson, *Algebraic Spaces*, Lecture Notes in Math. *203*, Springer Berlin-Heidelberg-New York, 1971.

[9] P. F. Kurchanov, *Elliptic curves of infinite rank over Γ-extensions*, Math. USSR Sbornik, *19*, 320–324 (1973).

[10] J. Manin, *The Tate height of points on an abelian variety. Its variants and applications*. Izv. Akad. Nauk SSSR Ser. Mat. *28* (1964), 1363–1390. (AMS Translations 59 (1966) 82–110).

[11] B. Mazur, *Rational points of abelian varieties with values in towers of number fields*, Inv. Math. *18* (1972) 183–266.

[12] D. Mumford, *Biextensions of formal groups*, in the *Proceedings of the Bombay Colloquium on Algebraic Geometry*, Tata Institute of Fundamental Research Studies in Mathematics 4, London, Oxford University Press 1968.

[13] D. Mumford, *A remark on Mordell's conjecture*, Amer. J. Math. 87 (1965) 1007–1016.

[14] J. P. Murre, *On contravariant functors from the category of preschemes over a field into the category of abelian groups*, Pub. Math. I. H. E. S. n° 23, 1964.

[15] A. Néron, *Quasi-fonctions et hauteurs sur les variétés abéliennes*, Annals of Math. *82*, n° 2, (1965) 249–331.

[16] A. Néron, *Hauteurs et fonctions thêta*. Rend. Sci. Mat. Milano *46* (1976) 111–135.

[17] A. Néron, *Fonctions thêta p-adiques*, Symposia Mathematica, *24* (1981) 315–345.

[18] A. Néron, *Fonctions thêta p-adiques et hauteurs p-adiques*, pp. 149–174, in Séminaire de theorie des nombres (Séminaire Delange-Pisot-Poitou), Paris 1980–81, Progress in Math., Vol 22, Birkhäuser Boston-Basel-Stuttgart (1982).

[19] P. Norman, *Theta Functions*. Handwritten manuscript.

[20] B. Perrin-Riou, *Descente infinie et hauteur p-adique sur les courbes elliptiques à multiplication complexe*. To appear in Inv. Math.

[21] M. Rosenblum, in collaboration with S. Abramov, *The Birch-Swinnerton-Dyer conjecture mod p*. Handwritten manuscript.

[22] P. Schneider, *p-adic height pairings I*, to appear in Inv. Math.

[23] J. Tate, *Variation of the canonical height of a point depending on a parameter*. To appear in Amer. J. Math.

[24] Ju. G. Zarhin, *Néron Pairing and Quasicharacters*, Izv. Akad. Nauk. SSSR Ser. Mat. Tom 36 (1972) No. 3, 497–509. (Math. USSR Izvestija, Vol. 6 (1972) No. 3, 491–503.)

[EGA I] A. Grothendieck, *Éléments de géométrie algébrigue*. Publications Mathématiques, I. H. E. S. *4* Paris (1961).

[SGA 2] A. Grothendieck, *Cohomologie locale des faisceaux cohérents et Théorèmes de Lefschetz locaux et globaux*, North-Holland Publishing Company - Amsterdam (1968).

[SGA 3] M. Demazure and A. Grothendieck et al, Séminaire de Géométrie Algébrique du Bois Marie 1962/64. *Schémas en Groupes* II. Lecture Notes in Mathematics. *152*, Springer, Berlin-Heidelberg-New York (1970).

[SGA 7 I] A. Grothendieck et al, Séminaire de Géometrie Algébrique du Bois Marie. 1967/69. *Groupes de Monodromie en Géométrie Algébrique*. Lecture Notes in Mathematics. *288*, Springer, Berlin-Heidelberg-New York (1972).

Received October 13, 1982

Supported by NSF

Professor Barry Mazur
Department of Mathematics
Harvard University
Cambridge, Massachusetts 02138

Professor John Tate
Department of Mathematics
Harvard University
Cambridge, Massachusetts 02138

The Action of an Automorphism of $\mathbf{C}$ On a Shimura Variety and its Special Points

J. S. Milne

To I.R. Shafarevich

In [8, pp. 222–223] Langlands made a very precise conjecture describing how an automorphism of $\mathbf{C}$ acts on a Shimura variety and its special points. The results of Milne-Shih [15], when combined with the result of Deligne [5], give a proof of the conjecture (including its supplement) for all Shimura varieties of abelian type (this class excludes only those varieties associated with groups having factors of exceptional type and most types D). Here the proof is extended to cover all Shimura varieties. As a consequence, one obtains a complete proof of Shimura's conjecture on the existence of canonical models. The main new ingredients in the proof are the results of Kazhdan [7] and the methods of Borovoi [2].

In the preprint [7], Kazhdan shows that the conjugate (by an automorphism of $\mathbf{C}$) of the quotient of a Hermitian symmetric domain by an arithmetic group is a variety of the same form. (For a precise statement of what we use from [7], see (3.2).) In sections 2 and 3 we apply this result to prove the following weak form of Langlands's conjecture:

(0.1) let $M^{o}(G, X^{+})$ be the connected Shimura variety defined by a simply-connected semi-simple algebraic group G and Hermitian symmetric domain X^{+}; then, for any automorphism τ of $\mathbf{C}$, there is a connected Shimura variety $M^{o}(G', X'^{+})$ for which there exist compatible isomorphisms

$$\varphi\colon \tau M^{o}(G, X^{+}) \longrightarrow M^{o}(G', X'^{+})$$
$$\psi\colon G_{\mathbf{A}^f} \longrightarrow G'_{\mathbf{A}^f}.G'_{\mathbf{A}^f}.$$

In [2] Borovoi shows that the analogue of (0.1) for non-connected Shimura varieties implies the existence of canonical models for all Shimura varieties. We adapt his methods to show, in section 4, 5, and 6, that (0.1) implies Langlands's conjecture for all connected Shimura varieties.

In the final section we review the main consequences of this result: Langlands's conjecture in its original form; the existence of canonical models in the sense of Deligne; Langlands's conjecture describing the action of complex conjugation on a Shimura variety with a real canonical model; the existence of canonical models in the sense of Shimura. An expository account of this, and related material, can be found in [11].

I am indebted to P. Deligne for several valuable conversations and, especially, for suggestions that led to the elimination of a hypothesis on the congruence nature of arithmetic subgroups in the statement of the main theorem.

Notations. The notations are the same as [4]. In particular, a reductive group G is connected with centre $Z(G)$. A superscript $+$ denotes a topological connected component, and $G(\mathbf{Q})_+$ denotes the inverse image of $G^{ad}(\mathbf{R})^+$ under $G(\mathbf{Q}) \to G^{ad}(\mathbf{R})$. The symbol $\mathbf{S}$ denotes $\mathrm{Res}_{\mathbf{C}/\mathbf{R}}\mathbf{G}_m$, and, for any homomorphism $h: \mathbf{S} \longrightarrow G$, μ_h denotes the restriction of $h_{\mathbf{C}}$ to the first factor in $\mathbf{S}_{\mathbf{C}} = \mathbf{G}_m \times \mathbf{G}_m$.

If (G, X) and (G', X') are pairs defining Shimura varieties, then a map $(G, X) \to (G', X')$ is a homomorphism $G \to G'$ carrying X into X'. An inclusion $(T, h) \to (G, X)$ will always mean that T is a maximal torus of G.

If G is a group over $\mathbf{Q}$, then $G_l = G_{\mathbf{Q}_l}$. By a homomorphism $G_{\mathbf{A}^f} \to G'_{\mathbf{A}^f}$ we mean a family of homomorphisms of algebraic groups $G_l \to G'_l$ whose product maps $G(\mathbf{A}^f)$ into $G'(\mathbf{A}^f)$. (As usual, $\mathbf{A}^f = (\varprojlim \mathbf{Z}/m\mathbf{Z}) \otimes \mathbf{Q}$.) The closure of $G(\mathbf{Q})$ in $G(\mathbf{A}^f)$ is denoted by $G(\mathbf{Q})^-$.

We say that a group G satisfies the Hasse principle for H^i if the map of Galois cohomology groups $H^i(\mathbf{Q}, G) \to \prod_l H^i(\mathbf{Q}_l, G)$ is injective, where l runs through all primes of $\mathbf{Q}$ including $l = \infty$.

If V is an algebraic variety over a field k, and $\tau: k \to K$ is an inclusion of fields, then τV denotes $V \otimes_{k,\tau} K = V \times_{\mathrm{spec}\, k} \mathrm{spec}\, K$.

The main definitions concerning connected Shimura varieties are reviewed in an Appendix.

§1. Statement of the First Theorem

To a pair (G, X^+) satisfying (C) (see the Appendix), a special point $h \in X^+$, and an automorphism τ of $\mathbf{C}$, Langlands [8] associates another pair $({}^{\tau}G, {}^{\tau}X^{+})$ also satisfying (C), a special point ${}^{\tau}h \in {}^{\tau}X^{+}$, and an isomorphism $\psi_\tau = (g \mapsto {}^{\tau}g)\colon G(\mathbb{A}^f) \longrightarrow {}^{\tau}G(\mathbb{A}^f)$. (See also [14]; in general, we shall use the definitions of [14] and [15] rather than [8].)

Theorem 1.1. *Assume that G is simply-connected; then, with the above notations, there exists an isomorphism*

$$\varphi_\tau\colon \tau M^{\circ}(G, X^{+}) \longrightarrow M^{\circ}({}^{\tau}G, {}^{\tau}X^{+})$$

such that

$$(a)\ \varphi_\tau(\tau[h]) = [{}^{\tau}h] \quad (\textit{for the particular special } h)$$
$$(b)\ \varphi_\tau(\tau(gx)) = {}^{\tau}g\varphi_\tau(x), \quad \textit{all } x \in M^{\circ}(G, X^{+}),\ g \in G(\mathbb{A}^f).$$

Remark 1.2. This is a weak form of part (a) of Conjecture C° [15, p. 340]. In §6 we shall see that it leads to a proof of the full conjecture.

Remark 1.3. The real approximation theorem [3, 0.4] shows that $G(\mathbb{Q})_{+}$ is dense in $G(\mathbb{R})_{+}$. Therefore, for any $x \in X^{+}$, $G(\mathbb{Q})_{+}x$ is (real) dense in X^{+}, and its image in $\Gamma \backslash X^{+}$ is Zariski dense. It follows that there is at most one map φ_τ satisfying the conditions of the theorem.

Remark 1.4. Let G be a semi-simple group over $\mathbb{Q}$, and let $i\colon T \to G$ be the inclusion of a maximal torus. Then $\mathrm{Aut}(G, i) = \overline{T} \overset{\mathrm{df}}{=} T/Z(G)$. Fix a finite Galois extension $L/\mathbb{Q}$ and consider triples (G', i', ψ) where $(G', T \overset{i'}{\to} G')$ is isomorphic to (G, i) over L, and ψ is an isomorphism $(G, i)_{\mathbb{A}^f} \overset{\approx}{\to} (G', i')_{\mathbb{A}^f}$ (i.e., an isomorphism $G_{\mathbb{A}^f} \to G'_{\mathbb{A}^f}$ carrying $i_{\mathbb{A}^f}$ into $i'_{\mathbb{A}^f}$). Given such a triple, choose an $a\colon (G, i)_L \overset{\approx}{\to} (G', i')_L$ and define $\beta = \beta(G', i', \psi) \in \overline{T}(\mathbb{A}^f_L)$ by the equation $\psi \circ \beta = a$. Let $\overline{\beta}$ be the image of β in $\overline{T}(\mathbb{A}^f_L)/\overline{T}(L)$. Then $(G', i', \psi) \mapsto \overline{\beta}(G', i', \psi)$ defines a one-to-one correspondence

$$\{\text{isomorphism classes of triples } (G', i', \psi)\} \leftrightarrow \left(\overline{T}(\mathbb{A}^f_L/\overline{T}(L)\right)^{\mathrm{Gal}(L/\mathbb{Q})}.$$

Consider now $(T, h) \overset{i}{\hookrightarrow} (G, X^{+})$. Using the element $\overline{\beta}(\tau, \mu_h)$ explicitly defined in [14, 3.18], one obtains from this correspondence a pair $({}^{\tau}G, {}^{\tau}i\colon T \hookrightarrow {}^{\tau}G)$ together with an isomorphism $\psi_\tau\colon G_{\mathbb{A}^f} \longrightarrow {}^{\tau}G_{\mathbb{A}^f}$ carrying

i into ${}^{\tau}i$. These are the objects in (1.1). The map ${}^{\tau}h\colon \mathbb{S} \to T \overset{{}^{\tau}i}{\hookrightarrow} {}^{\tau}G$ is that whose associated cocharacter is $\tau\mu_h$, and ${}^{\tau}X^+$ is the ${}^{\tau}G^{ad}(\mathbb{R})^+$-conjugacy class containing ${}^{\tau}h$.

One other fact we shall need concerns the class γ of $({}^{\tau}G, {}^{\tau}i)$ in $H^1(\mathbb{Q}, \overline{T})$ (equal to the image of $\overline{\beta}(\tau, \mu_h)$ under

$$(\overline{T}(\mathbb{A}_L^f)/\overline{T}(L))^{\mathrm{Gal}(L/\mathbb{Q})} \xrightarrow{d} H^1(\mathrm{Gal}(L/\mathbb{Q}), \overline{T}(L)) \to H^1(\mathbb{Q}, \overline{T}).$$

The existence of ψ shows that the image of γ in $H^1(\mathbb{Q}_l, \overline{T})$ is zero, for all finite l; its image in $H^1(\mathbb{R}, \overline{T})$ is represented by $\tau\mu(-1)/\mu(-1)$ [14, 3.14].

Remark 1.5. Theorem 1.1 (in fact, Conjecture C^o) is proved in ([5], [15]) for Shimura varieties of abelian type. Since we shall need to make use of this result for groups of type A, we outline the main steps in its proof. For pairs (G, X^+) with G the symplectic group and X^+ the Siegel upper half-space, (1.1) is shown in [15, 7.17] to be a consequence of a statement about abelian varieties of CM-type. This statement is proved in [5]. Let G be of type A, and suppose that G is almost simple over $\mathbb{Q}$. Then G can be embedded into a symplectic group ([4, 2.3.10]), and the following easy lemma can be applied.

Lemma 1.6. *Let (G, X^+) satisfy (C), and let $\overline{H}$ be a reductive subgroup of G^{ad}. Suppose that some $h \in X^+$ factors through $\overline{H}_{\mathbb{R}}$, and let X_H^+ be the $\overline{H}^{ad}(\mathbb{R})^+$-conjugacy class containing the composite h' of h with $\overline{H} \to \overline{H}^{ad}$. Assume that $\overline{H}^{ad}$ satisfies (C_3) and let H be the simply connected covering group of $\overline{H}^{ad}$. Then (H, X_H^+) satisfies (C), and there is an embedding $M^o(H, X_H^+) \hookrightarrow M^o(G, X^+)$ compatible with $H(\mathbb{A}^f) \hookrightarrow G(\mathbb{A}^f)$ under which $[h'] \mapsto [h]$. If h is special, so also is h', and if (1.1) holds for (G, X^+) and h, then it does also for (H, X_H^+) and h'.*

§2. Morphisms of Shimura Varieties

A morphism $\varphi\colon M^o(G, X^+) \longrightarrow M^o(G', X'^+)$ will be said to be finite and étale if, for any $\Gamma' \in \sum(G')$, there exists a $\Gamma \in \sum(G)$ such that

$$\varphi_{\Gamma',\Gamma}\colon \Gamma \setminus X^+ \longrightarrow \Gamma' \setminus X'^+$$

is finite and étale.

Proposition 2.1. *Let (G, X^+) and (G', X'^+) satisfy (C), and let $\psi\colon G_{\mathbf{A}^f} \longrightarrow G'_{\mathbf{A}^f}$ be an isomorphism such that $\psi(G(\mathbf{Q})^-_+) = G'(\mathbf{Q})^-_+$. For any finite étale morphism $\varphi\colon M^\circ(G, X^+) \longrightarrow M^\circ(G', X'^+)$ compatible with ψ, there exists an element $g \in G(\mathbf{Q})^-_+$ and an isomorphism $\psi_o\colon G \xrightarrow{\approx} G'$ such that $\varphi = M^\circ(\psi_o) \circ g$. In particular, φ is an isomorphism.*

Proof. Choose an $h \in X^+$ and write $\varphi([h]) = g'[h']$, some $h' \in X'^+$, $g' \in G'(\mathbf{Q})^-_+$ (see the Appendix). Let $g^{-1} = \psi^{-1}(g')$; then

$$\varphi \circ g\colon M^\circ(G, X^+) \longrightarrow M^\circ(G', X'^+)$$

and

$$\psi \circ \underline{ad}\, g\colon G_{\mathbf{A}^f} \longrightarrow G'_{\mathbf{A}^f}$$

satisfy the same conditions as φ and ψ,and $\varphi \circ g[h] = [h']$. It therefore suffices to prove the following proposition.

Proposition 2.2. *In addition to the hypotheses of (2.1), suppose there exist $h \in X^+$ and $h' \in X'^+$ such that $\varphi([h]) = [h']$. Then ψ is defined over $\mathbf{Q}$ and $\varphi = M^\circ(\psi)$.*

Proof. There exists a unique isomorphism $\tilde{\varphi}\colon X^+ \xrightarrow{\approx} X'^+$ lifting all $\varphi_{\Gamma',\Gamma}\colon \Gamma \setminus X^+ \longrightarrow \Gamma' \setminus X'^+$ and sending h to h'. Let $\tilde{\varphi}_*$ be the map $\alpha \mapsto \tilde{\varphi} \circ \alpha \circ \tilde{\varphi}^{-1}\colon \mathrm{Aut}(X^+) \to \mathrm{Aut}(X'^+)$. For any

$$\alpha \in G^{ad}(\mathbf{Q})^+ \subset \mathrm{Aut}(X^+),$$

$\tilde{\varphi}_*(\alpha)$ induces an automorphism of $M^\circ(G', X'^+)$; in particular, it lies in the commensurability group of any $\Gamma' \in \sum(G')$ and therefore belongs to $G'^{ad}(\mathbf{Q})$ (see [1, Thm. 2]). Consider a $q \in G(\mathbf{Q})_+$, and write q_∞ and q_f for its images in $G^{ad}(\mathbf{Q})_+$ and $G(\mathbf{Q})^-_+$. Then q_∞ and q_f define the same automorphism of $M^\circ(G, X^+)$, and so $\tilde{\varphi}_*(q_\infty)$ and $\psi(q_f)$ define the same automorphism of $M^\circ(G', X'^+)$. They therefore have the same image in $G'^{ad}(\mathbf{Q})^{+\wedge(\mathrm{rel}\, G')} = G'(\mathbf{Q})^-_+ *_{G'(\mathbf{Q})_+} G'^{ad}(\mathbf{Q})^+$. Therefore $\psi(q_f) \in G'(\mathbf{Q}_+)$ (and $\tilde{\varphi}_*(q_\infty) = \psi(q_f)$ in $G'^{ad}(\mathbf{Q})^+$). As $G(\mathbf{Q})_+$ is Zariski dense in G, we conclude that ψ is defined over $\mathbf{Q}$. Write ψ_o for ψ regarded as a $\mathbf{Q}$-rational map, and consider

$$M^\circ(\psi_o)^{-1} \circ \varphi\colon M^\circ(G, X^+) \longrightarrow M^\circ(G, X^+).$$

It remains to show that this map is the identity. We know that it is finite and étale, maps $[h]$ to $[\psi_o^{-1} \circ h']$, and commutes with the action of $G(\mathbf{Q})^-_+$. We have therefore to prove the proposition in the case that $G = G'$ and ψ is the identity map. The equality noted parenthetically in the last paragraph shows that in this case $\tilde{\varphi}_*(q) = q$ for $q \in G(\mathbf{Q})_+/Z(\mathbf{Q})$. The real approximation theorem [3, 0.4] states that $G(\mathbf{Q})_+$ is dense in $G(\mathbf{R})_+$, and so $\tilde{\varphi}_*$ is the identity map on $G^{ad}(\mathbf{R})^+$. As $\tilde{\varphi}_* = \underline{ad}\,\tilde{\varphi}$ this means that $\tilde{\varphi}$ centralizes $G^{ad}(\mathbf{R})^+$, which implies that $\tilde{\varphi} = id$ [18, II 2.6].

Corollary 2.3. *The map*

$$G(\mathbf{Q})^-_+ \mathop{*}_{G(\mathbf{Q})_+} G^{ad}(\mathbf{Q})^+ \to \mathrm{Aut}(M^o(G,X^+))$$

identifies $\mathrm{Aut}_{G(\mathbf{Q})_+}(M^o(G,X^+))$ *with* $\{g * \alpha \mid \underline{ad}\, g = \alpha^{-1} \text{ in } G^{ad}(\mathbf{A}^f)\}$.

Proof. An automorphism of $M^o(G,X^+)$ commuting with the action of $G(\mathbf{Q})_+$ commutes (by continuity) with the action of $G(\mathbf{Q})^-_+$. It can therefore be written $g \circ M^o(\alpha)$ for some $g \in G(\mathbf{Q})^-_+$ and $\alpha \in \mathrm{Aut}(G)$. In order for this map to commute with the action of $G(\mathbf{Q})_+$, α^{-1} and $\underline{ad}\, g$ must be equal. In particular α must be an inner automorphism, $\alpha \in G^{ad}(\mathbf{Q})$, and so the map is that defined by $g * \alpha$.

Corollary 2.4. *If $Z = Z(G)$ satisfies the Hasse principle for H^1, then $Aut_{G(\mathbf{Q})_+}(M^o(G,X^+)) = Z(\mathbf{A}^f) \cap G(\mathbf{Q})^-_+/Z(\mathbf{Q})$; for example, if G is an adjoint group,* $\mathrm{Aut}_{G(\mathbf{Q})_+}(M^o(G,X^+)) = 1$.

Proof. The hypothesis implies that if an element of $G^{ad}(\mathbf{Q})^+$ lifts to an element of $G(\mathbf{A}^f)$, then it lifts to an element of $G(\mathbf{Q})$.

Example 2.5. The last corollary applies to the Shimura varieties defined by simply connected groups without factors of type A_n, $n \geq 8$ (see (3.8) below). For these groups $G(\mathbf{Q})^-_+ = G(\mathbf{A}^f)$ and so

$$\mathrm{Aut}_{G(\mathbf{Q})_+}(M^o(G,X^+)) = Z(\mathbf{A}^f)/Z(\mathbf{Q}).$$

The propositions can also be used to compute the automorphism groups of non-connected Shimura varieties.

Corollary 2.6. *Let (G,X) satisfy [4, 2.1.1.1–2.1.1.3]; then the canonical map* $\left(G(\mathbf{A}^f)/Z(\mathbf{Q})^-\right) *_{G(\mathbf{Q})} G^{ad}(\mathbf{Q}) \to \operatorname{Aut}(M(G,X))$ *identifies*

$$\operatorname{Aut}_{G(\mathbf{A}^f)}(M(G,X))$$

with $\{g * \alpha \mid \underline{ad}(g) = \alpha^{-1} \text{ in } G^{ad}(\mathbf{A}^f)\}$.

Proof. Let $\varphi \in \operatorname{Aut}_{G(\mathbf{A}^f)}(M(G,X))$. Then there exists a $g \in G(\mathbf{A}^f)$ such that $g \circ \varphi$ maps $[h,1]$ to $[h',1]$ for some $h \in X^+$ and $h' \in X'^+$. Then $g \circ \varphi$ maps $M^o(G^{der}, X^+)$ into $M^o(G^{der}, X^+)$ and we can therefore apply 2.2.

Corollary 2.7. *Assume, in (2.6), that the centre Z of G satisfies the Hasse principle for H^1 for finite primes. Then*

$$\operatorname{Aut}_{G(\mathbf{A}^f)}(M(G,X)) = Z(\mathbf{A}^f)/Z(\mathbf{Q})^-.$$

Proof. This follows from (2.6) as (2.4) follows from (2.3).

§3. Proof of a Weak Form of (1.1)

This section is devoted to proving the following result.

Proposition 3.1. *Let (G,X^+) satisfy (C), and assume that G is simply connected; then, for any automorphism τ of $\mathbf{C}$, there is a pair (G',X'^+) satisfying (C) for which there exist compatible isomorphisms*

$$\varphi\colon \tau M^o(G,X^+) \longrightarrow M^o(G',X'^+)$$
$$\psi\colon G_{\mathbf{A}^f} \longrightarrow G'_{\mathbf{A}^f}.$$

We begin by recalling a theorem of Kazhdan. Let X^+ be a Hermitian symmetric domain, so that the identity component G of $\operatorname{Aut}(X^+)$ is a product of connected non-compact simple real Lie groups. For Γ an arithmetic subgroup of G, $\Gamma \backslash X^+$ carries a unique structure of an algebraic variety, and so $\tau(\Gamma \backslash X^+)$ is defined, $\tau \in \operatorname{Aut}(\mathbf{C})$.

Theorem 3.2 (Kazhdan). (a) *The universal covering space X'^+ of $\tau(\Gamma \backslash X^+)$ is a Hermitian symmetric domain.*

(b) *Let G' be the identity component of $\mathrm{Aut}(X'^+)$, and identify the fundamental group Γ' of $\tau(\Gamma \backslash X^+)$ with a subgroup of G'; then Γ' is a lattice in G'.*

Proof. The assumption that Γ is an arithmetic subgroup of G means that there exists a group G_1 over $\mathbf{Q}$ and a surjective homomorphism $f\colon G_1(\mathbf{R})^+ \longrightarrow G$ with compact kernel carrying an arithmetic subgroup of G_1 into a group commensurable with Γ. If G_1 is the symplectic group, then the theorem follows from the theory of moduli varieties of abelian varieties. If G_1 has no $\mathbf{Q}$-simple factor G_0 such that $G_{0\mathbf{R}}$ is of type E_6 or E_7 or has factors of both types $D^{\mathbf{R}}$ and $D^{\mathbf{H}}$, then the $\mathbf{Q}$-simple factors of G_1 can be embedded into symplectic groups, and this case follows from the last case. When $\Gamma \backslash X^+$ is compact (so that G_1 has $\mathbf{Q}$-rank zero), the theorem is proved in [6] (it also follows from Yau's theorem [21] on the existence of Einstein metrics). The remaining cases are treated in [7].

Remark 3.3. If Γ is irreducible (for example, if G is $\mathbf{Q}$-simple) then Γ' is also irreducible because otherwise $\tau(\Gamma\backslash X^+)$, and hence $\Gamma\backslash X^+$, would have a finite étale covering that was a product. Consequently, when $\mathrm{rank}_{\mathbf{R}}\, G' > 1$, Margulis's theorem [10, Thm. 1] shows that Γ' is arithmetic.

Let (G, X^+) be as in the statement of (3.1). In proving the proposition, we can assume that G is almost simple over $\mathbf{Q}$ and is not of type A (because when G is of type A we know much more — see (1.5)). This last assumption implies that G_l is not compact for any l.

Choose a compact open subgroup K of $G(\mathbf{A}^f)$ containing $Z(\mathbf{Q})$, and let $\Gamma = G(\mathbf{Q}) \cap K$ be the corresponding congruence subgroup. Then

$$M^{\circ}_K(G, X^+) = G(\mathbf{Q}) \backslash X^+ \times G(\mathbf{A}^f)/K = \Gamma \backslash X^+.$$

On applying (3.2), one obtains a Hermitian symmetric domain X_1^+, a real Lie group G' such that $G' = \mathrm{Aut}(X_1^+)^+$, and an irreducible lattice Γ' in G' such that $\tau M^{\circ}_K(G, X^+) = \Gamma' \backslash X_1^+$.

For any $g \in G(\mathbf{Q})$, let $\Gamma_g = \Gamma \cap g^{-1}\Gamma g$. There are two obvious maps $1, g\colon \Gamma_g \backslash X^+ \rightrightarrows \Gamma \backslash X^+$, namely the projection map and the projection map preceded by left multiplication by g. On applying τ, we obtain maps

$$\tau(1), \tau(g)\colon \tau(\Gamma_g \backslash X^+) \rightrightarrows \tau(\Gamma \backslash X^+) = \Gamma' \backslash X_1^+.$$

Choose $X_1^+ \to \tau(\Gamma_g \backslash X^+)$ so as to make the following diagram commute with the upper arrows, and choose $\tilde{g}$ to make it commute with the lower arrows:

$$\begin{array}{ccc} X_1^+ & \overset{id}{\underset{\tilde{g}}{\rightrightarrows}} & X_1^+ \\ \downarrow & & \downarrow \\ \tau(\Gamma_g \backslash X^+) & \overset{\tau(1)}{\underset{\tau(g)}{\rightrightarrows}} & \Gamma' \backslash X_1^+ \end{array} \qquad (3.3.1)$$

The double coset $\Gamma'\tilde{g}\Gamma' \subset \mathrm{Aut}(X_1^+)$ is well-defined, and we let

$$\Gamma_0 = \bigcup \Gamma'\tilde{g}\Gamma', \qquad g \in G(\mathbf{Q}).$$

Then Γ_0 is a subgroup of $\mathrm{Aut}(X_1^+)$ and is independent of the choice of K. (In [7] it is denoted by G^σ.)

There is a map $\gamma \mapsto \gamma_f\colon \Gamma_0 \longrightarrow G(\mathbf{A}^f)/Z(\mathbf{Q})$ that can be characterized as follows: for all $\Gamma = K \cap G(\mathbf{Q})$ (as above), and all $g \in G(\mathbf{Q}) \cap \gamma_f K$, the diagram (3.3.1) commutes with $\tilde{g}$ replaced by γ. We have therefore a canonical embedding

$$\gamma \mapsto (\gamma_\infty, \gamma_f)\colon \Gamma_0 \longrightarrow \mathrm{Aut}(X_1^+) \times G(\mathbf{A}^f)/Z(\mathbf{Q}).$$

Both γ_∞ and γ_f act on $\tau M^\circ(G, X^+)$, the first through its action on X_1^+ and the second through its action on $M^\circ(G, X^+)$. These actions are equal.

Lemma 3.4. *Regard* Γ_0 *as a subgroup of* $G(\mathbf{A}^f)/Z(\mathbf{Q})$.

(a) Γ_0 *is dense in* $G(\mathbf{A}^f)/Z(\mathbf{Q})$.
(b) $\Gamma_0 \cap Z(\mathbf{A}^f)/Z(\mathbf{Q}) = 1$.
(c) *For any compact open subgroup* K *of* $G(\mathbf{A}^f)$ *containing* $Z(\mathbf{Q})$,

$$\tau M_K^\circ(G, X^+) = (\Gamma_0 \cap K/Z(\mathbf{Q})) \backslash X_1^+ = \Gamma_0 \backslash X_1^+ \times G(\mathbf{A}^f)/K.$$

Proof. (a) it is clear from the definition of $\gamma \mapsto \gamma_f$ that, for any K, $\Gamma_0 \to G(\mathbf{A}^f)/Z(\mathbf{Q})K$ is surjective.

(b) Suppose $\gamma \in \Gamma_0$ is such that $\gamma_f \in Z(\mathbf{Q}^f)/Z(\mathbf{Q})$. Then the remark preceding the statement of the lemma shows that γ_∞ centralizes Γ_0 in G'. As Γ_0 has finite covolume in G', this implies that γ_∞ is in the centre of G' [17, 5.4, 5.18], and so $\gamma_\infty = 1$.

(c) We can assume that K is the group used in the construction of Γ_0. It is then clear that $\Gamma' = \Gamma_0 \cap K/Z(\mathbf{Q})$. The second equality follows from the first and (a).

Later we shall show that Γ_0 is contained in the identity component G' of $\mathrm{Aut}(X_1^+)$, but for the present we define $\Gamma_0^+ = \Gamma_0 \cap G'$.

We now fix an integral structure for G and define, for any finite set S of finite primes, $G(\mathbf{A}_S^f) = \prod_{l \in S} G(\mathbf{Q}_l) \times \prod_{l \notin S} G(\mathbf{Z}_l)$. Let $\Gamma_{0,S} = \Gamma_0 \cap G(\mathbf{A}_S^f)$; then $\Gamma_{0,S}^+ \stackrel{\mathrm{df}}{=} \Gamma_{0,S} \cap \Gamma_0^+$ can be regarded as a subgroup of $G'_S \stackrel{\mathrm{df}}{=} G' \times \prod_{l \in S} G_l$.

Lemma 3.5. *The group $\Gamma_{0,S}^+$ is an irreducible lattice in G'_S.*

Proof. It follows from (3.4c) that Γ_0 is a discrete subgroup of $G' \times G(\mathbf{A}^f)$, and therefore $\Gamma_{0,S}^+$ is a discrete subgroup of $G' \times G(\mathbf{A}_S^f)$. As $\prod_{l \notin S} G(\mathbf{Z}_l)$ is compact, the projection $G'_S \times \prod_{l \notin S} G(\mathbf{Z}_l) \to G'_S$ takes discrete groups to discrete groups [20, p. 4], and so $\Gamma_{0,S}^+$ is discrete in G'_S.

Let U be a compact open subgroup of $\prod_{l \in S} G_l$, and let

$$\Gamma = (U \times \prod_{l \notin S} G(\mathbf{Z}_l)) \cap \Gamma_0^+.$$

Then $U\Gamma_{0,S} \setminus G'_S = \Gamma' \setminus G'$, which carries an invariant finite measure. It follows that $\Gamma_{0,S}$ is of finite covolume in G'_S.

Our assumption that G is almost simple over $\mathbf{Q}$ implies that $\Gamma_{0,S}$ is irreducible (cf. 3.3).

Now assume that S is nonempty and sufficiently large that $\mathrm{rank}(G'_S) \geq 2$. Then Margulis's theorem [10, Thm. 7] shows that $\Gamma_{0,S}^+$ is arithmetic. More precisely, there is the following result.

Lemma 3.6. *There exists an algebraic group G_1 over $\mathbf{Q}$ and a map $\psi_S \colon G_{1S} \longrightarrow G'_S$, where $G_{1S} = G_{1\mathbf{R}} \times \prod_{l \in S} G_{1,l}$, having the following properties:*

(a) *there exists an S-arithmetic subgroup Γ_S in $G_1(\mathbf{Q})$ such that $\psi_S(\Gamma_S)$ is commensurable with $\Gamma_{0,S}^+$;*

(b) *write $\psi_S = \psi_\infty \times \prod_{l \in S} \psi_l$; then ψ_∞ is surjective with a compact kernel, and each ψ_l is an isomorphism.*

Moreover, (G_1, ψ_S) is uniquely determined by the conditions (a) and (b).

Proof. Margulis's theorem gives us a pair (G_1, ψ_S) satisfying (a) and such that ψ_S is surjective with a compact kernel. We can suppose that Γ_S is irreducible. Then G_1 is almost simple over $\mathbf{Q}$ and so cannot be of type A. Therefore G_{1l} does not have any compact factors and ψ_l must be an isomorphism.

Let Λ be a subgroup of $\Gamma_{0,S}$ of finite index, and let $A_\Lambda = \{(f_l) \in \prod_{l \in S} \Gamma(G_l, \mathcal{O}_{G_l}) \mid f_l(\lambda) \in \mathbf{Q},\ f_l(\lambda) = f_{l'}(\lambda), \text{ all } l, l' \in S, \lambda \in \Lambda\}$. Then, for all sufficiently small Λ, A_Λ is independent of Λ and $\operatorname{Spec} A_\Lambda = G_1$. This shows the uniqueness.

When we enlarge S, to S' say, then G_1 does not change and $\psi_{S'}|G_{1S} = \psi_S$. We therefore get a map

$$\psi = \psi_\infty \times \psi^f : G_{1\mathbf{R}} \times G_{1\mathbf{A}^f} \longrightarrow G' \times G_{\mathbf{A}^f}$$

such that for all finite S, $\psi^f(G_{1,S})$ is commensurable with $\Gamma_{0,S}$, where $G_{1,S} = G_1(\mathbf{Q}) \cap G_1(\mathbf{A}^f_S)$. Let G_1^{comp} be the product of the anisotropic factors of $G_{1\mathbf{R}}$, and consider

$$\begin{array}{ccc} G^1(\mathbf{Q}) & \hookrightarrow & (G_1/G_1^{comp})(\mathbf{R}) \times G_1(\mathbf{A}^f) \\ & & \downarrow \bar{\psi} \\ \Gamma_0 & \hookrightarrow & \operatorname{Aut}(X_1^+) \times G(\mathbf{A}^f)/Z(\mathbf{Q}). \end{array}$$

For any finite set S, $\tilde{\Gamma}_{0,S} \stackrel{\mathrm{df}}{=} \bar{\psi}^{-1}(\Gamma_{0,S})$ is commensurable with $G_1(\mathbf{Q})_S$. Ultimately we shall show that $\Gamma_0 = G_1(\mathbf{Q})/Z(\mathbf{Q})$, but we begin with a weaker result.

Lemma 3.7. *The group* $\Gamma_0 \subset \psi(G_1(\mathbf{Q})) \cdot Z(\mathbf{A}^f_S)$.

Proof. Let $\gamma \in \Gamma_0$, and let $\tilde{\gamma} \in \operatorname{Aut}(X_1^+) \times G_1(\mathbf{A}^f)$ map to γ regarded as an element of $\operatorname{Aut}(X_1^+) \times G(\mathbf{A}^f)/Z(\mathbf{Q})$. For large enough S, $\gamma \in \Gamma_{0,S}$, and $\underline{ad}\,\tilde{\gamma}$ maps $\tilde{\Gamma}_{0,S}$ into $\tilde{\Gamma}_{0,S}$.

Choose an irreducible representation $T: G_1^{ad} \hookrightarrow GL_n$ of G_1^{ad}, and consider the representation $\underline{ad}\,\tilde{\gamma} \circ T$ of $\tilde{\Gamma}_{0,S} \cap G_1(\mathbf{Q})$. The Zariski closure of $(\underline{ad}\,\tilde{\gamma} \circ T)(\tilde{\Gamma}_{0,S} \cap G_1(\mathbf{Q}))$ in GL_n is $T(G_1^{ad})$, and so [10, Thm. 8] can be applied to show that there is a morphism $\bar{\alpha}: G_1 \longrightarrow G_1^{ad}$ whose restriction to $\tilde{\Gamma}_{0,S}$ is $\underline{ad}\,\tilde{\gamma}$. Lift $\bar{\alpha}$ to an isomorphism $\alpha: G_1 \longrightarrow G_1$. Then α and $\underline{ad}\,\tilde{\gamma}$ agree on a subgroup of $\tilde{\Gamma}_{0,S}$ of finite index (therefore also on a subgroup of $G_1(\mathbf{Q})_S$ of finite index). As $\Gamma_{0,S}$ is dense in $\prod_{l \in S} G(\mathbf{Q}_l)/Z(\mathbf{Q})$, this shows that

$\alpha = \underline{ad}\,\gamma$ on $\prod G(\mathbf{Q}_l)/Z(\mathbf{Q})$. In particular, α is an inner automorphism, i.e., $\alpha \in G_1^{ad}(\mathbf{Q})$. Moreover, α has the property that it lifts to $G_1(\mathbf{Q}_l)$ for all $l \in S$, and hence for all l because we can extend S. The next lemma shows that this implies that α lifts to an element $\alpha_1 \in G_1(\mathbf{Q})$. This element α_1 has the same image as γ in $\mathrm{Aut}(X_1^+) \times G^{ad}(\mathbf{A}^f)$, which completes the proof.

Lemma 3.8. *Let G be a simply connected semi-simple group over a number field k, and let $Z = Z(G)$. Then $H^1(k, G) \to \prod_{v\ finite} H^1(k_v, G)$ is injective, provided G has no factors of type A_n, $n \geq 4$.*

Proof. We can assume G is absolutely almost simple. Then $Z(\bar{k}) = \mathbf{Z}/2\mathbf{Z} \times \mathbf{Z}/2\mathbf{Z}$ or $\mathbf{Z}/n\mathbf{Z}$, $n \leq 4$. If $Z(\bar{k}) = Z(k)$, then the result is obvious from class field theory. In any case, $Z(\bar{k}) = Z(L)$ for L a Galois extension of k with Galois group S_3 or $\mathbf{Z}/n\mathbf{Z}$, $n \leq 3$. The exact sequence

$$0 \to H^1(L/k, Z) \to H^1(k, Z) \to H^1(L, Z)$$

shows that is suffices to prove that $H^1(L/k, Z) \to \prod H^1(L_v/k_v, Z)$ is injective. In fact it suffices to do this with k replaced by the fixed field of a Sylow subgroup of $\mathrm{Gal}(L/k)$. But then the Galois group is cyclic, and the result is obvious.

Lemma 3.7 implies that Γ_0 and $G_1(\mathbf{Q})$ have the same image in $G^{ad}(\mathbf{A}^f) = G_1^{ad}(\mathbf{A}^f)$. If we form the quotient of $M^o(G, X^+)$ by the action of $Z(\mathbf{A}^f)$, we get $Z(\mathbf{A}^f) \backslash M^o(G, X^+) = M^o(G^{ad}, X^+)$. Therefore,

$$\begin{aligned} {}_\tau M^o(G^{ad}, X^+) &= \Gamma_0 Z(\mathbf{A}^f) \backslash X_1^+ \times G(\mathbf{A}^f) \\ &= G_1(\mathbf{Q}) Z_1(\mathbf{A}^f) \backslash X_1^+ \times G_1(\mathbf{A}^f) = M^o(G_1^{ad}, X_1^+). \end{aligned}$$

We have proved (3.1) with G replaced by G^{ad}. An argument of Borovoi (see 5.2a below; it is not necessary to assume there that G is simply connected) shows that the map $G^{ad}_{\mathbf{A}^f} \to G^{ad}_{1\mathbf{A}^f}$ defined by ψ identifies G_1^{ad} with an inner form of G^{ad}. Therefore $\psi^f \colon G_{\mathbf{A}^f} \to G_{1\mathbf{A}^f}$ has the same property, and so $\psi^f | Z(G)$ is defined over $\mathbf{Q}$, i.e., ψ^f identifies $Z \overset{\mathrm{df}}{=} Z(G)$ with $Z_1 \overset{\mathrm{df}}{=} Z(G_1)$.

From (3.4) we know that $\Gamma_0 \cap Z(\mathbf{A}^f)/Z(\mathbf{Q}) = 1$. Therefore any element g of $G_1(\mathbf{Q})/Z(\mathbf{Q})$ can be written uniquely as $g = \gamma \cdot z_g$, $\gamma \in \Gamma_0$,

$z_g \in Z(\mathbb{A}^f)/Z(\mathbb{Q})$. The map $g \mapsto z_g$ is a homomorphism

$$G_1(\mathbb{Q})/Z(\mathbb{Q}) \to Z(\mathbb{A}^f).$$

If we knew, as is conjectured, that $G_1(\mathbb{Q})/Z(\mathbb{Q})$ is simple, then this homomorphism would have to be zero, and we would have achieved our immediate goal of showing that $\Gamma_0 \supset G_1(\mathbb{Q})/Z(\mathbb{Q})$. Instead, we argue as follows.

Let F be a totally real Galois extension of $\mathbb{Q}$ with Galois group Δ; let $G_* = \mathrm{Res}_{F/\mathbb{Q}} G_F$ and let X_*^+ be such that there is an embedding

$$(G, X^+) \hookrightarrow (G_*, X_*^+).$$

Then Δ acts on G_* and X_*^+, and G is the unique subgroup of G_* such that $\prod_{\delta\in\Delta} G_F \overset{(\delta,\ldots)}{\longrightarrow} (G_*)_F$ is an isomorphism. The group Δ as continues to act when we make the above constructions for (G_*, X_*^+). In particular, Δ acts on G_{1*} and there is an inclusion $G_1 \quad \to G_{1*}$ that identifies G_{1*} with $\mathrm{Res}_{F/\mathbb{Q}} G_1$. We can conclude:

Lemma 3.9. *For any F as above, the diagram*

$$\begin{array}{ccc} g \mapsto z_g\colon G_1(\mathbb{Q})/Z(\mathbb{Q}) & \to & Z(\mathbb{A}^f)/Z(\mathbb{Q}) \\ \downarrow & & \downarrow \\ g \mapsto z_g\colon G_{1*}(\mathbb{Q})/Z_*(\mathbb{Q}) & \to & Z_*(\mathbb{A}^f)/Z_*(\mathbb{Q}) \end{array}$$

commutes, where $Z_* = \mathrm{Res}_{F/\mathbb{Q}} Z$.

Let $\gamma \in G_1(\mathbb{Q})$; we shall show that $\gamma \,(\mathrm{mod}\, Z(\mathbb{Q})) \in \Gamma_0$.

Lemma 3.10. *There exist fundamental maximal tori $T_i \subset G_1$, $i = 1, \ldots, k$, and elements $\gamma_i = T_i(\mathbb{Q})$ such that $\gamma = \gamma_1 \cdots \gamma_k$.*

Proof. Let U be the set of $g \in G_1(\mathbb{R})$ such that the centralizer of g is a compact maximal torus. Then, the usual argument using the Lie algebra, shows that U is open in $G_1(\mathbb{R})$. Moreover, U generates $G_1(\mathbb{R})$. Let $\gamma = \gamma_1' \cdots \gamma_k'$ with $\gamma_i' \in U$. According to the real approximation theorem, the set $G_1(\mathbb{Q}) \cap U$ is dense in U. We can therefore choose $\gamma_i \in G_1(\mathbb{Q}) \cap U$,

$i = 2, 3, \ldots, k$, so close to γ_i' that $\gamma_1 \stackrel{\text{df}}{=} \gamma(\gamma_2 \ldots \gamma_k)^{-1}$ also lies in U. As $\gamma_1 \in G_1(\mathbf{Q})$, the elements $\gamma_1, \ldots, \gamma_k$ fulfill the requirements of the lemma.

Thus we can assume $\gamma \in T(\mathbf{Q}) \subset G(\mathbf{Q})$, where T is a maximal torus such that $T(\mathbf{R})$ is compact. This last condition on T implies that T splits over a CM-field L, which can be chosen to be Galois over $\mathbf{Q}$. Let F be the maximal totally real subfield of L and let $G_* = \operatorname{Res}_{F/\mathbf{Q}} G$. The construction of Borovoi recalled below in §4, gives a reductive group H_α of type A_1 such that $T_* \subset H_\alpha \subset G_{1*}$. After possibly extending F we can assume no $(H_\alpha)_l$ is anisotropic. Consider

$$\begin{array}{ccc} \gamma \in G_1(\mathbf{Q})/Z(\mathbf{Q}) & \to & Z(\mathbf{A}^f)/Z(\mathbf{Q}) \\ \cap & & \cap \\ G_{1*}(\mathbf{Q})/Z_*(\mathbf{Q}) & \to & Z_*(\mathbf{A}^f)/Z_*(\mathbf{Q}). \end{array}$$

Note that $\gamma \in H_\alpha(\mathbf{Q})/Z_*(\mathbf{Q}) \cap H_\alpha(\mathbf{Q})$. A theorem of Platonov and Rapinčuk [16] shows that $H_\alpha(\mathbf{Q})$ has no non-central normal subgroup. Therefore, the lower map is zero on $H_\alpha(\mathbf{Q})/Z_*(\mathbf{Q}) \cap H_\alpha(\mathbf{Q})$ and so γ maps to zero. It therefore lies in Γ_0.

We have shown $\Gamma_0 \supset G_1(\mathbf{Q})/Z(\mathbf{Q})$. Therefore, for any compact open $K \subset G(\mathbf{A}^f)$ containing $Z(\mathbf{Q})$, we have a finite étale map

$$\Gamma_0 \backslash X_1^+ \times G(\mathbf{A}^f)/K \to G_1(\mathbf{Q}) \backslash X_1^+ \times G_1(\mathbf{A}^f)/K.$$

We therefore have a finite étale map

$$\tau M^\circ(G, X^+) \to M^\circ(G_1, X_1^+)$$

which is $G(\mathbf{A}^f)$-equivariant when we identify $G_{\mathbf{A}^f}$ with $G_{1\mathbf{A}^f}$ by means of ψ^f. When we apply τ^{-1} to this map, and repeat the whole of the above construction for (G_1, X_1^+) and τ^{-1}, we obtain maps

$$M^\circ(G, X^+) \to \tau^{-1} M^\circ(G_1, X_1^+) \to M^\circ(G_2, X_2^+).$$

The composite map satisfies the hypotheses of (2.1) and therefore is an isomorphism. This implies that the first map

$$M^\circ(G, X^+) \to \tau^{-1} M^\circ(G_1, X_1^+)$$

is an isomorphism and, on applying τ^{-1}, we get that

$$\tau M^\circ(G, X^+) \to M^\circ(G_1, X_1^+)$$

is an isomorpism; this is what we had to prove.

Corollary 3.11. *Suppose in (3.2) that Γ is a congruence group, i.e., that there exists a simply-connected semi-simple group G_1 over $\mathbf{Q}$ and a surjective homomorphism $f\colon G_1(\mathbf{R}) \longrightarrow G$ with compact kernel carrying a congruence subgroup of $G(\mathbf{Q})$ into a subgroup of Γ with finite index. Then $\tau(\Gamma\backslash X^+)$ is the quotient of a Hermitian symmetric domain by a congruence group.*

§4. Embedding Forms of SL_2 into G

In this section, we recall some results of Borovoi [2]. Let (G, X^+) satisfy (C), and let $(T, h) \subset (G, X)$. Throughout the section, we assume the following conditions hold:

(4.1a) $G = \operatorname{Res}_{F/\mathbf{Q}} G'$, where F is totally real and G' is absolutely almost simple;

(4.1b) the maximal torus $T' \subset G'$ such that $\operatorname{Res}_{F/\mathbf{Q}} T' = T$ splits over a quadratic, totally imaginary extension L of F.

Remark 4.2. The condition (b) implies that T', and any subtorus, is a product of one-dimensional tori, and therefore satisfies the Hasse principle for H^1.

As T'_L is split, we can write

$$\operatorname{Lie} G'_L = \operatorname{Lie} T'_L \bigoplus_{\alpha \in R} (\operatorname{Lie} G'_L)_\alpha$$

where $R = R(G'_{\mathbf{C}}, T'_{\mathbf{C}})$. An $\alpha \in R$ will be said to be *totally compact* if it is a compact root of $(G' \otimes_{F,\sigma} \mathbf{R})_{\mathbf{C}}$ for all embeddings $\sigma\colon F \hookrightarrow \mathbf{R}$. Let R^{ntc} denote the set of roots that are not totally compact. Then, for $\alpha \in R^{ntc}$,

$$\mathfrak{H}'_\alpha \overset{\mathrm{df}}{=} \operatorname{Lie} T'_L \oplus (\operatorname{Lie} G'_L)_\alpha \oplus (\operatorname{Lie} G'_L)_{-\alpha}$$

is defined over F, and we let H'_α be the corresponding connected subgroup of G'. Write $H_\alpha = \operatorname{Res}_{F/\mathbf{Q}} H'_\alpha$ and $Z_\alpha = Z(H_\alpha)$.

Proposition 4.3 (Borovoi, Oniščic). $Z(G) = \bigcap_\alpha Z_\alpha$, $(\alpha \in R^{ntc})$.

Proof. From the formula $[\mathfrak{p}, \mathfrak{p}] = \mathfrak{k}$, valid whenver $\mathfrak{g} = \mathfrak{k} \oplus \mathfrak{p}$ is the Cartan decomposition of a non-compact real Lie algebra, it follows that $R^{ntc} + R^{ntc} \supset R$. Thus

$$\bigcap_{\alpha \in R^{ntc}} Z(H'_\alpha) = \bigcap_{\alpha \in R^{ntc}} \mathrm{Ker}(\alpha) = \bigcap_{\alpha \in R} \mathrm{Ker}(\alpha) = Z(G'),$$

and the proposition follows by applying $\mathrm{Res}_{F/\mathbb{Q}}$.

Corollary 4.4. *Let $\overline{T} = T/Z$ and $\overline{Z}_\alpha = Z_\alpha/Z$ where $Z = Z(G)$. Regard $\overline{Z}_\alpha(\mathbb{A}^f)/\overline{Z}_\alpha(\mathbb{Q})$ as a subgroup of $\overline{T}(\mathbb{A}^f)/\overline{T}(\mathbb{Q})$. Then*

$$\bigcap_\alpha \overline{Z}_\alpha(\mathbb{A}^f)/\overline{Z}_\alpha(\mathbb{Q}) = 1, (\alpha \in R^{ntc}).$$

Proof. This follows easily from the fact that $\bigcap \overline{Z}_\alpha = 1$.

§5. Completion of the Proof of (1.1)

In this section, we use the methods of Borovoi [2] to deduce (1.1) from (3.1).

Let (G, X^+) satisfy (C), and consider $(T, h) \subset (G, X^+)$. For any $q \in T(\mathbb{Q})$, $q[h] = [h]$, and so there is a representation ρ_h of T on the tangent space $\mathfrak{t}_h$ to $M^\circ(G, X^+)$ at $[h]$.

Lemma 5.1 (cf. [2, 3.2, 3.3]). (a) $\rho_h \sim \oplus\alpha$, $\alpha \in R(G_{\mathbb{C}}, T_{\mathbb{C}})$, $\langle \alpha, \mu_h \rangle = 1$.

(b) *Suppose that (G, X'^+) also satisfies (C), and that*

$$(T', h') \subset (G, X'^+);$$

if $\rho_h \sim \rho_{h'}$, then $h = h'$ (and $X'^+ = X^+$).

Proof. (a) According to (C_1) there is a decomposition of $\mathfrak{g} = \mathrm{Lie}(G)$

$$\mathfrak{g}_{\mathbb{C}} = \mathfrak{g}^{o,o} \oplus \mathfrak{p}^+ \oplus \mathfrak{p}^-$$

such that $Ad\,\mu(z)$ acts trivially on $\mathfrak{g}^{o,o}$, as z on $\mathfrak{p}^+$, and as z^{-1} on $\mathfrak{p}^-$. Thus $\mathfrak{g}_\alpha \subset \mathfrak{p}^+$ if and only if $\alpha(\mu(z)) = z$, i.e., $\langle \alpha, \mu \rangle = 1$. The canonical maps $\mathfrak{p} \overset{\approx}{\to} \mathfrak{t}_h$ and $\mathfrak{p} \hookrightarrow \mathfrak{p}_{\mathbf{C}} = \mathfrak{p}^+ \oplus \mathfrak{p}^-$ induce a C-linear, equivariant isomorphism $\mathfrak{p}^+ \overset{\approx}{\to} \mathfrak{t}_h$ (see [4, 1.1.14]).

(b) It follows from [4, 1.2.7] that $X'^+ = X^+$. Since

$$\{\alpha \mid \langle \alpha, \mu_h \rangle = 1\} = \{\alpha \mid \langle \alpha, \mu_{h'} \rangle = 1\},$$

h and h' define the same Hodge filtration on Lie(G), but this implies that they are equal [4, p. 254].

Proposition 5.2 (Borovoi [2]). *Let (G, X^+), (G', X'^+), and τ, φ, and ψ be as in (3.1). Consider $i\colon (T, h) \hookrightarrow (G, X^+)$.*

(a) *There exists an inclusion $i'\colon T \hookrightarrow G'$ and a $g \in G(\mathbf{A}^f)$ such that $\psi \circ \underline{ad}(g) \circ i_{\mathbf{A}^f} = i'_{\mathbf{A}^f}$.*

(b) *There exists an $h' \in X'$ factoring through i' such that $\mu_{h'} = \tau\mu_h$ (as maps into T).*

(c) *The pair (G', j') is a form of (G, j); its class in $H^1(\mathbf{Q}, \overline{T})$, $\overline{T} \overset{\mathrm{df}}{=} T/Z(G)$, has image zero in $H^1(\mathbf{Q}_l, \overline{T})$ for all $l \neq \infty$ and is represented by $\tau\mu_h(-1)/\mu_h(-1)$ for $l = \infty$.*

Proof. (a) Let $\varphi(\tau[h]) = g'[h']$, $h' \in X'^+$, $g' \in G(\mathbf{A}^f)$ (see the Appendix). For any $t \in i(T(\mathbf{Q}))$, $t[h] = [h]$, and so $\psi(t)g'[h'] = g'[h']$, i.e.,

$$(g'^{-1}\psi(t)g')[h'] = [h'].$$

This implies that $g'^{-1}\psi(t)g' \in G'(\mathbf{Q})$. Let $g = \psi^{-1}(g'^{-1})$; then $\psi \circ \underline{ad}\, g \circ i\colon T_{\mathbf{A}^f} \longrightarrow G'_{\mathbf{A}^f}$ maps $T(\mathbf{Q})$ into $G'(\mathbf{Q})$. As $T(\mathbf{Q})$ is Zariski dense in T, this means that $\psi \circ \underline{ad}\, g \circ i$ is defined over $\mathbf{Q}$; we denote it by i'.

(b) Let h' be as in (a); then $\rho_{h'} = \tau\rho_h$ because $(\varphi \circ g)(\tau[h]) = [h']$ and $\varphi \circ g$ is $T(\mathbf{Q})$-equivariant. Therefore (b) follows from (5.1b).

(c) As $\psi \circ \underline{ad}\, g$ is an isomorphism $(G, i)_{\mathbf{A}^f} \overset{\approx}{\to} (G', i^p)_{\mathbf{A}^f}$, it is clear that the class of (G, i) in $H^1(\mathbf{Q}_l, \overline{T})$ is trivial for all finite l. The element $\underline{ad}\,\mu(-1)$ is a Cartan involution on G; therefore the class of $\mu(-1)$ in $H^1(\mathbf{R}, \overline{T})$ corresponds to the (unique) compact form of $G_{\mathbf{C}}$. Similarly, the class of $\tau\mu(-1)$ corresponds to the compact form of $G'_{\mathbf{C}} = G_{\mathbf{C}}$. The conclusion is now clear.

Remark 5.3. If in (5.2) we replace φ by $\varphi \circ g$ and ψ by $\psi \circ \underline{ad}(g)$, then φ and ψ are still compatible, but now $\varphi(\tau[h]) = [h']$ and $\psi \circ i = i'$.

Corollary 5.4. *Consider* $(T,h) \xrightarrow{i} (G,X^+)$, *and assume that* G *is simply-connected and that* $\overline{T} \stackrel{\text{df}}{=} T/Z(G)$ *satisfies the Hasse principle for* H^1. *Let* $(T,{}^\tau h) \xhookrightarrow{{}^\tau i} ({}^\tau G,{}^\tau X^+)$ *be as in* §1. *Then there exist isomorphisms*

$$\varphi\colon M^\circ(G,X^+) \longrightarrow M^\circ({}^\tau G,{}^\tau X^+)$$
$$\psi\colon G_{\mathbf{A}^f} \longrightarrow {}^\tau G_{\mathbf{A}^f}$$

such that

(a) $\varphi(\tau[h]) = [{}^\tau h]$;

(b) φ *and* ψ *are compatible;*

(c) $\psi \circ i_{\mathbf{A}^f} = {}^\tau i_{\mathbf{A}^f}$.

Proof. Let (G',X'^+) be as in (3.1), and let $i'\colon T' \hookrightarrow G'$ and $h' \in X'^+$ be as in (5.2a) and (5.2b). Choose φ and ψ as in (5.3). Then (5.2c) shows that there exists an isomorphism $(G',i') \approx ({}^\tau G,{}^\tau i)$ (cf. (1.4)), and (5.2b) shows that the isomorphisms carries h' into ${}^\tau h$. The corollary is now clear.

We now prove (1.1). Let (G,X^+), h, and τ be as in the statement of (1.1). We can assume that G is almost simple over $\mathbf{Q}$. Then $G = \operatorname{Res}_{F_1/\mathbf{Q}} G_1$ for some totally real field F_1 and absolutely almost-simple group G_1. Let $(T,h) \subset (G,X^+)$ and let T_1 be the maximal torus in G_1 such that $\operatorname{Res}_{F_1/\mathbf{Q}} T_1 = T$. As $T_{\mathbf{R}}$ is anisotropic, T_1 splits over a *CM*-field $L \supset F$. Let F be the maximal totally real subfield of L and let $G_* = \operatorname{Res}_{F/\mathbf{Q}} G'$ where $G' = G_{1F}$. Let h_* be the composite of h with the canonical inclusion $G \hookrightarrow G_*$. Then the $G_*(\mathbf{R})^+$-conjugacy class X_* containing h_*, together with G_*, satisfy (C), and (1.6) shows that it suffices to prove (1.1) for (G_*,X_*) and h_*. This allows us to assume that the original objects, $(T,h) \hookrightarrow (G,X^+)$, satisfy (4.1).

Let φ and ψ be isomorphisms as in (5.4). As both ψ and ψ_τ (see §1) are isomorphisms $(G,i)_{\mathbf{A}^f} \approx ({}^\tau G,{}^\tau i)_{\mathbf{A}^f}$, there exists a $t \in \overline{T}(\mathbf{A}^f)$ such that $\psi_\tau = \underline{ad}(t) \circ \psi$. To any non totally compact root α, there corresponds a subgroup H_α of G containing T (see §4). Let $H'_\alpha = H_\alpha^{der}$, and let X_α^+ be the $H_\alpha^{ad}(\mathbf{R})^+$-conjugacy class containing $h_\alpha \stackrel{\text{df}}{=} (\mathbf{S} \xrightarrow{h} H_\alpha \to H_\alpha^{ad})$. Then φ maps $\tau M^\circ(H'_\alpha,X_\alpha^+)$ into $M^\circ({}^\tau H_\alpha,{}^\tau X_\alpha^+)$ because it maps $\tau(g[h_\alpha])$ to $\underline{ad}\, t^{-1}(\psi_\tau(g))[{}^\tau h_\alpha]$ for all $g \in H'_\alpha(\mathbf{A}^f)$ and $H'_\alpha(\mathbf{A}^f).[h_\alpha]$ is dense in $M^\circ(H'_\alpha,X_\alpha^+)$. Since H'_α is of type A_1, we know (1.1) for it: there exists an isomorphism $\varphi_\tau\colon \tau M^\circ(H'_\alpha,X_\alpha^+) \longrightarrow M^\circ({}^\tau H'_\alpha,{}^\tau X_\alpha^+)$, compatible with ψ_τ, and taking $\tau[h_\alpha]$ to $[^\tau h_\alpha]$. The map

$$\tau^{-1}(\varphi_\tau^{-1} \circ \varphi)\colon M^\circ(H'_\alpha, X^+_\alpha) \longrightarrow M^\circ(H'_\alpha, X^+_\alpha)$$

fixes $[h_\alpha]$ and is compatible with $\underline{ad}\,t$: $H'_\alpha(\mathbf{A}^f) \longrightarrow H'_\alpha(\mathbf{A}^f)$. Therefore (2.2) shows that $\underline{ad}\,t$, regarded as element of $H^{ad}_\alpha(\mathbf{A}^f)$, lies in $H^{ad}_\alpha(\mathbf{Q})$, i.e., $t \in \overline{Z}_\alpha(\mathbf{A}^f)/\overline{Z}_\alpha(\mathbf{Q})$. Now (4.4) shows that $t \in \overline{T}(\mathbf{Q})$, and so $\varphi \circ t$ and $\underline{ad}(t) \circ \psi$ fulfill the requirements of the theorem.

§6. Compatibility of the Maps φ_τ for Different Special Points

Let (G, X^+) satisfy (C) and let $(T, h) \subset (G, X^+)$. Then Langlands's constructions lead to the definition of a map

$$\psi_\tau = (g \mapsto {}^\tau g)\colon G^{ad}(\mathbf{Q})^{+\wedge}(\mathrm{rel}\, G) \longrightarrow {}^\tau G^{ad}(\mathbf{Q})^{+\wedge}(\mathrm{rel}\, {}^\tau G)$$

that is compatible with the maps of the same name, $G(\mathbf{A}^f) \to {}^\tau G(\mathbf{A}^f)$, $G^{ad}(\mathbf{A}^f) \to {}^\tau G^{ad}(\mathbf{A}^f)$ (see [14, §8]).

Proposition 6.1. *Assume that G is simply connected, and let φ_τ be an isomorphism $\tau M^\circ(G, X^+) \to M^\circ({}^\tau G, {}^\tau X^+)$ with the properties (a) and (b) of (1.1). Then (1.1b) holds for all $g \in G^{ad}(\mathbf{Q})^{+\wedge}(\mathrm{rel}\, G)$.*

Proof. Let $G_* = \mathrm{Res}_{F/\mathbf{Q}}(G_F)$ for some totally real field F, and let h_* be the composite of h with $G \hookrightarrow G_*$. Then there is a commutative diagram

$$\begin{array}{ccc} \tau M^\circ(G, X^+) & \overset{\varphi_\tau}{\underset{\approx}{\to}} & M^\circ({}^\tau G, {}^\tau X^+) \\ \downarrow \cap & & \downarrow \cap \\ \tau M^\circ(G_*, X^+_*) & \overset{\varphi_\tau}{\underset{\approx}{\to}} & M^\circ({}^\tau G_*, {}^\tau X^+_*). \end{array}$$

$(({}^\tau G, {}^\tau X^+)$ and $({}^\tau G_*, {}^\tau X^+_*)$ defined using h and h_*). This shows that φ_τ (for G) is compatible with the action of g for all $g \in G_*(\mathbf{A}^f)$ (any F) and all $g \in H_\alpha(\mathbf{Q})^+$ (any $H_\alpha \subset G_*$; see §4), but these elements generate a dense subgroup of $G^{ad}(\mathbf{Q})^{+\wedge}(\mathrm{rel}\, G)$.

Remark 6.2. Theorem 1.1 and (6.1) prove part (a) of Conjecture C° (see [14, p. 340]) for simply connected groups.

We shall now need to consider two special points h and h' for a given (G, X^+). To distinguish the objects ${}^\tau G, {}^\tau X^+, \varphi_\tau, \ldots$ constructed relative to h from the similar objects constructed relative to h', we shall write the former ${}^{\tau,h}G, {}^{\tau,h}X^+, \varphi_{\tau,h}, \ldots$.

Let $\overline{\beta}(\tau, h)$ and $\overline{\beta}(\tau, h')$ be the elements of $\left(G^{ad}(\mathbb{A}_L^f)/G^{ad}(L)\right)^{\mathrm{Gal}(L/\mathbb{Q})}$ corresponding to h and h' respectively (see (1.4)). The image of $\overline{\beta}(\tau, h)$ and $\overline{\beta}(\tau, h')$ in $H^1(\mathbb{Q}, G^{ad})$ are trivial at the finite primes and are equal at the infinite primes (see [15, p. 315–316]). As G^{ad} satisfies the Hasse principal for H^1 ([9, VII. 6]) this shows that $\overline{\beta}(\tau, h)$ and $\overline{\beta}(\tau, h')$ have the same image in $H^1(\mathbb{Q}, G^{ad})$ and therefore $B \overset{\mathrm{df}}{=} \overline{\beta}(\tau, h')\overline{\beta}(\tau, h)^{-1}$ lies in $G^{ad}(\mathbb{A}^f)/G^{ad}(\mathbb{Q})$. Moreover, there is an isomorphism $f\colon {}^{\tau,h}G \longrightarrow {}^{\tau,h'}G$ such that

$$f_{\mathbb{A}^f} = \psi_{\tau,h'} \circ B \circ \psi_{\tau,h}^{-1}\colon {}^{\tau,h}G(\mathbb{A}^f) \longrightarrow {}^{\tau,h'}G(\mathbb{A}^f).$$

Define

$$\varphi^\circ(\tau; h', h) : M^\circ({}^{\tau,h}G, {}^{\tau,h}X^+) \overset{\approx}{\rightarrow} M^\circ({}^{\tau,h'}G, {}^{\tau,h'}X^+)$$

to be $({}^{\tau,h'}B)^{-1} \circ M^\circ(f)$. It carries the action of ${}^{\tau,h}g$, $g \in G(\mathbb{A}^f)$, into the action of ${}^{\tau,h'}g$. (See [15, p. 312–318].)

Theorem 6.3. *Assume that G is simply connected; then for any special $h, h' \in X^+$, the diagram*

$$\begin{array}{ccc}
\tau M^\circ(G, X^+) & \xrightarrow{\varphi_{\tau,h}} & M^\circ({}^{\tau,h}G, {}^{\tau,h}X^+) \\
 & \searrow^{\varphi_{\tau,h'}} & \downarrow{\scriptstyle \varphi(\tau;h',h)} \\
 & & M^\circ({}^{\tau,h'}G, {}^{\tau,h'}X^+)
\end{array}$$

commutes.

Proof. We first prove this under the assumption that $h' = \underline{ad}\, q \circ h, q \in G^{ad}(\mathbb{Q})^+$. In this case $B = q$ (see [15, 9.1]) so that $M^\circ(f) = ({}^{\tau,h'}q) \circ \phi^\circ(\tau; h', h) = \phi^\circ(\tau; h', h) \circ ({}^{\tau,h}q)$.

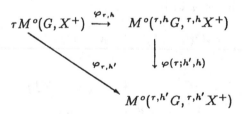

Consider the diagram

$$\begin{array}{ccc}
\tau M^o(G,X^+) & \xrightarrow{\phi_{\tau,h}} & M^o({}^{\tau,h}G, {}^{\tau,h}X^+) \\
\downarrow \tau(q) & & \downarrow {}^{\tau,h}q \\
\tau M^o(G,X^+) & \xrightarrow{\phi_{\tau,h}} & M^o({}^{\tau,h}G, {}^{\tau,h}X^+) \\
& \searrow \varphi_{\tau,h'} & \downarrow \overset{o}{\phi}(\tau;h',h) \\
& & M^o({}^{\tau,h'}G, {}^{\tau,h'}X^+)
\end{array}
\qquad \Big\downarrow M^o(f)$$

The upper square commutes because of (6.1). The outside of the diagram commutes because both maps send $[h]$ to $[{}^{\tau}h'] = [f \circ {}^{\tau}h]$ and the action of $\tau(g)$ to that of ${}^{\tau,h'}g$. Thus the lower triangle commutes.

Before proving the general case, we need some lemmas.

Lemma 6.4. *Consider an inclusion $i\colon (G,X) \hookrightarrow (G',X')$. Theorem (6.3) holds for h and h' (as elements of X) if and only if it holds for $i \circ h$ and $i \circ h'$.*

Proof. Since i induces an embedding $M^o(G,X^+) \hookrightarrow M^o(G',X'^+)$ and the maps $\phi_{\tau,i\circ h}$, $\phi_{\tau,i\circ h'}$, and $\phi(\tau; i\circ h', i\circ h)$ restrict to $\phi_{\tau,h}$, $\phi_{\tau,h'}$, and $\phi(\tau;h',h)$ on $M^o(G,X^+)$, the sufficiency is clear. The necessity follows from the facts that $G'(\mathbf{Q})_+[i\circ h]$ is dense in $M^o(G',X'^+)$ and $\phi_{\tau,i\circ h'}$ and $\phi(\tau; i\circ h', i\circ h)\circ\phi_{\tau,i\circ h}$ have the same behaviour with respect to the Hecke operators.

Lemma 6.5. *If (6.3) is true for the pairs (h,h') and (h',h''), then it is also true for the pair (h,h'').*

Proof. This is immediate, since $\phi(\tau;h'',h')\circ\phi(\tau;h',h) = \phi(\tau;h'',h)$.

Lemma 6.6 (Borovoi). *Let G be a simple noncompact group over* $\mathbf{R}$ *and let X be a $G(\mathbf{R})$-conjugacy class of homomorphisms $\mathbf{S} \to G$ satisfying (C_1) and (C_2). Suppose $h, h' \in X$ factor through the same maximal torus $T \subset G$. Then $h' = h$ of $h' = h^{-1}$.*

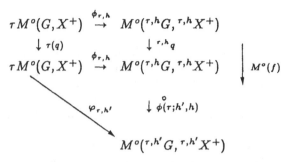

Proof. Let K_∞ be the centralizer of $h(\mathbf{S})$. Then [4, 1.2.7], K_∞ is a maximal connected compact subgroup of G. The pair (G,T) determines the set of compact roots in $R(G_{\mathbf{C}}, T_{\mathbf{C}})$, which determines K_∞. The centre Z of K_∞ is 1-dimensional (loc. cit.), and h defines an isomorphism $\mathbf{S}/\mathbf{G}_m \xrightarrow{\approx} Z$; it is therefore determined up to sign.

We now complete the proof of (6.3). As usual, we can assume that G is almost simple over $\mathbf{Q}$ and therefore that $G = \mathrm{Res}_{F/\mathbf{Q}} G_1$, where F is totally real and G_1 is absolutely almost simple. Let $(T,h) \subset (G,X^+)$ and $(T',h') \subset (G,X^+)$, and let T_1, $T_1' \subset G_1$ be such that $\mathrm{Res}_{F/\mathbf{Q}} T_1 = T$ and $\mathrm{Res}_{F/\mathbf{Q}} T_1' = T'$. There exists a CM-field L splitting both T_1 and T_1'. After replacing G with $G_* = \mathrm{Res}_{F'/\mathbf{Q}} G_{1,F'}$, where F' is the maximal totally real subfield of L, and using (6.4), we can assume that L is a quadratic extension of F. As $T_{1,L}$ and $T'_{1,L}$ are split, there exists a $\beta \in G_1(L)$ such that $\beta T_1 \beta^{-1} = T_1'$. For each real prime v: $F \hookrightarrow \mathbf{R}$ of F, choose an extension (also denoted by v) of v to L and write H_v for $H \otimes_{F,v} \mathbf{R}$, any F-group H. As $T_{1,v}$ and $T'_{1,v}$ are compact, there exists a $\gamma_v \in G_v(\mathbf{R})$ such that $\gamma_v T_{1,v} \gamma_v^{-1} = T'_{1,v}$. Let $c_\sigma = \beta^{-1} \cdot \sigma\beta$, where σ generates $\mathrm{Gal}(L/F)$. Then $c_\sigma \in N(L)$, where N is the normalizer of T, and so it defines a class $c \in H^1(L/F, N)$. As $\gamma_v^{-1} \cdot v(\beta) \in N(\mathbf{C})$ and

$$v(c_\sigma) = \left(\gamma_v^{-1} \cdot v(\beta)\right)^{-1} \cdot \iota\left(\gamma_v^{-1} \cdot v(\beta)\right),$$

where ι denotes complex conjugation, we see that c maps to 1 in $H^1(L_v/F_v, N)$. Let ω_v be the image of $\gamma_v^{-1} \cdot v(\beta)$ in $W(L_v)$, where $W = N/T$. The image ω of c_σ in $W(L)$ is ι because $v(\omega) = \omega_v^{-1} \cdot \iota\omega_v$ and ι acts trivially on $W(L_v)$ (see [15, p. 307]). Thus $c_\sigma \in T(L)$ and

$$c \in \mathrm{Ker}\Big(H^1(L/F,T) \to \bigoplus_v H^1(L_v/F_v, N)\Big).$$

The following diagram is useful:

$$\begin{array}{ccccccc}
N(F) & \to & W(F) & \to & H^1(L/F,T) & \to & H^1(L/F,N) \\
\downarrow & & \downarrow & & \downarrow & & \downarrow \\
\bigoplus_v N(F_v) & \to & \bigoplus_v W(F_v) & \to & \bigoplus_v H^1(L_v/F_v,T) & \to & \bigoplus_v H^1(L_v/F_v,N)
\end{array}$$

We now prove (6.3) by induction on $l = \sum_v l(\omega_v)$, where $l(\omega_v)$ is the length of ω_v as an element of $W(\mathbf{C})$. Suppose first that $l = 0$. Then $\gamma_v^{-1} \cdot v(\beta) \in T(L_v)$ and so c maps to 1 in $H^1(L_v/F_v, T)$ for all v. Note that

$T \approx U^r$ some r, where U is the unique one-dimensional non-split F-torus split by L, and therefore $H^1(L/F,T) \approx (F^\times/NL^\times)^r$. The penultimate assertion shows that c is represented by a family $(c_1,\ldots,c_r)$ of totally positive elements of $F^\times$. After adjoining $\sqrt{c_i}$ to F, $i=1,\ldots,r$, we can assume $c=1$. Then $c_\sigma = t^{-1}\cdot\sigma t$ some $t\in T(L)$ and so, after replacing β with βt^{-1}, we can assume it lies in $G_1(F)$. Regard β as an element of $G(\mathbf{Q})$. Lemma (6.5) and the first part of the proof show that we need only prove the theorem for $\underline{ad}\,\beta\circ h$ and h'. This means that we can assume that h and h' factor through the same torus. But then they must be equal because, in the context of (6.6), h^{-1} does not lie in the same connected component as h.

Finally, suppose $l(\omega_{v_o})\neq 0$, say $\omega_{v_o} = s_\alpha\omega'_{v_o}$ with $l(\omega'_{v_o}) < l(\omega_{v_o})$ and s_α the reflection corresponding to the root α. If α is compact at v, then s_α lifts to $\gamma_\alpha\in N(F_v)$ (see [15, p. 308]) and we can replace γ_{v_o} with $\gamma_{v_o}\gamma_\alpha^{-1}$. Then ω_{v_o} is replaced with ω'_{v_o} and we can apply the induction hypothesis. Suppose therefore that α is not compact at v and define $H'_\alpha\subset G_1$, $H'_\alpha\supset T_1$, H'_α of type A_1, as in §4. Let H_α be the derived group of H'_α and let $T_\alpha = H_\alpha\cap T$. Then $T_\alpha\approx U$, which implies that $H^1(L/F,T_\alpha)\to\bigoplus_v H^1(L_v/F_v,T_\alpha)$ is surjective. Choose $c_\alpha\in H^1(L/F,T_\alpha)$ mapping to (c_v) where $c_v=1$ for $v\neq v_o$ and $c_{v_o}=\delta(s_\alpha)$ where δ is the boundary map

$$W_\alpha(F_{v_o})\to H^1(L_{v_o}/F_{v_o},T_\alpha),\qquad W_\alpha = N_\alpha/T_\alpha,$$

$N_\alpha = \mathrm{Norm}(T_\alpha)$. Then c_α maps to 1 in $\bigoplus H^1(L_v/F_v,N_\alpha)$. The Hasse principle therefore shows that c_α splits in $H^1(L/F,H_\alpha):(c_\alpha)_\sigma = g^{-1}\cdot\sigma g$, $g\in H_\alpha(L)$. Lift s_α to $n_\alpha\in N_\alpha(L)$; then $v_o((c_\alpha)_\sigma) = n_\alpha^{-1}\cdot\iota n_\alpha$ and so $gn_\alpha^{-1}\in H_\alpha(\mathbf{R})$. Since we know the theorem for $\mathrm{Res}_{F/\mathbf{Q}}H_\alpha$, (6.4) allows us to replace (T,h) with $\big(\underline{ad}\,g\circ T,\underline{ad}(gn_\alpha^{-1})\circ h\big)$. This replaces β with βg^{-1}, γ_{v_o} with $\gamma_{v_o}n_\alpha g^{-1}$, γ_v with $\gamma_v g^{-1}$, $v\neq v_o$, ω_{v_o} with $gs_\alpha^{-1}\omega_{v_o}g^{-1} = g\omega'_{v_o}g^{-1}$, and ω_v with $g\omega_v g^{-1}$, $v\neq v_o$. Thus $\sum l(\omega_v)$ is diminished, and we can apply the induction hypothesis.

§7. Conclusions

We have shown that Conjecture C^o of [15, p. 340–341] is true for (G,X^+) whenever G is simply connected. As is remarked in [15, 9.6], this implies the general case.

Theorem 7.1. *The conjecture of Langlands [8, p. 232–233] (see also [15, p. 311]) is true for all Shimura varieties.*

Proof. In [15, 9.4] it is shown that this conjecture is equivalent to Conjecture C^o.

Theorem 7.2. *Canonical models (in the sense of [4, 2.2.5]) exist for all Shimura varieties.*

Proof. This is a consequence of (7.1) (see [15, §7]).

Theorem 7.3. *The conjecture of Langlands describing the action of complex conjugation on a Shimura variety having a real canonical model [8, p. 234] is true.*

Proof. This again follows from (7.1)

Theorem 7.4. *The main theorems of [13], viz. (4.6) and (4.9), are true for all Shimura varieties.*

Proof. They are proved in [13] under the assumption that G is classical and the canonical model exists, but the first assumption is only used to simplify the proof of [13, 1.3], and we can instead deduce this theorem from Proposition 2.1 above.

Remark 7.5. Theorem 7.4 gives a definitive answer to the question of Shimura [19, p. 347].

Remark 7.6. For Shimura varieties of Abelian type, (7.1), (7.2), and (7.3) were first proved in ([5], [15]), [4], and [12] respectively.

Appendix

We say that (G, X^+) satisfies (C) if G is a semi-simple group over $\mathbf{Q}$ and X^+ is a $G(\mathbf{R})^+$-conjugacy class of maps $\mathbf{S} \to G^{ad}_{\mathbf{R}}$ for which the following hold:

(C_1) for all $h \in X^+$, the Hodge structure on $\mathrm{Lie}(G_{\mathbf{R}})$ defined by h is of type $\{(-1,1),(0,0),(1,-1)\}$;

(C_2) $\underline{ad}\, h(i)$ is a Cartan involution on $G^{ad}_{\mathbf{R}}$;

(C_3) G^{ad} has no non-trivial factors defined over $\mathbf{Q}$ that are anisotropic over $\mathbf{R}$.

Such a (G,X^+) defines a connected Shimura variety $M^o(G,X^+)$. The topology $\tau(G)$ on $G^{ad}(\mathbf{Q})$ is that for which the images of the congruence subgroups of $G(\mathbf{Q})$ form a fundamental system of neighbourhoods of the identity, and

$$M^o(G,X^+) = \varprojlim \Gamma \backslash X^+$$

where the limit is over the set $\sum(G)$ of torsion-free arithmetic subgroups of $G^{ad}(\mathbf{Q})$ that are open relative to the topology $\tau(G)$. For $h \in X^+$, $[h]$ and $[h]_\Gamma$ denote the images of h in $M^o(G,X^+)$ and ${}_\Gamma M^o(G,X^+) \overset{\mathrm{df}}{=} \Gamma \backslash X^+$.

Any $\alpha \in G^{ad}(\mathbf{Q})^+$ acts on $M^o(G,X^+)$ by transport of structure: $\alpha[h]_\Gamma = [\alpha \circ h]_{\alpha(\Gamma)}$. Any $g \in G(\mathbf{Q})^{-}_{+}$ acts as follows: Let $\Gamma \in \sum(G)$ and let K be a compact open subgroup of $G(\mathbf{A}^f)$ such that Γ contains the image of $K \cap G(\mathbf{Q})_+$; then $g \in qK$ some $q \in G(\mathbf{Q})_+$, and $g[h]_\Gamma \overset{\mathrm{df}}{=} [\underline{ad}\, q \circ h]_{q\Gamma q^{-1}}$. These actions combine to give an action of $G(\mathbf{Q})^{-}_{+} \times G^{ad}(\mathbf{Q})^+$ (semi-direct product for the obvious action of $G^{ad}(\mathbf{Q})^+$ on $G(\mathbf{Q})^{-}_{+}$). The map $q \mapsto (q, \underline{ad}\, q^{-1})$ identifies $G(\mathbf{Q})_+$ with a normal subgroup of the product, and the quotient

$$G(\mathbf{Q})^{-}_{+} *_{G(\mathbf{Q})_+} G^{ad}(\mathbf{Q})^+ \overset{\mathrm{df}}{=} G(\mathbf{Q})^{-}_{+} \times G^{ad}(\mathbf{Q})^+ / G(\mathbf{Q})_+$$

continues to act on $M^o(G,X^+)$. The completion of $G^{ad}(\mathbf{Q})^+$ for the topology $\tau(G)$, $G^{ad}(\mathbf{Q})^{+\wedge}(\mathrm{rel}\, G)$, is equal to $G(\mathbf{Q})^{-}_{+} *_{G(\mathbf{Q})_+} G^{ad}(\mathbf{Q})^+$, and this identification is compatible with the actions of the groups on $M^o(G,X^+)$ (see [4, 2.1.6.2]).

Any $x \in M^o(G,X^+)$ can be written $x = g[h]$ for some $g \in G(\mathbf{Q})^{-}_{+}$ and $h \in X^+$. For suppose $x_\Gamma = [h]_\Gamma$; then, for any $\Gamma_1 \subset \Gamma$, $x_{\Gamma_1} = \gamma_{\Gamma_1}[h]$ some $\gamma_{\Gamma_1} \in \Gamma$; let $\gamma = \lim_{\Gamma_1 \to 1} \gamma_{\Gamma_1}$, and let $\gamma = g * \alpha$; then $x = \gamma[h] = g[\alpha(h)]$.

If G is simply connected, then $G(\mathbf{Q})^{-}_{+} = G(\mathbf{A}^f)$; moreover

$${}_\Gamma M^o(G,X^+) = \Gamma \backslash X^+ = G(\mathbf{Q}) \backslash X^+ \times G(\mathbf{A}^f)/K$$

if K is a compact open subgroup of $G(\mathbf{A}^f)$ containing $Z(\mathbf{Q})$, and Γ is the image of $K \cap G(\mathbf{Q})$ in $G^{ad}(\mathbf{Q})^+$.

References

[1] A. Borel, Density and maximality of arithmetic subgroups. J. Reine Angew. Math. *224* (1966), 78–89.

[2] M. Borovoi, Canonical models of Shimura varieties. Handwritten notes dated 26/5/81.

[3] P. Deligne, Travaux de Shimura. Sém. Bourbaki Février 71, Exposé 389, Lecture Notes in Math., 244, Springer, Berlin, 1971.

[4] P. Deligne, Variétés de Shimura: interpretation modulaire, et techniques de construction de modéles canoniques. Proc. Symp. Pure Math., A.M.S., *33* (1979), part 2, 247–290.

[5] P. Deligne, Motifs et groupes de Taniyama. Lecture Notes in Math., 900, Springer, Berlin, 1982, pp. 261–279.

[6] D. Kazhdan, On arithmetic varieties. *Lie Groups and Their Representations*. Budapest, 1971, pp. 151–217.

[7] D. Kazhdan, On arithmetic varieties, II. Preprint.

[8] R. Langlands, Automorphic representations, Shimura varieties, and motives. Ein Märchen. Proc. Symp. Pure Math., A.M.S. *33* (1979), part 2, 205–246.

[9] R. Langlands, Les débuts d'une formule des traces stables. Publ. Math. Univ. Paris VII. (To appear).

[10] G. Margulis, Discrete groups of motions of manifolds of nonpositive curvature. Amer. Math. Soc. Transl. *109* (1977), 33–45.

[11] J. Milne, The arithmetic of automorphic functions. In preparation.

[12] J. Milne and K-y. Shih, The action of complex conjugation on a Shimura variety. Annals of Math., *113* (1981), 569–599.

[13] J. Milne and K-y. Shih, Automorphism groups of Shimura varieties and reciprocity laws. Amer. J. Math., *103* (1981), 911–935.

[14] J. Milne and K-y. Shih, Langlands's construction of the Taniyama group. Lecture Notes in Math., 900, Springer, Berlin, 1982, 229–260.

[15] J. Milne and K-y. Shih, Conjugates of Shimura varieties. Lecture Notes in Math., 900, Springer, Berlin, 1982, pp. 280–356.

[16] V. Platonov and A. Rapinčuk, On the group of rational points of three-dimensional groups. Soviet Math. Dokl. *20* (1979), 693–697.

[17] M. Ragunathan, *Discrete Subgroups of Lie Groups*. Erg. Math. *68*, Springer, Berlin, 1972.

[18] I. Satake, *Algebraic Structures of Symmetric Domains*. Publ. Math. Soc. Japan 14, Princeton University Press, Princeton, 1980.

[19] G. Shimura, On arithmetic automorphic functions. Actes, Congrès Intern. Math. (1970) Tom 2, 343–348.

[20] G. Shimura, *Arithmetic Theory of Automorphic Functions*. Publ. Math. Soc. Japan 11, Princeton University Press, Princeton, 1971.

[21] S-T. Yau, On the Ricci curvature of a compact Kähler manifold and the complex Monge-Ampère equation, I. Comm. Pure Appl. Math., *31* (1978), 339–411.

Received April 29, 1982

This work was completed while the author was at The Institute for Advanced Study and was supported in part by NSF grant MCS 8103365.

Professor J. S. Milne
Department of Mathematics
University of Michigan
Ann Arbor, Michigan 48109

The Torelli Theorem for Ordinary K3 Surfaces over Finite Fields

Niels O. Nygaard

To I.R. Shafarevich

Introduction

Shafarevich's and Piatetski-Shapiro's proof of the Torelli theorem for K3 surfaces over $\mathbf{C}$ [13] is one of the most beautiful proofs in complex algebraic geometry.

In one of its several formulations the theorem states the following: Let $(X, \mathcal{L})$ and $(X', \mathcal{L}')$ be polarized surfaces over $\mathbf{C}$ and assume that ϕ: $H^2(X', \mathbf{Z}) \longrightarrow H^2(X, \mathbf{Z})$ is an isomorphism compatible with the cup-product pairings and the Hodge structures. Assume further that ϕ maps the cohomology class of $\mathcal{L}'$ to that of $\mathcal{L}$ then $(X, \mathcal{L}) \simeq (X', \mathcal{L}')$ and in fact ϕ is induced by an isomorphism between the polarized surfaces.

One can ask whether an analogous theorem holds in positive characteristics, i.e., whether there are linear algebra data associated to a polarized K3 surface defined over a field of characteristic $p > 0$ such that these data uniquely determine the isomorphism class of the K3 surface. Let me explain why there seems to be hope for such a theorem: Combining the Torelli theorem and Kulikov's result, that the period map is proper, one shows that the moduli space of polarized K3 surfaces of some fixed degree, say d, (here the term "K3 surface" has to be broadened slightly to include certain singular surfaces) is isomorphic to the period space. The period space is a disjoint union of Shimura varieties $Sh_K(G, \mu)$ where $G = 0(2, 19)$. Assume that we can construct a moduli space over Spec $\mathbf{Z}[\frac{1}{n}]$, n a suitable integer, for polarized K3 surfaces of degree d. Then we have in fact constructed a model for the period space and one can show that this model is canonical in the sense of Shimura varieties. Let p be a prime not dividing n and consider the reduction mod p of the moduli space. We then get a variety over $\mathbf{F}_p$ which is on the one hand a moduli space for polarized K3 surfaces of degree d in characteristic p and on the other hand the reduction of the canonical model of a Shimura variety. Langlands has conjectured a purely group theoretical (or linear algebraic) description of the set of $\overline{\mathbf{F}}_p$ points of the reduction of a Shimura variety [9]. In the present case it then makes sense to use the term "periods" for the elements of this set and to view Langlands' conjecture as a conjectural Torelli theorem for polarized K3 surfaces over $\overline{\mathbf{F}}_p$. One instance that indicates that this makes sense is the supersingular case. Here Langlands' conjecture predicts that a supersingular K3 surface is uniquely determined by its crystalline cohomology. This was conjectured by Ogus in [11] and recently proved by him [12] as a consequence of Rudakov's and Shafarevich's results on degenerations of K3 surfaces in characteristic p [14].

1. Review of the Canonical Lifting and the Kuga-Satake-Deligne Abelian Variety

In this section we recall some of the results of our previous paper [10] and of Deligne's paper [4].

Let k be a perfect field of characteristic p.

Definition 1.1. A K3 surface X_0/k is said to be *ordinary* if the following equivalent conditions are satisfied.

(i) The formal Brauer group $\hat{Br}_{X_0}$ has height 1.

(ii) The frobenius F: $H^2(O_{X_0}) \longrightarrow H^2(O_{X_0})$ is bijective.

(iii) The Hodge polygon and the Newton polygon of $H^2_{\mathrm{crys}}(X_0/W)$ coincide.

Let A be an artinian local ring with residue field k and let $X/\operatorname{Spec} A$ be a lifting of the K3 surface $X_0/\operatorname{Spec} k$.

By Artin-Mazur[2] the enlarged formal Brauer group ψ_{X_A} defines a p-divisible group on $\operatorname{Spec} A$ lifting $\psi_{X_0}/\operatorname{Spec} k$.

In [10] we proved the following theorem.

Theorem 1.2. *Let $X_0/\operatorname{Spec} k$ be an ordinary K3 surface. The map*

$$\{\text{Iso. classes of liftings } X/\operatorname{Spec} A\} \to \{\text{Iso. classes of liftings } G/\operatorname{Spec} A\}$$

defined by

$$X/\operatorname{Spec} A \mapsto \psi_X/\operatorname{Spec} A$$

is a functorial isomorphism.

Proof. [10] theorem 1.3.

Since height one groups are rigid, there is precisely one lifting G_A^0 of $Br_{X_0}^{\wedge}$ to $\operatorname{Spec} A$. Similarly étale groups are rigid so there is precisely one lifting G_A^{et} of $\psi_{X_0}^{et}$ to $\operatorname{Spec} A$. It follows that if G is any lifting of ψ_{X_0} to $\operatorname{Spec} A$ we have an extension

$$0 \to G_A^0 \to G \to G_A^{et} \to 0$$

lifting the extension

$$0 \to Br_{X_0}^{\wedge} \to \psi_{X_0} \to \psi_{X_0}^{et} \to 0.$$

In particular we can consider the lifting of ψ_{X_0} defining the trivial extension $G = G_A^0 \times G_A^{et}$. By the theorem there exists a unique lifting $X_{can,A}/\operatorname{Spec} A$ such that $\psi_{X_{can,A}} = G_A^0 \times G_A^{et}$. Now let $A = W_n$ and put $X_n = X_{can,W_n}$ then we get a proper flat formal scheme $\{X_n\}/\operatorname{Spf} W$.

Theorem 1.3. *The formal scheme $\{X_n\}/\operatorname{Spf} W$ is algebraizable and defines a K3 surface $X_{can}/\operatorname{Spec} W$. Furthermore X_{can} has the property that any line bundle on X_0 lifts uniquely to X_{can}.*

Proof. [10] proposition 1.6.

Next recall from [4] the following facts: Let $(X, \mathcal{L})/\operatorname{Spec} V$ be a polarized K3 surface where V is a discrete valuation ring with residue field k a finite extension of $\mathbf{F}_p$ and fraction field K a finite extension of $\mathbf{Q}_p$.

Choose an isomorphism $\overline{K} \simeq \mathbf{C}$ and base extend X to a K3 surface $X_{\mathbf{C}}/\operatorname{Spec} \mathbf{C}$.

We let $P^2(X_{\mathbf{C}}, \mathbf{Z}(1))$ denote the primitive part of the cohomology, i.e., $\{cl(\mathcal{L})\}^{\perp}$.

Theorem 1.4. *(Kuga-Satake-Deligne). There exists a finite extension K'/K and an abelian variety $A/\operatorname{Spec} K'$ with the following properties.*

(i) *A has complex multiplications by the even Clifford algebra $C = C^+P^2(X_{\mathbf{C}}, \mathbf{Z}(1))$ of the bilinear form on $P^2(X_{\mathbf{C}}, \mathbf{Z}(1))$.*

(ii) *$H^1(A_{\mathbf{C}}, \mathbf{Z})$ is isomorphic to $C^+P^2(X_{\mathbf{C}}, \mathbf{Z}(1))$ as a right C-module, and left multipliction defines an isomorphism of algebras*

$$\mu\colon C^+P^2(X_{\mathbf{C}}, \mathbf{Z}(1)) \longrightarrow \operatorname{End}_{C^+}(H^1(A_{\mathbf{C}}, \mathbf{Z}))$$

(iii) *μ is an isomorphism of Hodge structures.*

(iv) *Tensoring with $\mathbf{Z}_l$ and identifying $\mathbf{Z}_l$-cohomology with l-adic étale cohomology μ induces an isomorphism of étale sheaves on $\operatorname{Spec} K'$*

$$\mu_l\colon C^+P^2_{et}f_*\mathbf{Z}_l(1) \longrightarrow \operatorname{End}_C(R^1_{et}g_*\mathbf{Z}_l)$$

where $f\colon X_{K'} \longrightarrow \operatorname{Spec} K'$ and $g\colon A \longrightarrow \operatorname{Spec} K'$ are the structure maps.

(v) *A has good reduction.*

Assume now that $(X_0, \mathcal{L}_0)$ is a polarized ordinary K3 surface over $\mathbf{F}_q$ and consider the canonical lifting $X_{can}/\operatorname{Spec} W$ with the unique polarization $\mathcal{L}$ lifting $\mathcal{L}_0$. Applying the above theorem to $(X_{can}, \mathcal{L})$ we get an abelian variety A over a finite extension of $\mathbf{Q}_p$.

Proposition 1.5. *The reduction A_0 of A (which is an abelian variety by (v)) is ordinary.*

Proof. [10] Proposition 2.5.

The key result of [10] is the following:

Theorem 1.6. *A is isogeneous to the canonical lifting A_{can} of A_0, in the sense of ordinary abelian varieties.*

It follows from theorem 1.6 that we have isomorphisms

a) $H^1(A_{\mathbf{C}}, \mathbf{Q}) \simeq H^1(A_{can,\mathbf{C}}, \mathbf{Q})$,
b) $H^1_{et}(\overline{A}, \mathbf{Q}_l) \simeq H^1(\overline{A}_{can}, \mathbf{Q}_l)$ all l,
c) $\mathrm{End}(A_{\mathbf{C}}) \otimes \mathbf{Q} \simeq \mathrm{End}(A_{can,\mathbf{C}}) \otimes \mathbf{Q}$,

where a) is an isomorphism of rational Hodge structures and b) is an isomorphism of Galois modules.

The canonical lifting of A_0 is characterized by the fact that the frobenius lifts to A_{can}. It follows that we get an isogeny $\sigma_{A_{can}} \in \mathrm{End}(A_{can,\mathbf{C}}) \otimes \mathbf{Q}$ and hence by c) above an isogeny $\sigma_A \in \mathrm{End}(A_{\mathbf{C}}) \otimes \mathbf{Q}$. Since the frobenius on A_0 commutes with the action of C it follows that $\sigma_A \in \mathrm{End}_C(A_{\mathbf{C}}) \otimes \mathbf{Q}$. The isogeny σ_A induces an automorphism of $H^1(A_{\mathbf{C}}, \mathbf{Q})$ compatible with the Hodge structure and the right C-module structure hence an element $\sigma_A \in \mathrm{End}_C(H^1(A_{\mathbf{C}}, \mathbf{Q}))$.

We define an algebra automorphism σ_{X_0} of $C^+P^2(X_{can,\mathbf{C}}, \mathbf{Q}(1))$ by

$$x \mapsto \mu^{-1}(\sigma_A) \cdot x \cdot \mu^{-1}(\sigma_A^{-1}).$$

In [10] it is proved that σ_X induces an automorphism

$$F_{X_0} \colon P^2(X_{can,\mathbf{C}}, \mathbf{Q}(1)) \longrightarrow P^2(X_{can,\mathbf{C}}, \mathbf{Q}(1))$$

compatible with the cup-product and the Hodge structure and such that F_{X_0} is identified with the geometric frobenius under the canonical isomorphism

$$P^2(X_{can,\mathbf{C}}, \mathbf{Q}(1)) \otimes \mathbf{Q}_l \simeq P^2_{et}(\overline{X}_0, \mathbf{Q}_l(1)).$$

2. The Torelli Theorem

We fix as before an isomorphism $\overline{\mathbf{Q}}_p \simeq \mathbf{C}$.

Let $(X_0, \mathcal{L}_0)$ be a polarized ordinary K3 surface over $\mathbf{F}_q$, $q = p^n$. Define $M(X_0) = H^2(X_{can,\mathbf{C}}, \mathbf{Z}(1))$ and let $v \in M(X_0)$ be the cohomology class of the lifting of $\mathcal{L}_0$. We extend F_{X_0} to $M(X_0) \otimes \mathbf{Q}$ by putting $F_{X_0}(v) = v$.

Theorem 2.1. *Let $(X_0, \mathcal{L}_0)$ and $(X'_0, \mathcal{L}'_0)$ be polarized ordinary* K3 *surfaces over $\mathbf{F}_q$. Assume that ϕ: $M(X'_0) \longrightarrow M(X_0)$ is an isomorphism* *compatible with the cup-product pairings and such that $\phi(v') = v$. Assume further that there is an integer m such that $\phi \cdot F^m_{X'_0} = F^m_{X_0} \cdot \phi$ then $(X_0, \mathcal{L}_0)$ and $(X'_0, \mathcal{L}'_0)$ are isomorphic over $\overline{\mathbf{F}}_q$.*

Proof. Enlarging $\mathbf{F}_q$ if necessary, we may assume

$$\phi \cdot F_{X'_0} = F_{X_0} \cdot \phi.$$

We shall prove that the map

$$\phi\colon H^2(X'_{can,\mathbf{C}}, \mathbf{Z}(1)) \longrightarrow H^2(X_{can,\mathbf{C}}, \mathbf{Z}(1))$$

is compatible with the Hodge structures. Since $\phi(v') = v$, it is clear that we get an isomorphism

$$\phi\colon P^2(X'_{can,\mathbf{C}}, \mathbf{Z}(1)) \longrightarrow P^2(X_{can,\mathbf{C}}, \mathbf{Z}(1))$$

and hence an isomorphism of algebras

$$\phi\colon C^+P^2(X'_{can,\mathbf{C}}, \mathbf{Z}(1)) \longrightarrow C^+P^2(X_{can,\mathbf{C}}, \mathbf{Z}(1)).$$

We have isomorphisms of bimodules

$$C^+P^2(X'_{can,\mathbf{C}}, \mathbf{Z}(1)) \simeq H^1(A'_{\mathbf{C}}, \mathbf{Z})$$
$$C^+P^2(X_{can,\mathbf{C}}, \mathbf{Z}(1)) \simeq H^1(A_{\mathbf{C}}, \mathbf{Z}),$$

hence we have an isomorphism

$$\phi\colon H^1(A'_{\mathbf{C}}, \mathbf{Z}) \xrightarrow{\sim} H^1(A_{\mathbf{C}}, \mathbf{Z}).$$

The frobenius maps on $H^1(A'_{\mathbf{C}}, \mathbf{Q})$ and $H^1(A_{\mathbf{C}}, \mathbf{Q})$ are maps of right modules, hence are given by left multiplication by elements

$$\sigma_{A'} \in C^+P^2(X'_{can,\mathbf{C}}, \mathbf{Q}(1))^*$$

and

$$\sigma_A \in C^+P^2(X_{can,\mathbf{C}}, \mathbf{Q}(1))^*.$$

The frobenius maps on $C^+P^2(X'_{can,\mathbf{C}}, \mathbf{Q}(1))$ and $C^+P^2(X_{can,\mathbf{C}}, \mathbf{Q}(1))$ induced by $F_{X'_0}$ and F_{X_0} are given by

$$\sigma_{X'_0}(x) = \sigma_{A'} \cdot x \cdot \sigma_{A'}^{-1}$$
$$\sigma_{X_0}(y) = \sigma_A \cdot y \cdot \sigma_A^{-1}.$$

It is clear from the assumptions that $\sigma_{X_0} \cdot \phi = \phi \cdot \sigma_{X'_0}$, so we get, for $y \in C^+P^2(X_{can,\mathbf{C}}, \mathbf{Q}(1))$,

$$\begin{aligned}\phi(\sigma_{A'}) \cdot y \cdot \phi(\sigma_{A'}^{-1}) &= \phi(\sigma_{A'} \cdot \phi^{-1}(y) \cdot \sigma_{A'}^{-1}) \\ &= \phi(\sigma_{X'_0}(\phi^{-1}(y))) \\ &= \sigma_{X_0} \cdot \phi(\phi^{-1}(y)) \\ &= \sigma_A \cdot y \cdot \sigma_A^{-1}.\end{aligned}$$

This shows that $\sigma_A^{-1} \cdot \phi(\sigma_{A'})$ is in the center of $C^+P^2(X_{can,\mathbf{C}}, \mathbf{Q}(1))$, which is a central simple algebra over $\mathbf{Q}$, hence $\sigma_A^{-1} \cdot \phi(\sigma_{A'}) = \lambda \in \mathbf{Q}$ and $\phi(\sigma_{A'}) = \lambda \cdot \sigma_A$.

Assume that $x \in H^1(A'_{\mathbf{C}}, \mathbf{C})$ is an eigenvector for frobenius corresponding to an eigenvalue $\mu \in \mathbf{C}$, i.e., we have $\sigma_{A'} \cdot x = \mu \cdot x$ then

$$\lambda \cdot \sigma_A \cdot \phi(x) = \phi(\sigma_{A'}) \cdot \phi(x) = \phi(\sigma_{A'} \cdot x) = \mu\phi(x).$$

It follows that $\phi(x)$ is an eigenvector for frobenius on $H^1(A_{\mathbf{C}}, \mathbf{C})$ corresponding to the eigenvalue $\lambda^{-1} \cdot \mu$.

Assume that the reductions of A and A' are defined over $\mathbf{F}_{p^r}$. Then all the eigenvalues of frobenius have absolute value $p^{r/2}$ by the Riemann hypothesis. In particular $|\mu| = p^{r/2}$ and $|\lambda^{-1}\mu| = p^{r/2}$ so $|\lambda| = 1$. Since $\lambda \in \mathbf{Q}$ we have $\lambda = \pm 1$, so $\phi(\sigma_{A'}^2) = \sigma_A^2$, and hence passing to a quadratic extension of the groundfield we can assume $\phi(\sigma_{A'}) = \sigma_A$. It follows that the isomorphism

$$\phi: H^1(A'_{\mathbf{C}}, \mathbf{Q}) \longrightarrow H^1(A_{\mathbf{C}}, \mathbf{Q})$$

is compatible with the frobenius maps.

The reduction map

$$\mathrm{Hom}(A_{can}, A'_{can}) \to \mathrm{Hom}(A_0, A'_0)$$

is a bijection. By theorem 1.6 we have

$$\mathrm{Hom}(A_{can}, A'_{can}) \otimes \mathbf{Q} \simeq \mathrm{Hom}(A, A') \otimes \mathbf{Q}.$$

Hence the reduction map

$$\mathrm{Hom}(A, A') \otimes \mathbf{Q} \to \mathrm{Hom}(A_0, A'_0) \otimes \mathbf{Q}$$

is bijective.

Let $\mathrm{Hom}_\sigma(H^1(A'_{\mathbf{C}}, \mathbf{Q}), H^1(A_{\mathbf{C}}, \mathbf{Q}))$ denote the set of homomorphisms $\psi\colon H^1(A'_{\mathbf{C}}, \mathbf{Q}) \longrightarrow H^1(A_{\mathbf{C}}, \mathbf{Q})$ satisfying $\psi\,\sigma_{A'} = \sigma_A\,\psi$.

It is clear that the obvious map

$$\mathrm{Hom}(A, A') \otimes \mathbf{Q} \to \mathrm{Hom}(H^1(A'_{\mathbf{C}}, \mathbf{Q}), H^1(A_{\mathbf{C}}, \mathbf{Q}))$$

maps into $\mathrm{Hom}_\sigma(H^1(A'_{\mathbf{C}}, \mathbf{Q}), H^1(A_{\mathbf{C}}, \mathbf{Q}))$, since this is true after tensoring with $\mathbf{Q}_l$ and identifying

$$H^1(A'_{\mathbf{C}}, \mathbf{Q}_l) \otimes \mathbf{Q}_l = H^1_{et}(\overline{A}'_0, \mathbf{Q}_l),$$
$$H^1(A_{\mathbf{C}}, \mathbf{Q}_l) \otimes \mathbf{Q}_l = H^1_{et}(\overline{A}_0, \mathbf{Q}_l).$$

By a theorem of Tate [15] the obvious map

$$\mathrm{Hom}(A_0, A'_0) \otimes \mathbf{Q} \to \mathrm{Hom}_\sigma(H^1(\overline{A}'_0, \mathbf{Q}_l), H^1(\overline{A}_0, \mathbf{Q}_l))$$

is a bijection. It follows from this and the previous remarks that

$$\mathrm{Hom}(A, A') \otimes \mathbf{Q} \to \mathrm{Hom}_\sigma(H^1(A'_{\mathbf{C}}, \mathbf{Q}), H^1(A_{\mathbf{C}}, \mathbf{Q}))$$

is a bijection.

By the above $\phi \in \mathrm{Hom}_\sigma(H^1(A'_{\mathbf{C}}, \mathbf{Q}), H^1(A_{\mathbf{C}}, \mathbf{Q}))$, hence is induced by an element of $\mathrm{Hom}(A, A') \otimes \mathbf{Q}$, and so, being induced by a geometric morphism, ϕ is compatible with the Hodge structures.

Now the commutative diagram

$$\begin{array}{ccc}
C^+P^2(X'_{can,\mathbf{C}}, \mathbf{Q}(1)) & \overset{\mu}{\simeq} & \mathrm{End}_C(H^1(A'_{\mathbf{C}}, \mathbf{Q})) \\
\downarrow \phi & & \downarrow x \mapsto \phi\cdot x\cdot\phi^{-1} \\
C^+P^2(X_{can,\mathbf{C}}, \mathbf{Q}(1)) & \overset{\mu}{\simeq} & \mathrm{End}_C(H^1(A_{\mathbf{C}}, \mathbf{Q}))
\end{array}$$

and the fact that μ is an isomorphism of Hodge structures implies that

$$\phi\colon C^+P^2(X'_{can,\mathbf{C}}, \mathbf{Q}(1)) \longrightarrow C^+P^2(X_{can,\mathbf{C}}, \mathbf{Q}(1))$$

is compatible with the Hodge structures. It follows from [10] lemma 3.2 that we have injective maps of Hodge structures

$$P^2(X'_{can,\mathbf{C}}, \mathbf{Q}(1)) \to C^+P^2(X'_{can,\mathbf{C}}, \mathbf{Q}(1))$$

and

$$P^2(X_{can,\mathbf{C}}, \mathbf{Q}(1)) \to C^+P^2(X_{can,\mathbf{C}}, \mathbf{Q}(1))$$

and a commutative diagram

$$\begin{array}{ccc} P^2(X'_{can,\mathbf{C}}, \mathbf{Q}(1)) & \to & C^+P^2(X'_{can,\mathbf{C}}, \mathbf{Q}(1)) \\ \downarrow \phi & & \downarrow \phi \\ P^2(X_{can,\mathbf{C}}, \mathbf{Q}(1)) & \to & C^+P^2(X_{can,\mathbf{C}}, \mathbf{Q}(1)), \end{array}$$

so ϕ: $P^2(X'_{can,\mathbf{C}}, \mathbf{Q}(1)) \longrightarrow P^2(X_{can,\mathbf{C}}, \mathbf{Q}(1))$ is compatible with the Hodge structures. Since v' and v have Hodge type $(0, 0)$ it follows that also

$$\phi\colon H^2(X'_{can,\mathbf{C}}, \mathbf{Q}(1)) \longrightarrow H^2(X_{can,\mathbf{C}}, \mathbf{Q}(1))$$

is compatible with the Hodge structures.

Now we apply the Torelli theorem over $\mathbf{C}$ to deduce that there is an isomorphism

$$\psi\colon (X_{can,\mathbf{C}}, \mathcal{L}) \simeq (X'_{can,\mathbf{C}}, \mathcal{L}').$$

This isomorphim is defined over a finite extension $K/\mathbf{Q}_p$, which we can choose large enough to contain $W(\mathbf{F}_q)$. Let R be the integral closure of $W(\mathbf{F}_q)$ in K. Then $(X_{can}, \mathcal{L})_R$ and $(X'_{can}, \mathcal{L})_R$ are polarized surfaces over Spec R whose generic fibers are isomorphic as polarized surfaces. By the Mumford-Matsusaka theorem [16] this implies that $(X_{can}, \mathcal{L})_R$ and $(X'_{can}, \mathcal{L}')_R$ are isomorphic. The residue field of R is a finite extension of $\mathbf{F}_q$ and so it follows that $(X_0, \mathcal{L}_0)$ and $(X'_0, \mathcal{L}'_0)$ become isomorphic over a finite extension of $\mathbf{F}_q$.

References

[1] Artin, M., *Supersingular K3 surfaces.* Ann. Sc. Éc. Norm. Sup. 4e série, t. 7, 543–568, (1974).

[2] Artin, M., Mazur, B., *Formal groups arising from algebraic varieties.* Ann. Sc. Éc. Norm. Sup. 4e série, t. 10, 87–132, (1977).

[3] Berthelot, P., *Cohomologie cristalline des schemas des caractéristique $p > 0$.* Lecture Notes in Math. no. 407, Springer Verlag. Berlin, Heidelberg, New York, (1974).

[4] Deligne, P., *La conjecture de Weil pour les surfaces K3.* Inv. Math. 15, 206–226, (1972).

[5] Deligne, P., *Variétés abéliennes ordinaires sur un corps fini.* Inv. Math. 8, 238–243, (1968).

[6] Deligne, P., Illusie, L., *Cristaux ordinaires et coordonées canoniques.* In Surfaces Algebrique, Seminaire Orsay, 1976-78. Lecture Notes in Math. no. 868, Springer Verlag. Berlin, Heidelberg, New York, (1981).

[7] Katz, N., *Serre-Tate local moduli of ordinary abelian varieties.* In Surfaces Algebrique, Seminaire Orsay, 1976-78. Lecture Notes in Math. no. 868, Springer Verlag. Berlin, Heidelberg, New York, (1981).

[8] Kuga, M., Satake, I., *Abelian varieties associated to polarized K3 surfaces.* Math. Ann. 169, 239-242, (1967).

[9] Langlands, R., *Some contemporary problems with origins in the Jugendtraum* (*Hilbert's problem* 12). In Mathematical developments arising from Hilbert problems, Symp. in Pure Math. vol. XXVIII, part 2. American Math. Soc. Providence, Rhode Island, (1976).

[10] Nygaard, N., *The Tate conjecture for ordinary K3 surfaces over finite fields.* To appear in Inv. Math.

[11] Ogus, A., *Supersingular K3 crystals.* Journeés de Géometrié Algébrique. Rennes 1978, Astérisque no. 64, 3–86, (1978).

[12] Ogus, A., *A crystalline Torelli theorem for supersingular K3 surfaces.* This volume.

[13] Shafarevich, I., Piatetski-Shapiro, I. *A Torelli theorem for surfaces of type K3.* Math. U.S.S.R. Izvestija 5, 547-588, (1971).

[14] Shafarevich, I., Rudakov, A., *Degeneration of K3 surfaces over fields of finite characteristic.* Preprint.

[15] Tate, J., *Endomorphisms of abelian varieties over finite fields.* Inv. Math. no. 2, 134–144, (1966).

[16] Mumford, D., Matsusaka, *Two fundamental theorems on deformations of polarized algebraic varieties.* Amer. Jour. of Math. 86, (1964).

Received August 10, 1982

Professor Niels O. Nygaard
Department of Mathematics
University of Chicago
Chicago, Illinois 60637

Real Points on Shimura Curves

A. P. Ogg

To I.R. Shafarevich

Let B be a quaternion algebra over $\mathbf{Q}$, i.e., a central simple alegbra of dimension 4 over $\mathbf{Q}$. We assume that B is indefinite, and fix an identification of $B_\infty = B \otimes \mathbf{R}$ with $M_2(\mathbf{R})$. The discriminant D of B is then the product of an even number of distinct primes. The completion $B_p = B \otimes \mathbf{Q}_p$ is a skew field if $p \mid D$ and is isomorphic to $M_2(\mathbf{Q}_p)$ if $p \nmid D$, and $D = 1$ if and only if B is isomorphic to $M_2(\mathbf{Q})$, i.e., B is not a skew field.

Let $\alpha \mapsto \alpha'$ denote the *canonical involution* of B; thus $(\alpha\beta)' = \beta'\alpha'$, and the *trace* $s(\alpha) = \alpha + \alpha'$ and the *norm* $n(\alpha) = \alpha\alpha'$ are in $\mathbf{Q}$ for any $\alpha \in B$. If B or one of its completions is isomorphic to a matrix algebra $M_2(K)$, then $\begin{pmatrix} a & b \\ c & d \end{pmatrix}' = \begin{pmatrix} d & -b \\ -c & a \end{pmatrix}$, and the trace and norm correspond to the trace and determinant of a two-by-two matrix. An element α of B is an *integer* if $s(\alpha)$ and $n(\alpha)$ are in $\mathbf{Z}$. An *order* in B is a subring O of B, consisting of integers and of rank 4 over $\mathbf{Z}$. The corresponding local orders are written $O_p = O \otimes \mathbf{Z}_p$, and we have $O = B \cap_p O_p$.

Given an integer $F \geq 1$ which is relatively prime to D, an *Eichler order* of level F is an order O such that the local order O_p is maximal if $p \nmid F$, and if $p \mid F$, there is an isomorphism of B_p onto $M_2(O_p)$ carrying O_p onto $\left\{\begin{pmatrix} a & b \\ Fc & d \end{pmatrix} : a, b, c, d \in \mathbf{Z}_p\right\}$. Thus an Eichler order of level 1 is a maximal order. All orders in this paper will be Eichler orders. Eichler has shown [3,4]:

1) There is only one Eichler order of a given level F, up to conjugation, i.e., the general one is $\alpha O \alpha^{-1}$, where $\alpha \in B^\times$.

2) O contains a unit of norm -1.

The group $O^{(1)}$ of units of norm 1 in O is thus of index 2 in $O^\times$. It is a discrete subgroup of $SL(2,\mathbf{R})$ and so acts on the upper half-plane $\mathfrak{H}$. Since $O^{(1)}$ depends only on D and F, up to conjugation, we have a well-defined Riemann surface $O^{(1)} \backslash \mathfrak{H}$. If $D > 1$, this Riemann surface is compact and is the *Shimura curve* $S = S_{D,F}$. If $D = 1$, then S denotes the usual compactification by adding cusps, i.e., S is the ordinary modular curve $X_0(F)$. According to Shimura, S has a canonical model defined over $\mathbf{Q}$.

If $D > 1$, then $S(\mathbf{Q})$ is empty since $S(\mathbf{R})$ is empty, as shown by Shimura and Shih; the proof will be recalled below. If $D = 1$, then $S(\mathbf{Q})$ is known by the work by Barry Mazur. The next problem along these lines is to determine, in the case $D > 1$, the rational points on a quotient $S/(w)$ of S by an *Atkin-Lehner involution;* these involutions will be described in detail in §2 below. A rational point on such a quotient corresponds to a pair of conjugate points on S over some (imaginary) quadratic number field, but of course not all quadratic points on S need arise in this manner. For example, if S is hyperelliptic over $\mathbf{Q}$, or more generally, if $S/(w)$ is isomorphic over $\mathbf{Q}$ to the projective line, then there are infinitely many such quadratic points. In §5, we complete the program of determining the values of (D, F) for which S is hyperelliptic, which was done in [12] for $D = 1$ and in [10] for $F = 1$, and determine for which of these values S is hyperelliptic over $\mathbf{R}$ or $\mathbf{Q}$.

The main part of the paper is devoted to the study of the real points on $S/(w)$; it is hoped that this will be of use later on in the study of the rational points. If $w = w(m)$ is associated to an exact divisor m of DF (see §2 for the notation), then real points on $S^{(m)} = S/(w(m))$ will in general arise from certain embeddings of $\sqrt{m}$ into $\mathcal{O}$, and the number $\nu(m)$ of classes of such embeddings is given by Eichler's results on embeddings, which we review in §1. Then $S^{(m)}(\mathbf{R})$ consists of $\nu(m)/2$ disjoint circles, if there are no cusps or elliptic fixed points of order 2; otherwise this formula must be suitably modified.

The basis results on the arithmetic of quaternions which are used in this paper can be found in the papers of Eichler or in the recent lecture notes [18] of Vignéras. In $\mathbf{Z}$ or $\mathbf{Z}_p$ we write $a \mid b$ if a divides b, $a \parallel b$ if a divides b exactly, i.e., if a divides b and a is relatively prime to b/a, and $a \sim b$ if a and b are associates.

It is a pleasure to thank Y. Ihara for several helpful conversations.

1. Eichler's Embedding Theorem.

Wet B be a quaternion algebra over $\mathbf{Q}$ and let L be a 2-dimensional algebra over $\mathbf{Q}$. The canonical involution of B and the non-trivial automorphism of L will both be denoted by $\alpha \mapsto \alpha'$. Consider an embedding $\varphi: L \longrightarrow B$. If $D > 1$, i.e., if B is a skew field, then L is a quadratic field, and the local algebra $L_p = L \otimes \mathbf{Q}_p$ is necessarily a field when p divides D. Thus the primes p which ramify in B cannot split in L; this is a necessary and sufficient condition for L to be embeddable in B. If $L = \mathbf{Q}(\alpha)$, and $s = \alpha + \alpha'$, $n = \alpha\alpha'$, then to give an embedding φ is to give an element of B with trace s and norm n. We assume that $x^2 - sx + n = 0$ has two distinct roots, i.e., if L is not a field (possible only if $D = 1$), then this polynomial has two distinct rational roots.

Let O be a fixed Eichler order of level F in B. Given an embedding $\varphi: L \longrightarrow B$, its *order* is $R = \varphi^{-1}(O)$, an order in L; in this situation one often says that φ is an *optimal embedding* of R into O. If $\gamma \in O^{\times}$, then we have an *equivalent embedding* $\varphi_\gamma: L \to B$, by $\varphi_\gamma(\alpha) = \gamma\varphi(\alpha)\gamma^{-1}$; it has the same order R as φ. If $R = \mathbf{Z}[\alpha]$, and $s = \alpha + \alpha'$, $n = \alpha\alpha'$, then to give an optimal embedding of R into O is to give a *primitive* element β of O with trace s and norm n; β is *imprimitive* if $\beta - a \in pO$ for some $a \in \mathbf{Z}$ and some prime p. If $\gamma \in O^{\times}$, then the conjugate element $\beta_1 = \gamma\beta\gamma^{-1}$ (often written $\beta_1 \sim \beta$) defines an equivalent embedding. Let $\nu(R, O)$ denote the number of inequivalent optimal embeddings of R into O.

Similarly, in the local situation, we have optimal embeddings of R_p into O_p, and two of them are equivalent if conjugate by an element of $O_p^{\times}$. Let $\nu_p(R, O)$ denote the number of inequivalent optimal embeddings of R_p into O_p. Then $\nu_p(R, O) = 1$ if $p \not| DF$. (We can take $O_p = M_2(\mathbf{Z}_p)$, and $\left(\begin{smallmatrix} 0 & -1 \\ n & s \end{smallmatrix}\right)$ is certainly a primitive element with given norm and trace. Conversely, let $\left(\begin{smallmatrix} a & b \\ c & d \end{smallmatrix}\right) = \alpha$ be such an element. We change bases in $\mathbf{Z}_p^2$ by $e_2 = \left(\begin{smallmatrix} x \\ y \end{smallmatrix}\right)$ and $e_1 = \alpha' e_2$, so $\alpha e_1 = n e_2$ and $e_1 + \alpha e_2 = s e_2$, as desired. This works if $\gamma = (e_1\ e_2) = \left(\begin{smallmatrix} dx - by & x \\ ay - cx & y \end{smallmatrix}\right) \in O_p^{\times}$, i.e., if we can find $x, y \in \mathbf{Z}_p$ with $cx^2 + (d - a)xy - by^2 \in \mathbf{Z}_p^{\times}$. This is possible unless p divides $c, d - a$, and b, in which case α is not primitive.)

Theorem 1 (Eichler). *Let $h(R)$ be the class number of R. Then*

$$\nu(R, O) = h(R) \prod_{p | DF} \nu_p(R, O).$$

More precisely, suppose we are given for each $p \mid DF$ an equivalence class of optimal embeddings of R_p into O_p. Then there are exactly $h(R)$ inequivalent optimal embeddings of R into O which are in the given local classes.

Remark. 1) Here we use the fact that the class number $h(O)$ is 1; in general, when B is not necessarily indefinite, the left side would be a sum over $h(O)$ classes. The result is not stated in exactly this form in [5], but the proof is the same.

2) Let R_1 be the principal order in L, i.e., the order consisting of all integers, and let $R_f = \mathbf{Z} + fR_1$ be the order of conductor f. Their class numbers are connected by Dedekind's formula

$$h(R_f)/h(R_1) = f\left(\prod_{p|f}\left(1 - \left(\frac{L}{p}\right)/p\right)\right)/(R_1^\times : R_f^\times),$$

where $\left(\frac{L}{p}\right)$ is the Legendre symbol, if L is a field, and otherwise $\left(\frac{L}{p}\right) = 1$. In that last case, say $L = \mathbf{Q} + \mathbf{Q} = \{\left(\begin{smallmatrix} a & 0 \\ 0 & d\end{smallmatrix}\right) : a, d \in \mathbf{Q}\}$, so $R_1 = \mathbf{Z} + \mathbf{Z}$ has four units ± 1, $\pm\epsilon$, where $1 = \left(\begin{smallmatrix} 1 & 0 \\ 0 & 1\end{smallmatrix}\right)$, $\epsilon = \left(\begin{smallmatrix} 1 & 0 \\ 0 & -1\end{smallmatrix}\right)$, and $h(R_1) = 1$; then $h(R_f) = \begin{cases} \varphi(f)/2 & \text{if } f \geq 3 \\ 1 & \text{if } f \leq 2\end{cases}$.

3) The *generalized Legendre symbol* for the order $R = R_f$ is

$$\left(\frac{R}{p}\right) = \begin{cases} \left(\frac{L}{p}\right) & (\text{if } p \nmid f) \\ 1 & (\text{if } p \mid f). \end{cases}$$

The formulas for the local factors $\nu_p(R, O)$ are then:

Theorem 2. *Let f be the conductor of R and let F be the level of O. Then $\nu_p = \nu_p(R, O)$ is given below, according to various cases; divisibilities are to be understood as in $\mathbf{Z}_p$, and ψ_p is the multiplicative function with $\psi_p(p^k) = p^k(1 + 1/p)$ and $\psi_p(n) = 1$ if $p \nmid n$.*

i) *If $p \mid D$, then $\nu_p = 1 - \left(\frac{R}{p}\right)$.*

ii) *If $p \parallel F$, then $\nu_p = 1 + \left(\frac{R}{p}\right)$.*

iii) *Suppose $p^2 \mid F$.*

a) *If $(pf)^2 \mid F$, then* $\nu_p = \begin{cases} 2\psi_p(f) & (\textit{if } \left(\frac{L}{p}\right) = 1) \\ 0 & (\textit{otherwise}). \end{cases}$

b) *If* $pf^2 \parallel F$ *(say* $f \sim p^k$*), then* $\nu_p = \begin{cases} 2\psi_p(f) & (\text{if } \left(\frac{L}{p}\right) = 1) \\ p^k & (\text{if } \left(\frac{L}{p}\right) = 0). \\ 0 & (\text{if } \left(\frac{L}{p}\right) = -1) \end{cases}$

c) *If* $f^2 \parallel F, f \sim p^k$*, then* $\nu_p = p^{k-1}(p + 1 + \left(\frac{L}{p}\right))$.

d) *If* $pF \mid f^2$*, then* $\nu_p = \begin{cases} p^k + p^{k-1} & (\text{if } F \sim p^{2k}) \\ 2p^k & (\text{if } F \sim p^{2k+1}) \end{cases}$.

Proof (sketch). Parts i) and ii) are very familiar, so let us assume that $p^2 \mid F$ and $O_p = \left\{\left(\begin{smallmatrix} a & b \\ Fc & d \end{smallmatrix}\right) : a, b, c, d \in \mathbb{Z}_p\right\}$. Then $R_p = \mathbb{Z}_p[\alpha]$, where we assume that $s = s(\alpha)$ and $n = n(\alpha)$ satisfy

$$\begin{cases} n = 0 & \text{and } f \parallel s \\ n \sim pf^2 & \text{and } pf \mid s \\ n \sim f^2 & \text{and } f \mid s \end{cases} \qquad \text{if} \left(\frac{L}{p}\right) = \begin{cases} 1 \\ 0 \\ -1. \end{cases}$$

Thus we take n as small as possible. If $\nu_p \geq 1$, then we can write $R_p = \mathbb{Z}_p[\beta]$ with $\beta = \left(\begin{smallmatrix} a & b \\ Fc & 0 \end{smallmatrix}\right)$, so $F|n(\beta)|n$. Conversely, if $F \mid n$, then $\left(\begin{smallmatrix} s & -1 \\ n & 0 \end{smallmatrix}\right)$ is primitive in O_p with the desired norm and trace. Thus $\nu_p \geq 1$ if and only if $F \mid n$, so $\nu_p = 0$ exactly in the cases listed in the statement of the theorem, and we may assume that $F \mid n$ from now on.

Note that conjugating $\left(\begin{smallmatrix} a & b \\ Fc & d \end{smallmatrix}\right)$ by an element of $O_p^\times$ leaves $a(\text{mod } F)$ and $b(\text{mod } F)$ unchanged.

Suppose that $p \nmid f$, so we are in case a), and $\left(\frac{L}{p}\right) = 1$, $n = 0$. Then $p \nmid s$, and $\left(\begin{smallmatrix} s & 0 \\ 0 & 0 \end{smallmatrix}\right)$ and $\left(\begin{smallmatrix} 0 & 0 \\ 0 & s \end{smallmatrix}\right)$ are two solutions, so $\nu_p \geq 2$. Conversely, let $\alpha = \left(\begin{smallmatrix} a & b \\ Fc & d \end{smallmatrix}\right)$ have trace s and norm 0. Then either

$$p \nmid a, \quad \gamma = (e_1\, e_2) = \begin{pmatrix} a & -b \\ Fc & d \end{pmatrix} \in O_p^\times,$$

and $\alpha e_2 = 0 = \alpha' e_1$, so $\alpha \sim \left(\begin{smallmatrix} s & 0 \\ 0 & 0 \end{smallmatrix}\right)$, or $p \nmid d$ and $\alpha \sim \left(\begin{smallmatrix} 0 & 0 \\ 0 & s \end{smallmatrix}\right)$. Thus $\nu_p = 2$ in this case, in agreement with a).

Suppose $p \mid f$ from now on. Then $p \mid s$, so if $\alpha = \left(\begin{smallmatrix} a & b \\ Fc & d \end{smallmatrix}\right)$ is primitive in O_p with norm n and trace s, then p divides both a and d and so p

cannot divide both b and c. If $p \not| b$, let d_0 be a standard representative of d modulo F. Then $\gamma = \begin{pmatrix} b_0 & 0 \\ d-d_0 & 1 \end{pmatrix} \in O_{p'}^{\times}$ and $\gamma^{-1}\alpha\gamma = \begin{pmatrix} s-d_0 & 1 \\ Fc_0 & d_0 \end{pmatrix} = \alpha(d_0) \sim \alpha$. Conversely, any d_0 with $d_0(s-d_0) \equiv 0 \pmod F$ leads to a solution $\alpha(d_0)$, since $F \mid n$. Similarly, if $p \mid b$ but $p \not| c$, we find that $\alpha \sim \beta(a_0) = \begin{pmatrix} a_0 & b_0 \\ F & s-a_0 \end{pmatrix}$, with b_0 divisible by p and a_0 a standard representative of a modulo F. Thus ν_p is the number of solutions $x \pmod F$ of $x(s-x) \equiv 0 \pmod F$ plus the number of solutions $y \pmod F$ of $y(y-s) \equiv n \pmod{pF}$. In case a), for example, where $n = 0$ and $f \parallel s$, and $(pf)^2 \mid F$, say $f \sim p^k$, the first equation has as solutions $x \equiv 0, s \pmod{F/p^k}$, which is $2p^k$ solutions $\pmod F$, and the second has as solutions $y \equiv 0, s \pmod{F/p^{k-1}}$, or $2p^{k-1}$ solution $\pmod F$. The other cases are similar.

2. The Atkin-Lehner Group.

As always, let O be an Eichler order of level F in the quaternion algebra B of discriminant D. As shown by Eichler, the group of (non-zero fractional two-sided) ideals of O is as follows. We have the obvious ideals xO for $x \in O^{\times}$. In addition, if $m \parallel DF$, we have an ideal $I = I(m)$, non-obvious if $m \neq 1$, with $I^2 = mO$. To define it, it is enough to describe the local completion I_p at any prime p. We take $I_p = O_p$ if $p \not| m$. If $p \mid m$ and $p \mid D$, then I_p is the ideal of all non-units in O_p. If $p^k \parallel m$ and F, and we identify O_p with $\left\{\begin{pmatrix} a & b \\ p^k c & d \end{pmatrix} : a, b, c, d \in \mathbf{Z}_p\right\}$, then we identify I_p with $\begin{pmatrix} 0 & 1 \\ p^k & 0 \end{pmatrix} O_p = O_p \begin{pmatrix} 0 & 1 \\ p^k & 0 \end{pmatrix}$. Note that

$$s(I(m)) \subset m\mathbf{Z}. \tag{1}$$

The quotient group of all ideals modulo the obvious ideals is generated by the $I(m)$ and so is isomorphic to C_2^r, where r is the number of prime factors of DF and C_2 is the group of order 2.

We can write $I(m) = \mu O = O\mu$, where $\mu \in O$ has norm m, using Eichler's results that any ideal (left, right, or two-sided) of O is principal and O has a unit of norm -1. Then $O^{\times} = \mu O^{\times}\mu^{-1}$, since $O = \mu O \mu^{-1}$, and also $O^{(1)} = \mu O^{(1)}\mu^{-1}$. Hence μ defines an automorphism $w(m)$ of S with $w(m)^2 = 1$, called the *Atkin-Lehner involution* associated to m; $w(m) \neq 1$ if $m \neq 1$. These involutions from the *Atkin-Lehner group*

$$W = \{w(m) : m \parallel DF\} \simeq C_2^r. \tag{2}$$

The genus $g^{(m)}$ of $S^{(m)} = S/(w(m))$ is related to the genus g of S by

$$g^{(m)} = (g+1)/2 - e(m)/4 \qquad (m \neq 1), \tag{3}$$

where $e(m)$ is the number of fixed points of $w(m)$ on S. Let $P \in S$ be a fixed point. Suppose that P is represented by $z \in \mathfrak{H}$. (Fixed cusps occur only when $m = 4$; cf. [12, Prop. 3].) Then $\mu(z) = \gamma(z)$, where $\gamma \in \mathcal{O}^{(1)}$. We replace μ by $\gamma\mu$ and assume that $\mu(z) = z$; we assume also that $s(\mu) \geq 0$, replacing μ by $-\mu$ if necessary. Then $\mathbf{Q}(\mu)$ is an imaginary quadratic number-field. Since μ' generates the same ideal $I = I'$, we have $\mu' = \epsilon\mu$, where $\epsilon \in \mathcal{O}^{(1)} \cap \mathbf{Q}(\mu)$. In general, $\epsilon = -1$ so $\mu^2 = -m$; the other possibilities are $\epsilon = 1 + i = 1 + \zeta_4$, if $m = 2$, and $\epsilon = 1 - \zeta_3$, if $m = 3$; here and in the following ζ_n denotes a primitive n-th root of 1. The order $R = \mathbf{Q}(\mu) \cap \mathcal{O}$ contains $\mathbf{Z}[\mu]$; if it is larger, then we claim that $m \equiv 3 \pmod 4$ and $R = \mathbf{Z}[(1+\mu)/2]$. (Assume that $\mu^2 = -m$ and let $\alpha = (a + b\mu)/2 \in \mathcal{O}$, where $a, b, \in \mathbf{Q}$. Then $a = s(\alpha) \in \mathbf{Z}$, and $-bm = s(\alpha\mu) \in s(I(m)) \subset m \cdot \mathbf{Z}$, by (1), so $b \in \mathbf{Z}$. Since $n(\alpha) = (a^2 + mb^2)/4 \in \mathbf{Z}$, we have $a \equiv b \pmod 2$ if $m \equiv 3 \pmod 4$, and otherwise a and b are even, using the fact that μ is primitive in $\mathcal{O}$, since it generates the ideal $I(m)$, in the case $m \equiv 0 \pmod 4$).

Thus, to $z \in \mathfrak{H}$ representing a fixed point $p \in S$ of $w(m)$, we have associated two optimal embeddings of R into $\mathcal{O}$, corresponding to μ and μ' (with trace ≥ 0). Consider an equivalent embedding, defined by $\gamma\mu\gamma^{-1}$, where $\gamma \in \mathcal{O}^{\times}$. If $n(\gamma) = 1$, then $\gamma\mu\gamma^{-1}$ fixes $\gamma(z)$, which also represents P. Suppose then that $n(\gamma) = -1$. Then $\gamma\mu\gamma^{-1}$ fixes $\gamma(\bar{z}) \in \mathfrak{H}$, which represents the complex conjugate point $\overline{P} \in S$, according to the real structure of S defined in [17], and discussed more fully in the next section. If $P = \overline{P}$, i.e., (changing the notation) if $\bar{z} = \gamma(z)$, where $\gamma \in \mathcal{O}^{\times}$ with $n(\gamma) = -1$, then $\bar{z}$ and hence z is fixed by $\gamma\mu\gamma^{-1}$, so $\gamma\mu\gamma^{-1} = \mu'$. (We do not have $\gamma\mu\gamma^{-1} = \mu$, for then γ would be a unit in R and have norm 1.) Conversely, suppose that $\mu' = \gamma\mu\gamma^{-1}$, where $\gamma \in \mathcal{O}^{\times}$. Then μ' fixes $\gamma(z)$, so $\gamma(z) = z$ or $\bar{z}$; actually $\gamma(z) = \bar{z}$, i.e., $P = \overline{P}$, since if $\gamma(z) = z$, then $\gamma \in \mathcal{O}(\mu)$ and $\gamma\mu\gamma^{-1} = \mu$ and not μ'. Thus $P = \overline{P}$ if and only if $\mu \sim \mu'$, and the number of fixed points of $w(m)$ on S is the number of inequivalent optimal embeddings of R into $\mathcal{O}$, such that $\mu\mathcal{O} = I(m)$. Now any two such μ are locally equivalent at any prime p dividing m (by Theorem 2 and its proof). By Theorem 1, then, we have (expect in the case $D = 1$ and $m = 4$)

$$e(m) = \sum_R h(R) \prod_{p \nmid m} \nu_p(R, \mathcal{O}), \tag{4}$$

where R ranges over $\mathbf{Z}[\sqrt{-m}]$, and also $\mathbf{Z}[(1+\sqrt{-m})/2]$ when $m \equiv 3 \pmod 4$, and $Z[i]$ when $m = 2$.

The elements of W are rational automorphisms of S, i.e., they are defined over $\mathbf{Q}$, and so they act on $S(K)$ for any field K.

3. Real Points, Especially when $D > 1$.

Let φ: $\mathfrak{H} \longrightarrow S$ be the natural map. According to Shimura, the real structure of S, i.e., the action of complex conjugation on S, satisfies

$$\overline{\varphi(z)} = \varphi(\epsilon(\bar{z})), \tag{5}$$

where ϵ is any unit of norm -1 in $\mathcal{O}$.

Given $m \parallel DF$, let $P = \varphi(z)$ satisfy $\overline{P} = w(m)P$, i.e., $\mu(z) = \gamma\epsilon(\bar{z})$, where $\gamma \in \mathcal{O}^{(1)}, n(\mu) = m$, and $\mu\mathcal{O} = I(m)$, in the notation of §2. Then $\bar{z} = \beta(z)$, where $\beta = \epsilon^{-1}\gamma^{-1}\mu$ has norm $-m$ and $\beta\mathcal{O} = I(m)$. Writing $\beta = \left(\begin{smallmatrix} a & b \\ c & d \end{smallmatrix}\right) \in M_2(\mathbf{R})$, we have $\bar{z} = (az+b)/(cz+d)$ or $az + b = c|z|^2 + d\bar{z}$; comparing imaginary parts, we have $s(\beta) = a+d = 0$, so $\beta^2 = -\beta\beta' = m$. This is not possible if m is a square and $D > 1$, so B is a skew field; in particular, taking $m = 1$, this proves [17] that $S(\mathbf{R})$ is empty when $D > 1$.

In any case, let φ_m: $\mathfrak{H} \longrightarrow S^{(m)} = S/(w(m))$ be the natural map, and put

$$U(m) = \{\beta \in \mathcal{O} : \beta\mathcal{O} = I(m), \quad s(\beta) = 0, \quad \beta^2 = m\}, \tag{6}$$

and

$$Z_\beta = \{z \in \mathfrak{H} : \quad \beta(z) = \bar{z}\}, \tag{7}$$

for $\beta \in U(m)$. Then $X_\beta = \varphi_m(Z_\beta)$ is contained in $S^{(m)}(\mathbf{R})$, and these sets cover $S^{(m)}(\mathbf{R})$ if $D > 1$; for $D = 1$, we must add in $S(\mathbf{R})/(w(m))$. Note that $Z_\beta : c(x^2+y^2) - 2ax = b$ is a geodesic arc in $\mathfrak{H}$; if $c = 0$ it is a vertical line and if $c \neq 0$ it is the circle $(cx-a)^2 + (cy)^2 = a^2 + bc = m$. If $D = 1$ (so we take $a, b, c, d \in \mathbf{Z}$) and m is a square in $\mathbf{Z}$, then Z_β begins and ends at a cusp; otherwise, X_β is closed.

Lemma. *If $\beta_1 = \gamma\beta\gamma^{-1}$, where $\gamma \in \mathcal{O}^\times$, then $X_{\beta_1} = X_\beta$.*

Proof. If $n(\gamma) = 1$, then $Z_{\beta_1} = \gamma(Z_\beta)$ has the same image as Z_β in S and hence in $S^{(m)}$. Suppose that $n(\gamma) = -1$. Then $\alpha = \gamma\beta$ defines $w(m)$, so $Z_{\beta_1} = \alpha(Z_\beta)$ has the same image as Z_β in $S^{(m)}$.

Let $\beta \in U(m)$, $m \neq 1$. The corresponding order $R = \mathbf{Q}(\beta) \cap O$ is either $\mathbf{Z}[\beta]$ or $\mathbf{Z}[(1+\beta)/2]$, the second possibility occurring only if $m \equiv 1 \pmod 4$. The proof of this is the same as that of the corresponding fact in §2, replacing $-m$ by m. Thus our sets X_β are indexed by the equivalence classes of certain optimal embeddings. Clearly $X_\beta = X_{-\beta}$, so the number of sets is $\nu(m)/2$, where $\nu(m)$ is the number of inequivalent elements of $U(m)$, provided that β and $-\beta$ are never conjugate under $O^\times$. In the general case, write $\nu(m) = \nu_1(m) + 2\nu_2(m)$, where we have ν_1 inequivalent $\beta \in U(m)$ with $\beta \sim -\beta$, and ν_2 inequivalent pairs $\pm\beta$ with $\beta \not\sim -\beta$. The number of sets X_β is thus

$$\nu_1(m) + \nu_2(m) = \big(\nu(m) + \nu_1(m)\big)/2. \tag{8}$$

The number $\nu(m)$ is given by Eichler's embedding theorem; as above, the local factor in the formula is 1 at a prime dividing m, since the ideal and hence the local embedding is fixed, so we get:

Proposition 1. *$\nu(m) = \sum_R h(R) \prod_{p \nmid m} \nu_p(R, O)$, for $m \neq 1$, where R ranges over $\mathbf{Z}[\sqrt{m}]$, and also $\mathbf{Z}[(1+\sqrt{m})/2]$ if $m \equiv 1 \pmod 4$. (Notation: Even if m is a square, $\alpha = \sqrt{m}$ is an element of B satisfying $s(\alpha) = 0$ and $\alpha^2 = m$.)*

The number $\nu(m)$ is easily computed, so at least we know in all cases whether $S^{(m)}(\mathbf{R})$ is empty or not. Now $S^{(m)}(\mathbf{R})$ is a real 1-manifold, i.e., a union of disjoint circles. If $D > 1$ and if $O^{(1)}$ has no elliptic fixed points of order 2 with real image in $S^{(m)}$, then X_β is a circle and $\beta \not\sim -\beta$, for all $\beta \in U(m)$, as we shall show, so the number of components of $S^{(m)}(\mathbf{R})$ is

$$\#(m) = \nu(m)/2 \tag{9}$$

in this case. This formula will have to be modified in the exceptional cases.

Suppose that $\beta \in U(m)$ satisfies $-\beta = \gamma\beta\gamma^{-1}$, where $\gamma \in O^\times$. Then $\beta = \gamma^2\beta\gamma^{-2}$, so γ^2 commutes with β and γ (which generate B) and so lies in the center $\mathbf{Q}$ of B. Thus $\gamma^2 = \pm 1$, and $\gamma^2 = -1$ if B is a skew field. In either case O contains the $\mathbf{Z}$-module $\langle 1, \beta, \gamma, \beta\gamma \rangle_{\mathbf{Z}}$ generated by anticommuting elements β, γ with $\beta^2 = m$ and $\gamma^2 = \pm 1$, so the (reduced) discriminant DF of $\mathcal{O}$ divides the discriminant $4m$ of this module; thus $m \parallel DF \mid 4m$. Suppose that $\gamma^2 = -1$. Then γ is an elliptic element of $\mathcal{O}^{(1)}$, of order 4, with a unique fixed point z_0 in $\mathfrak{H}$. Since $\beta\gamma = -\gamma\beta$, we see that $\beta(z_0)$ is also fixed by γ, and in the lower half-plane (since $n(\beta) < 0$), so $\beta(z_0) = \bar{z}_0$. Thus $z_0 \in Z_\beta$ and $\varphi_m(z_0)$ is a real image of an elliptic fixed point of order 2, or a real "E-point" for short. Note that $\beta_1 = \beta\gamma$ is also in $U(m)$, and $z_0 \in Z_{\beta_1}$. The sets Z_β and Z_{β_1} meet at a right angle at z_0 (say because $\beta_1 = (1+\gamma)^{-1}\beta(1+\gamma)$, and $1+\gamma$ acts as a rotation through 90°), so X_β and X_{β_1} meet at $P_0 = \varphi_m(z_0)$ at an angle of 180°, as they should. It may happen that $\beta \sim \beta_1$, in which case X_β is a circle, i.e., a component of $S^{(m)}(\mathbf{R})$; we discuss this possibility later. Conversely, let z_0 be an elliptic fixed point of order 2 mapping to a real point P_0 on $S^{(m)}$. Then $z_0 \in z_\beta$, where $\beta \in U(m)$, and z_0 is fixed by $\gamma \in \mathcal{O}$, where $\gamma^2 = -1$. Then $\beta\gamma(z_0) = \bar{z}_0$, so $\beta_1 = \beta\gamma$ has trace 0 (as shown early in this section), i.e., β and γ anticommute and we are in the situation just discussed. Thus $-\beta = \gamma\beta\gamma^{-1}$, with $\gamma^2 = -1$, if and only if an end of X_β is an E-point.

The number of elliptic fixed points of order 2 on S is $\nu(\mathbf{Z}[i], \mathcal{O})$. If this number is $\neq 0$, then $4 \not| F$, and the number is either 2^r (if $2 \not| DF$) or 2^{r-1} (if $2 \parallel DF$), where r is the number of prime factors of DF. In either case, the E-points form a single orbit under W, with $(w(2))$ as stabilizer when $2 \parallel DF$. In particular, they all look the same, on S or on $S^{(m)}$.

Consider now the second case; $\beta\gamma = -\gamma\beta$, with $\gamma^2 = 1$, possible only if $D = 1$. Then $n(\gamma) = -1$ (otherwise $\gamma = \gamma'$), so $\gamma' = -\gamma$. Then $\alpha = \beta\gamma$ defines $w(m)$ and $\alpha^2 = -m$, so α has a fixed point $z \in \mathfrak{H}$. Since α anticommutes with β and with γ, it fixed $\beta(z)$ and $\gamma(z)$, which are therefore $\bar{z}$, so $z \in Z_\beta \cap Z_\gamma$. Thus X_β and $X_\gamma = \varphi_m(Z_\gamma)$ meet at a real fixed point $\varphi_m(z)$ of $w(m)$. Conversely, let $P = \varphi(z) \in S$ be a real fixed point of $w(m)$, not a cusp; we can assume that $\alpha(z) = z$ and $\beta(z) = \bar{z}$, where α defines $w(m)$ and $\beta \in U(m)$. We assume that $m > 3$ and hence $\alpha^2 = -m$. (If $m = 2$ or 3, then F divides 8 or 12 and $S = X_0(F)$ has genus 0, and the real locus of any quotient of S is a single circle.) Write $\alpha = \beta\gamma$, where $\gamma \in \mathcal{O}^\times$ and $n(\gamma) = -1$; then $\gamma(z) = \bar{z}$ and $s(\gamma) = 0$, so $-\alpha = \alpha' = \gamma'\beta' = \gamma\beta = -\beta\gamma$, and we are back in the situation just discussed. This proves the first two parts of:

Proposition 2. *Let $\beta \in U(m)$. The possibilities for $\beta \sim -\beta$ are as follows.*

i) *If* $-\beta = \gamma\beta\gamma^{-1}$, *where* $\gamma^2 = -1$, *then one end of* $X_\beta = \varphi_m(Z_\beta)$ *is a real E-point, and every real E-point arises in this way.*

ii) *If* $-\beta = \gamma\beta\gamma^{-1}$, *where* $\gamma^2 = 1$ *(necessarily* $D = 1$*), then one end of* X_β *is a fixed point of* $w(m)$*, and every real fixed point of* $w(m)$ *arises in this way.*

iii) *The two ends of* X_β *are of opposite type (as in i) and ii)) if and only if* $R = \mathcal{O} \cap \mathbf{Q}(\beta)$ *contains a unit of norm* -1.

Proof of iii). If $-\beta = \gamma\beta\gamma^{-1} = \gamma_1\beta\gamma_1^{-1}$, then $\gamma_1 = \gamma\rho$, where $\rho \in \mathcal{O}^\times$ and ρ commutes with β, i.e., $\rho \in R^\times$, and conversely. The type changes only if $n(\rho) = -1$.

Now suppose that X_{β_1} meets X_{β_2} in a non-cusp, where $\beta_i \in U(m)$ and β_2 is not equivalent to $\pm\beta_1$; we can suppose that there is a $z \in \mathfrak{H}$ with $z \in Z_{\beta_1} \cap Z_{\beta_2}$. Then $\beta_2 = \beta_1\gamma$, where $\gamma \in \mathcal{O}^\times$ and $n(\gamma) = 1$, and $\gamma(z) = z$. Thus z is an elliptic fixed point of order $e = 2$ or 3, and we can suppose that γ is of order $2e = 4$ or 6. We have $\beta_2 = \beta_1\gamma = \gamma'\beta_1$, since $\beta_i' = -\beta_i$. If $e = 3$, then $\gamma\beta_1\gamma^{-1} = \beta_1\gamma^{-2} = -\beta_1\gamma = -\beta_2$, contrary to the assumption. Thus $e = 2$, so we are in the situation i) of Proposition 2:

Proposition 3. *If the sets* X_{β_1} *and* X_{β_2} *meet in a non-cusp* P, *and* $\beta_1 \not\sim \pm\beta_2$, *then* P *is a real E-point.*

We have seen that real E-points can occur only if $m \mid DF \mid 2m$, and of course $\nu(m) > 0$ and $\nu(\mathbf{Z}[i], \mathcal{O}) > 0$. Let us show next that these cases do occur, at least if m is not a square (automatic if $D > 1$). Suppose first that DF is odd; then $m = DF$ and $m \equiv 1 \pmod 4$, since D is the product of an even number of primes, each $\equiv 3 \pmod 4$, and $p \equiv 1 \pmod 4$ if $p \mid F$ (we are assuming that $i \in \mathcal{O}$). Put $R_1 = \mathbf{Z}[(1+\sqrt{m})/2]$ and $R_2 = \mathbf{Z}[\sqrt{m}]$. We realize B in standard form as

$$B = \left\{ \begin{pmatrix} a & b \\ -b' & a' \end{pmatrix} : a, b, \in \mathbf{Q}(\sqrt{m}) \right\}.$$

(A prime p ramifies in B if and only if $p \mid m$ and $\left(\frac{-4}{p}\right) = -1$, i.e., $p \mid D$, so this is the quaternion algebra of discriminant D.) Put $\beta = \begin{pmatrix} \sqrt{m} & 0 \\ 0 & -\sqrt{m} \end{pmatrix}$ and $\gamma = \begin{pmatrix} 0 & 1 \\ -1 & 0 \end{pmatrix}$. Then $\mathcal{O} = \left\{ \begin{pmatrix} a & b \\ -b' & a' \end{pmatrix} : a, b \in R_1 \right\}$ is an order in B of discriminant DF, and an Eichler order of level F. (This needs to be verified only at primes p dividing F. Then $p \equiv 1 \pmod 4$, so $i \in \mathbf{Z}_p$. We have an isomorphism of O_p onto the standard order $\left\{\left(\begin{smallmatrix} a & b \\ Fc & d \end{smallmatrix}\right) : a, b, c, d, \in \mathbf{Z}_p\right\}$, with $\gamma \mapsto \left(\begin{smallmatrix} i & 0 \\ 0 & -i \end{smallmatrix}\right)$ and $\beta \mapsto \left(\begin{smallmatrix} 0 & 1 \\ m & 0 \end{smallmatrix}\right)$.) Thus we do have real E-points in this case. Furthermore, $\mathbf{Q}(\beta) \cap O \simeq R_1$, while $\beta_1 = \beta\gamma = \left(\begin{smallmatrix} 0 & \sqrt{m} \\ \sqrt{m} & 0 \end{smallmatrix}\right)$ satisfies $\mathbf{Q}(\beta_1) \cap O \simeq R_2$, so β and β_1 are certainly not equivalent. Finally, we claim that if $D > 1$ and we are in this situation, then the component of $S^{(m)}(\mathbf{R})$ containing these sets is

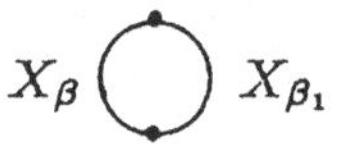

i.e., there are only two E-points on the component. This is true because the stability group of the component in $W/(w(m))$ is an elementary abelian 2-group acting freely and transitively on the E-points on the component, and so it is of order 2. Thus the number of components of $S^{(m)}(\mathbf{R})$ is again $\#(m) = \nu(m)/2$, as in (9), although the reason has changed; in this case we have 2^{r-2} special components as just discussed, plus circles X_β where $\beta \not\sim -\beta$.

Thus we are reduced to the case $DF = 2t$, where t is odd, and $m = t$ or $2t$. If $p \mid t$, then $p \equiv 3 \pmod 4$ if $p \mid D$ and $p \equiv 1 \pmod 4$ if $p \mid F$. We assume that t is not a square (automatic if $D > 1$). Then we can realize B as $\left\{\left(\begin{smallmatrix} a & b \\ -b' & a' \end{smallmatrix}\right) : a, b, \in \mathbf{Q}(\sqrt{t})\right\}$. (The odd ramified primes are those dividing t and $\equiv 3 \pmod 4$, i.e., those dividing D.) The order $\overline{O} = \left\{\left(\begin{smallmatrix} a & b \\ -b' & a' \end{smallmatrix}\right) : a, b \in \mathbf{Z}[(\sqrt{t}]\right\}$ has discriminant $4t$ and the desired completion $\overline{O}_p = O_p$ at any odd prime p; the proof is the same as above. Write $\overline{O} = \langle 1, \gamma, \rho, \gamma\rho \rangle_{\mathbf{Z}}$, where $\gamma = \left(\begin{smallmatrix} \sqrt{t} & 0 \\ 0 & -\sqrt{t} \end{smallmatrix}\right)$, $\rho = \left(\begin{smallmatrix} 0 & 1 \\ -1 & 0 \end{smallmatrix}\right)$, $\gamma\rho = \left(\begin{smallmatrix} 0 & \sqrt{t} \\ \sqrt{t} & 0 \end{smallmatrix}\right)$; our Eichler order will contain $\overline{O}$ with index 2. If $2 \mid D$, so O_2 is to be the maximal order, then we just add in $(1 + \rho)(1 + \gamma)/2$, which has trace 1 and norm $(1 - t)/2 \in \mathbf{Z}$. This also works when $2 \mid F$, as follows. Write $t = b^2 + c^2$, which is possible, and map $\overline{O}_2$ into the standard order

$O_2 = \left\{\begin{pmatrix} a & b \\ 2c & d \end{pmatrix} : a,b,c,d \in \mathbf{Z}_2\right\}$ by $\rho \mapsto \begin{pmatrix} 1 & -1 \\ 2 & -1 \end{pmatrix}$ and $\gamma \mapsto \begin{pmatrix} c-b & b \\ 2c & b-c \end{pmatrix}$, hence $\rho\gamma = -\gamma\rho \mapsto \begin{pmatrix} -b-c & c \\ -2b & b+c \end{pmatrix}$. Since

$$1 + \rho + \gamma + \rho\gamma \mapsto 2\begin{pmatrix} 1-b & (b+c-1)/2 \\ -b-c+1 & b \end{pmatrix} \in 2O_2,$$

we have what we want. Note that in either case we have

$$R = O \cap \mathbf{Q}(\gamma) = \mathbf{Z}[\gamma] \simeq \mathbf{Z}[\sqrt{t}\,].$$

Let us treat first the case $m = 2t = DF$. Here we take $\beta = (1+\rho)\gamma = \gamma(1-\rho)$, so $\beta^2 = m$, $\rho^2 = -1$, and $\beta\rho = -\rho\beta$. Thus there is a real E-point at an end of X_β, and we need to know whether $\rho\beta$ is equivalent to β or not. Now $(\rho+1)^{-1}\beta(\rho+1) = \gamma(\rho+1) = \gamma(1-\rho)\rho = \beta\rho$, so the general solution $\alpha \in B^\times$ of $\alpha\beta\alpha^{-1} = \beta\rho$ is $(\rho+1)\alpha = x + y\beta \in \mathbf{Q}(\beta)$. Here $\alpha \in O^\times$ if and only if $x + y\beta$ is an element of $R = O \cap \mathbf{Q}(\beta) = \mathbf{Z}[\beta]$ of norm 2, i.e., $\beta \sim \beta\rho$ if and only if $x^2 - my^2 = \pm 2$ is solvable with $x, y \in \mathbf{Z}$. If this is not solvable, and $D > 1$, then we have $\#(m) = \nu(m)/2$ as in the previous case. It it is solvable, then we have 2^{r-2} circles X_β with $\beta \sim -\beta$, plus circles X_β with $\beta \not\sim -\beta$, so $\#(m) = (\nu(m) + 2^{r-2})/2$, with $\nu(m) = h(4m) = h(\mathbf{Z}[\sqrt{m}\,])$.

Suppose now that $m = t$ and take $\beta = \gamma$; as noted above, we have $R = O \cap \mathbf{Q}(\beta) = \mathbf{Z}[\beta]$. We have only to decide whether β and $\beta\rho$ are equivalent or not. Again $(\rho+1)^{-1}\beta(\rho+1) = \beta\rho$, so we ask again whether there is an $\alpha \in O^\times$ with $(\rho+1)\alpha = x + y\beta$, i.e., whether $x^2 - my^2 = \pm 2$ is solvable with $x, y \in \mathbf{Z}$ or not.

Thus we have proved:

Theorem 3. *Let $D > 1$ and $m \parallel DF$, where m is not a square ($S^{(m)}(\mathbf{R})$ is empty if m is a squre). Let $\nu(m)$ be the number defined in Proposition 1. Then the number $\#(m)$ of components of $S^{(m)}(\mathbf{R})$ is $\nu(m)/2$, unless $\nu(m) > 0, i \in O, DF = 2t$ with t odd, $m = t$ or $2t$, and $x^2 - my^2 = \pm 2$ is solvable with $x, y \in \mathbf{Z}$, in which case $\#(m) = (\nu(m) + 2^{r-2})/2$, where r is the number of distinct prime factors of DF.*

Example. Consider the case $F = 1$, $D = 2p$, $m = p$ (a prime). Then $S^{(p)}(\mathbf{R})$ is empty if and only if $\sqrt{p} \notin B$, i.e., $p \equiv 1 (\mathrm{mod}\, 8)$. If $p \equiv 5 (\mathrm{mod}\, 8)$, then $\#(p) = \nu(p)/2 = h(p)$. If $p \equiv 3 (\mathrm{mod}\, 4)$, then $i \in O$ and $x^2 - py^2 = \pm 2$

is solvable, so $\#(p) = (1 + h(4p))/2$. The fixed points of $w(2)$ on S are as follows:

i) two points in $\mathbf{Q}(i)$, if $p \equiv 3 \pmod 4$;
ii) two points in $\mathbf{Q}(\sqrt{-2})$, if $p \equiv 5, 7 \pmod 8$.

(Cf. Shimura [16] for the rationality.) The images in $S^{(p)}$ are then rational over $\mathbf{Q}$, so we see that $S^{(p)}(\mathbf{Q})$ is non-empty (with at least two elements when $p \equiv 7 \pmod 8$) if $S^{(p)}(\mathbf{R})$ is.

4. Real Points when $D = 1$.

In this section, $S = X_0(F)$ is the ordinary modular curve of level F; $B = M_2(\mathbf{Q})$ and $\mathcal{O} = \left\{\begin{pmatrix} a & b \\ Fc & d \end{pmatrix} : a, b, c, d \in \mathbf{Z}\right\}$. Let $\varphi\colon \hat{\mathfrak{H}} \longrightarrow S$ be the natural map, where $\hat{\mathfrak{H}}$ denotes the upper half-plane $\mathfrak{H}$ together with cusps. We recall the description (cf.[11]) of the cusps of S. For each positive divisor y of F we have $\varphi(t)$ cusps x/y, where $t = (y, F/y)$ and x is taken modulo t (and prime to t). These cusps are conjugate over $\mathbf{Q}$, and their field of rationality is $\mathbf{Q}(\zeta_t)$, $\zeta_t = e^{2\pi i/t}$, so they are real (or rational) only if $t = 1$ or 2. Thus we have 2^r rational cusps with $t = 1$, forming a single orbit under W; cf. [12] for the action of W on the cusps. If $4 \mid F$, then the rational cusps with $t = 2$ also form a single orbit under W, consisting of 2^r elements if $8 \mid F$, and of 2^{r-1} elements (stabilized by $w(2)$) if $4 \parallel F$.

The set of real points of S is the union of the sets $Y_\epsilon = \varphi(Z_\epsilon)$, where Z_ϵ is the geodesic arc $\epsilon(z) = \bar{z}$ in $\hat{\mathfrak{H}}$ and $\epsilon \in \mathcal{O}$ satisfies $s(\epsilon) = 0$ and $n(\epsilon) = -1$, or $\epsilon \in U(1)$. Thus Y_ϵ is an arc in $S(\mathbf{R})$, beginning and ending at a cusp and passing through no other cusps. We do not care whether $\varphi\colon Z_\epsilon \longrightarrow Y_\epsilon$ is one-one or not, although that is the case when $F > 1$ for the "fundamental arc" (cf. [9]) $Z_{\left(\begin{smallmatrix} 1 & 0 \\ 0 & -1 \end{smallmatrix}\right)} = \{iy : y \geq 0\}$, which lies in a fundamental domain for $\Gamma_0(F) = \mathcal{O}^{(1)}$.

Let us determine the component Y^∞ of $S(\mathbf{R})$ containing the cusp $\infty = 1/F$. It contains $Y_{\left(\begin{smallmatrix} 1 & 0 \\ 0 & -1 \end{smallmatrix}\right)}$, connecting ∞ to $0 = w(F)(\infty)$, so it is fixed by $w(F)$. Taking $\epsilon = \left(\begin{smallmatrix} 1 & 0 \\ 0 & -1 \end{smallmatrix}\right)$, the set Z_ϵ is the line $Re(z) = -1/2$, connecting ∞ to $-1/2$. If F is odd (and > 1, say), then $-1/2$ is the same as the cusp 0, so Y^∞ consists of these two arcs and contains just two cusps. Thus $S(\mathbf{R})$ has 2^{r-1} components, a single orbit under W, with stabilizer group $(w(F))$, when F is odd (and > 1). On each component, $w(F)$ is a "reflection," with two fixed points.

Let us assume from now on that the genus of S is > 0; otherwise $S(\mathbf{R})$ consists of one circle.

Let F be even (and > 2). We have connected 0 to ∞ to $1/2$, and hence also 0 to $2/F$, applying $w(F)$; we express this by writing $(F/2, 1, F, 2)$, the cusp $1/y$ being listed by its denominator y. If $2 \parallel F$, then $w(2)$ sends F to $F/2$ and 1 to 2 (cf. [12]), so we have connected up 2 and $F/2$, and Y^∞ contains four cusps. Thus we have in this case 2^{r-2} components, each containing four cusps, and forming a single orbit under W. The stabilizer of a component is $(w(2), w(F))$, with $w(2)$ a rotation and $w(F)$ and $w(F/2)$ reflections. Suppose now that $4 \mid F$. Then the matrix $\alpha = \begin{pmatrix} 1 & 1/2 \\ 0 & 1 \end{pmatrix}$ normalizes $\mathcal{O}^{(1)}$ and so defines an involution of S, defined over $\mathbf{Q}$, fixing the cusp $\infty (y = F)$ and interchanging the cusps $y = 1$ and $y = 2$. If $8 \mid F$, then the cusp $y = F/2$ is fixed by α, so $F/2$ and 2 are connected and Y^∞ contains four cusps. Thus we have 2^{r-1} components, a single orbit under W, with $(w(F))$ as stabilizer, when $8 \mid F$. Suppose finally that $4 \parallel F$, say $F = 4F'$, where F' is odd (and > 1). Then α sends the cusp $y = 2F'$ to the cusp $y = F'$, so we have $(2F', 1, F, 2, F')$; $w(4)$ sends $(1, F)$ to $(4, F')$, so F' is connected to 4, and $w(F)$ sends $(2, F')$ to $(2F', 4)$, closing the circle. Thus we have 2^{r-2} components, a single orbit under W with stabilizer $(w(4), w(F))$, with each component containing six cusps, when $4 \parallel F$. In this last case, $w(F)$ and $w(4)$ act as reflections, the fixed points of $w(4)$ being cusps (with $t = 2$), while $w(F')$ is a rotation. Thus:

Theorem 4. *Let $S = X_0(F)$, $F > 4$. Then the components of $S(\mathbf{R})$ form a single orbit under W, and are described by the following table:*

Case	#Components	#Cusps per component	stabilizer in W
$2 \nmid F$	2^{r-1}	2	$(w(F))$
$2 \parallel F$	2^{r-2}	4	$(w(F), w(2))$
$4 \parallel F$	2^{r-2}	6	$(w(F), w(4))$
$8 \mid F$	2^{r-1}	4	$(w(F))$

In all cases, $w(F)$ acts on a component as a reflection, the two fixed points being non-cusps. If $2 \parallel F$, then $w(F/2)$ is a reflection (with non-cusps as *fixed points) and* $w(2)$ *is a rotation, and if* $4 \parallel F$, *then* $w(F/4)$ *is a rotation and* $w(4)$ *is a reflection with cusps (with* $t = 2$*) as fixed points.*

Now let $m \parallel F, m > 1$, and let $\varphi_m \colon \hat{\mathfrak{H}} \longrightarrow S^{(m)}$ be the natural map. Then $S^{(m)}(\mathbf{R})$ consists of the old part $S(\mathbf{R})/(w(m))$ plus the new part:

$$S^{(m)}(\mathbf{R}) = \bigcup_{\epsilon \in U(l)} X_\epsilon \cup \bigcup_{\beta \in U(m)} X_\beta, \tag{10}$$

where $X_\epsilon = \varphi_m(Z_\epsilon) = Y_\epsilon/(w(m))$ is known, by Theorem 4. We continue to assume that S has genus > 0, to avoid trivial special cases. Propositions 2 and 3 are still available, as we made no assumption about D being > 1 in that part of §3; recall that these propositions can apply only if $F \mid 4m$.

We have noted that Z_β begins and ends at a cusp if and only if $m = n^2$ is a square. This can also be seen by the rule in [12]: let P be the cusp of S represented by x/y, where as usual $y \mid F$ and x is taken modulo $t = (y, F/y)$. Write $y = y'y''$ and $t = t't''$, where y' and t' divide m and y'' and t'' divide F/m. Then $P^0 = w(m)(P)$ is represented by x^0/y^0, where $y^0 = (m/y') \cdot y''$, $t^0 = t$, and $x^0 \equiv -x \pmod{t'}$, $x^0 \equiv x \pmod{t''}$. On the other hand, the complex conjugate $\overline{P}$ is represented by $-x/y$ (cf.[11]), so $\overline{P} = P^0$ if and only if $y'^2 = m$ and $x \equiv -x \pmod{t''}$, i.e., $t'' = 1$ or 2.

Let us take care of the cases $m = 2, 4$. If $m = 2$, then $S(\mathbf{R})$ consists of 2^{r-2} circles on which $w(2)$ acts as a rotation, so $S(\mathbf{R})/(w(2))$ is also 2^{r-2} circles. The situation of Propositions 2 and 3 does not apply (for $F \mid 4m = 8$ implies $g = 0$, which was excluded), so the new part consists of $\nu(2)/2 = \nu(\mathbf{Z}[\sqrt{2}], \mathcal{O})/2$ circles X_β. Thus:

$$\#(2) = 2^{r-2} + \nu(2)/2. \tag{11}$$

Suppose $m = 4$, which is the only case in which $w(m)$ has fixed cusps; here $S(\mathbf{R})/(w(4))$ consists of 2^{r-2} half-circles, with end points at the cusps with $t = 2$. The discussion above shows that the sets $X_\beta, \beta \in U(4)$, will end at cusps with $t = 2$, so we have at most 2^{r-2} components. Actually, it is 2^{r-2}; this is seen from the action of W, or directly: taking $\beta = \begin{pmatrix} 2(1+F') & 2+F' \\ -4F' & -2(1+F') \end{pmatrix}$, where $F = 4F'$, the arc Z_β runs from $-1/2$ to $(2+F')/(2F')$ and so X_β connects the cusps with $y = 2$ and $y = 2F'$, and closes off the component containing the cusp ∞. Thus:

$$\#(4) = 2^{r-2}. \tag{12}$$

We suppose that $m \neq 2, 4$ from now on.

Let us suppose next that m is not a square, in which case the situation is somewhat similar to that when $D > 1$. According to Theorem 4, $w(m)$ has real fixed points only if $m = F$ or $F/2$, in which cases $S(R)/(w(m))$ consists of 2^{r-1} or 2^{r-2} half-circles.

Taking now $m = F$ or $F/2$, let us determine the component X^∞ of $S^{(m)}(\mathbf{R})$ containing the cusp ∞. It contains X_ϵ, going from ∞ to a fixed point of $w(m)$, where it meets X_β; here $\epsilon \in U(1)$, $\beta \in U(m)$, and $\epsilon\beta = -\beta\epsilon$. Specifically, if $m = F$ we can take $\epsilon = \left(\begin{smallmatrix} 1 & 0 \\ 0 & -1 \end{smallmatrix}\right)$, $\beta = \left(\begin{smallmatrix} 0 & 1 \\ m & 0 \end{smallmatrix}\right)$, and if $2m = F$ we can take $\epsilon = \left(\begin{smallmatrix} 1 & 1 \\ 0 & -1 \end{smallmatrix}\right)$, $\beta = \left(\begin{smallmatrix} m & (m-1)/2 \\ -2m & -m \end{smallmatrix}\right)$; in either case $R = \mathcal{O} \cap \mathbf{Q}(\beta) = \mathbf{Z}[\beta] \simeq \mathbf{Z}[\sqrt{m}]$. The general solution of $\epsilon_1\beta = -\beta\epsilon_1$, $\epsilon_1 \in \mathcal{O}^\times$, is $\epsilon_1 = \epsilon\rho$, where $\rho \in R^\times$. If $\rho = \gamma^2$, $\gamma \in R$, then

$$\epsilon\rho = \epsilon\gamma^2 = \gamma'\epsilon\gamma = \pm\gamma^{-1}\epsilon\gamma \sim \pm\epsilon.$$

Thus we are interested in ρ only modulo squares and take $\rho = a + b\beta$ to be a base unit in R. Suppose $-1 = n(\rho) = a^2 - mb^2$, so the other end of X_β is an E-point, by Proposition 2. The next arc is X_{β_1}, where $\beta_1 = \beta\epsilon_1 = \beta\epsilon\rho \in U(m)$, going from the E-point to a fixed point of $w(m)$, where it meets X_{ϵ_2},

$$\epsilon_2 = \epsilon_1\rho_1 = \epsilon_1(a + b\beta_1) = a\epsilon_1 - b\rho_1\epsilon_1 = a\epsilon\rho + b\beta = a^2\epsilon + ab\epsilon\beta + b\beta.$$

If $m = F$, we have $\epsilon_2 = \left(\begin{smallmatrix} a^2 & b(1+a) \\ bm(1-a) & -a^2 \end{smallmatrix}\right)$, fixing the cusp $b/(1-a)$. The denominator of this cusp is divisible by no odd prime divisor of m (since $a^2 - mb^2 = -1$), so it has $y = 1$ or 2 and is connected to ∞ on $S(\mathbf{R})$, by Theorem 4, and we are back on the half-circle with which we began. For $F = 2m$, we have $\epsilon_2 = \left(\begin{smallmatrix} a^2+bm(1-a) & a^2-ab(m+1)/2+b(m-1)/2 \\ 2bm(a-1) & -a^2-bm(1-a) \end{smallmatrix}\right)$, fixing the cusp $(1 - a + b)/2(a - 1)$, again connected to ∞ on $S(\mathbf{R})$. Thus, if $n(\rho) = -1$, we have

$X^\infty =$ 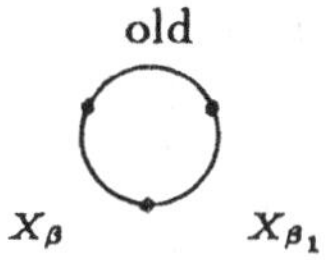

and the formula $\#(m) = \nu(m)/2$ holds, as we have 2^{r-1} (if $2 \nmid F$) or 2^{r-2} (if $2 \parallel F$) such components, a single orbit under W, plus ordinary components X_β with $\beta \not\sim -\beta$. Suppose now that $n(\rho) = 1$, so the other end of X_β is also a fixed point of $w(m)$, where X_β meets X_{ϵ_1}; $\epsilon_1 = \epsilon\rho$ as before. If $m = F$, then $\epsilon_1 = \begin{pmatrix} a & b \\ -mb & -a \end{pmatrix}$ fixes the cusps $x_1 = -b(1+a)$ and $x_2 = b/(1-a)$, which may or may not be connected to ∞ on $S(\mathbf{R})$. (E.g., if $m = 119$, then $\rho = 120 + 11\beta$ and $x_2 = -11/119 \sim \infty$, while if $m = 21$, then $\rho = 55 + 12\beta$ and $x_1 = -3/14 \sim 1/7$ is on the other component of $S(\mathbf{R})$.) By Theorem 4, the test is as follows. Let y_1 resp. y_2 be the greatest common divisor of m and the denominator of x_1 resp. x_2. Then x_i is connected to ∞ on $S(\mathbf{R})$ if and only if y_1 or y_2 is a power of 2. If $F = 2m$, then $\epsilon_1 = \begin{pmatrix} a-bm & a-b(m+1)/2 \\ 2bm & -a+bm \end{pmatrix}$, fixing the cusps $(1-a+b)/2(a-1)$, $(b-a-1)/2(a+1)$, which again may or may not be connected to ∞ on $S(\mathbf{R})$. If the fixed cusps of ϵ_1 are connected to ∞ on $S(\mathbf{R})$, then we have one such component of $S^{(m)}(\mathbf{R})$ corresponding to each component of $S(\mathbf{R})$, with only one class in $U(m)$ associated to such components. In the contrary case, X^∞ contains two or more half-circles from $S(\mathbf{R})/(w(m))$; actually the number is exactly two, since the elementary abelian 2-group W is operating transitively. In this case, we have two clases from $U(m)$ per component, and so no modification here of the expected formula $\#(m) = \nu(m)/2$.

As to the E-points, they have been taken into account if $n(\rho) = -1$, so assume that $n(\rho) = 1$ and $i \in O$. By the argument of §3 (where we assumed that m is not a square, but not that $D > 1$), $w(F)$ acts as complex conjugation on the elliptic fixed points of order 2 on S; if $2 \parallel F$, then so does $w(F/2)$, since $w(2)$ leaves them fixed. Thus the E-points are real if $m = F$ or $F/2$. The argument preceding Theorem 3 (deciding whether the components containing E-points consist of one or two arcs) is still valid, so we get, for $m = F$ or $F/2$:

$$(13) \qquad \#(m) = \begin{cases} \nu(m)/2 & (\text{if } a^2 - mb^2 = -1) \\ (\nu(m) + A + B)/2 & (\text{if } a^2 - mb^2 = 1), \end{cases}$$

where A is the number of components of $S(\mathbf{R})$ if the fixed cusps of ϵ_1 (above) are connected to ∞ there, and otherwise $A = 0$, and $B = 2^{r-2}$ if $2 \mid F$, $i \in O$, and $x^2 - my^2 = \pm 2$ is solvable, and otherwise $B = 0$.

The remaining case where $w(m)$ stabilizes a component of $S(\mathbf{R})$ is $m = F/4$. In this case $S(\mathbf{R})/(w(m))$ consists of 2^{r-2} circles, by Theorem 4, and there are no real E-points, so we have

$$(14) \qquad \#(m) = 2^{r-2} + \nu(m)/2 \quad (F = 4m).$$

In the remaining cases ($m \neq 2, 4, F, F/2, F/4$), $S(\mathbf{R})/(w(m))$ consists of half as many circles as does $S(\mathbf{R})$, and we also have $\nu(m)/2$ circles X_β, $\beta \not\sim -\beta$. Thus we have:

Theorem 5. *Let $S = X_0(F)$ have genus > 0; let $m \parallel F$, with m not of the form $n^2, n \geq 3$. Then the number $\#(m)$ of components of $S^{(m)}(\mathbf{R})$ is given by formulas (11) to (14), if $m = 2, 4, F, F/2, F/4$, and otherwise $\#(m) = \big(\#(1) + \nu(m)\big)/2$.*

Finally, we have the case $m = n^2, n \geq 3$; then each Z_β begins and ends at a cusp, so each X_β is an arc with a cusp at least one end. The general element of $U(m)$ is $\beta = \begin{pmatrix} n^2a & b \\ n^2c & -n^2a \end{pmatrix}$, where $F \mid n^2c$, i.e., $d = F/m$ divides c, and $bc = 1 - a^2n^2$; then β fixes the cusps $(\pm 1 + an)/nc$. According to [11], for example, the standard form for the cusp represented by a given rational number is as follows: write the rational number in lowest terms as r/ys, where $y \mid F, y > 0$; let $t = (y, F/y)$. We now take r and s modulo t and act on the left by $\begin{pmatrix} s & 0 \\ 0 & s^{-1} \end{pmatrix}$, getting rs/y. In our case we have $t = n$ or $2n$; since x/y and $-x/y$ are interchanged by $w(m)$ and by complex conjugation, we have for each such y exactly $\varphi(n)/2$ real cusps x/y on $S^{(m)}$, with $x \in G = (Z/nZ)^\times/\pm 1$. We call these the new cusps on $S^{(m)}(\mathbf{R})$; the old cusps are on $S(\mathbf{R})/(w(m))$. Changing the notation, the set of all new cusps on $S^{(m)}(\mathbf{R})$ is the set of all

$$P(x, y) = x/yn \tag{15}$$

where $x \in G$ and $y > 0$ is a divisor of $d = F/m$ with $(y, d/y) = 1$ or 2. The field of rationality of $P(x, y)$ is the real subfield of $\mathbf{Q}(e^{2\pi i/n})$. In general, $P(x, y)$ is an end of two arcs X_β and so has two neighbors, for which we need a formula; using the action of W, we may assume that $y = 1$ or 2 with no essential loss of generality.

Given an element of G, we represent it by $x \in \mathbf{Z}$, coprime to $2n$. We can solve $cx = -1 + an$, with $c, a \in \mathbf{Z}$ and $d \mid c$, getting an element β of $U(m)$, with $b = -x(1 + an)$. One end is $(-1 + an)/nc = x/n = P(x, 1)$ and the other is $(1 + an)/nc$. If n is even (this is the easiest case), then this fraction is already in lowest terms, so $(1 + an)/nc = (1 + an)/(n \cdot d \cdot c/d) = P(x', d)$, where $x' = c/d = 1/dx \in G$. Thus $P(1/dx, d)$ is the only neighbor of $P(x, 1)$, and more generally $P(1/dx, d/y)$ is the only neighbor of $P(x, y)$, when n is even. If $d > 1$ (and $2 \mid n$), then each new component contains exactly two cusps and we are done, so suppose that $d = 1$; write now $P(x) = P(x, 1)$. Now an exception to the general rule occurs only when

$x = x^{-1}$ in G; then the corresponding component may be a loop X_β, with only one cusp, or it may be connected to the old part. For example, $\beta = \left(\begin{smallmatrix} 0 & 1 \\ m & 0 \end{smallmatrix}\right)$ fixes $P(1)$ and anticommutes with $\epsilon = \left(\begin{smallmatrix} 1 & 0 \\ 0 & -1 \end{smallmatrix}\right)$, which fixes ∞, so $P(1)$ is on the component X^∞ containing ∞. Also, $\beta_1 = \left(\begin{smallmatrix} n^2 & 1-n \\ n^2(1+n) & -n^2 \end{smallmatrix}\right)$ fixes $P(1)$ and anticommutes with $\epsilon_1 = \left(\begin{smallmatrix} 1-n^2/2 & -1+n/2 \\ -n^2(n/2+1) & -1+n^2/2 \end{smallmatrix}\right)$, which fixes $1/(2+n)$, i.e., the cusp $1/2$, which is connected to ∞ on the old part. Thus, in either case, $P(x)$ is the only new cusp on its component when $x^2 = 1$ in G, so we have:

Theorem 6.1. *Let $F = dm$, where $m = n^2$, with n even, $n \geq 4$ (and $m \,\|\, F$).*

i) *If $d > 1$, then each new component of $S^{(m)}(\mathbf{R})$ contains exactly two cusps, $P(x, y)$ and $P(1/xd, d/y)$, and*

$$\#(m) = \#(1)/2 + 2^{s-2}\varphi(n),$$

where s is the number of prime factors of d.

ii) *If $d = 1$, then each component of $S^{(m)}(\mathbf{R})$ contains either two new cusps, $P(x)$ and $P(x^{-1})$, or one new cusp $P(x)$, if $x^2 = 1$ in G. Thus*

$$\#(m) = A/2 + \varphi(n)/4,$$

where A is the number of solutions of $x^2 = 1$ in G.

Remark. Then

$$A = \begin{cases} 2^r & (\text{if } 8 \mid n) \\ 2^{r-2} & (\text{if } 2 \,\|\, n \text{ and } x^2 \equiv -1 \pmod{n} \text{ is not solvable}) \\ 2^{r-1} & (\text{otherwise}). \end{cases}$$

Suppose n odd, $n \geq 3$, from now on; we retain the notation of the beginning of the paragraph preceding Theorem 6.1. A neighbor of $P(x, 1)$ is $(1 + an)/nc$. This fraction must be put into lowest terms; note that $(nc, 1 + an)$ is a power of 2. If d is odd (i.e., if F is odd), we can choose c to be either odd or even; taking c odd we have $(1 + an)/(n \cdot d \cdot c/d) = P(1/xd, d)$, and taking c even we have $P(1/4xd, d)$. Thus (applying $w(y)$), the two neighbors of $P(x, y)$ are $P(1/xd, d/y)$ and $P(1/4xd, d/y)$. A general component, with no cusp equal to one of its neighbors, looks like

(16) $\ldots, P(x,y),\ P(1/xd, d/y),\ P(xd/4, y),\ P(4/xd, d/y),\ P(xd/16, y), \ldots$

and its length is $2k$, i.e., it contains $2k$ new cusps, where k is the order of 4 in G. Exceptions can occur only if $d = 1$; then $P(x)$ is equal to one of its neighbors if $x^2 = 1$ or $4x^2 = 1$ in G. If so, then $P(x)$ is connected to an E-point or to a fixed point of $w(m)$. Now $P(1)$ is connected to ∞, by the same proof as above (this proof did not need n to be even), and ∞ is connected to $P(1/2)$, since $\beta = \begin{pmatrix} m & -(m-1)/2 \\ 2m & -m \end{pmatrix}$ fixes $(1+n)/2n = P(1/2)$ and anticommutes with $\epsilon = \begin{pmatrix} 1 & -1 \\ 0 & -1 \end{pmatrix}$, which fixes ∞. Thus the component X^∞ contains

(17) $\quad \ldots, P(2^3),\ P(2^{-3}),\ P(2),\ P(2^{-1}),\ \infty,\ P(1),\ P(4^{-1}),\ P(4), \ldots.$

Similarly, whenever $\pm x$ satisfies $x^2 \equiv 1 \pmod n$, then an old cusp lies between $P(x)$ and $P(x/2)$. Since there are 2^{r-1} old cusps and 2^{r-1} such solutions $\pm x$, the correspondence is one-one. Similarly, if $i \in \mathcal{O}$, i.e., each prime divisor of n is $\equiv 1 \pmod 4$, then we have 2^{r-1} solutions $\pm x$ of $x^2 \equiv -1 \pmod n$, and an E-point lies between $P(x)$ and $P(x/2)$. To see this, write $n = c^2 + c_1^2$, where $(c, c_1) = 1$ and (say) c is odd, c_1 is even; then $1 = ac_1 + a_1 c$ for suitable $a, a_1 \in \mathbf{Z}$. Put $x = ac - a_1 c_1$. Then $xc + c_1 = an^2$ and $-xc_1 + c = a_1 n^2$. Put $b = a_1 - ax$, $b_1 = a + a_1 x$. Then $bc = 1 - a^2 n^2$ and $b_1 c_1 = 1 - a_1^2 n^2$, so $\beta = \begin{pmatrix} an^2 & b \\ cn^2 & -an^2 \end{pmatrix}$ and $\beta_1 = \begin{pmatrix} a_1 n^2 & -b_1 \\ -c_1 n^2 & -a_1 n^2 \end{pmatrix}$ are in $U(m)$. They anticommute, since $\epsilon = \beta\beta_1 n^{-2} = \begin{pmatrix} x & -n^{-2}(1+x^2) \\ n^2 & -x \end{pmatrix}$ is in $U(-1)$. Thus Z_β meets Z_{β_1} at the fixed point of ϵ, so X_β meets X_{β_1} in an E-point, and we claim that the cusps at the other ends are $P(x)$ and $P(x/2)$. To see this, write $n = (u^2 + v^2)/2$, $c = uv$, $c_1 = (u^2 - v^2)/2$, where u and v are coprime odd integers. Then $1 + an = ca_1 + a(n + c_1) = a_1 uv + au^2$, and $(1 + an)/c = (a_1 v + au)/v$ in lowest terms; since

$$v(a_1 v + au) \equiv -a_1 c_1 + ac \equiv x \pmod n,$$

this is $P(x)$. Similarly,

$$1 + a_1 n = ac_1 + a_1(n + c) = a(u^2 - v^2)/2 + a_1(u+v)^2/2,$$

$$(1 + a_1 n)/c_1 = \big(a(u-v)/2 + a_1(u+v)/2\big)/\big((u-v)/2\big)$$

in lowest terms, and $((u-v)/2)(a(u-v)/2+a_1(u+v)/2) \equiv x/2 \pmod n$, so we have $P(x/2)$.

Thus, whenever $x^2 = 1$ in G, an old cusp or an E-point lies between $P(x)$ and $P(x/2)$. Now let l be the order of 2 in G. If l is odd, then the component of $P(x)$ has length l (contains l new cusps), where $x^2 = 1$ in G. In fact, it contains exactly the new cusps $P(x \cdot 2^i)$, while ordinary components have length $2l = 2k$, as we have seen. If $l = 2k$ is even, then ordinary components have order $2k$ (k is the order of 4 in G); so do the special components, since $P(x/2)$ is on the component if $x^2 = 1$ in G, and again our component contains exactly the new cusps $P(x \cdot 2^i)$. Thus:

Theorem 6.2. *Let $F = dm$ be odd, where $m = n^2, n \geq 3$. Let k resp. l be the order of 4 resp. 2 in $G = (\mathbf{Z}/n\mathbf{Z})^\times/(\pm 1)$.*

i) *If $d > 1$, then each new component of $S^{(m)}(\mathbf{R})$ has length $2k$, so*

$$\#(m) = \#(1)/2 + 2^{s-2}\varphi(n)/k,$$

where s is the number of prime factors of d.

ii) *If $d = 1$ and $l = 2k$ is even, then $S^{(m)}(\mathbf{R})$ has $\varphi(n)/4k$ components, each of length $2k$.*

iii) *If $d = 1$ and $l = k$ is odd, and A is the number of solutions of $x^2 = 1$ in G, then $S^{(m)}(\mathbf{R})$ has A special components of length k, plus ordinary components of length $2k$. Thus*

$$\#(m) = A/2 + \varphi(n)/4k, \quad \text{and } A = \begin{cases} 2^{r-1} (\text{if } i \not\in O) \\ 2^r \quad (\text{if } i \in O) \end{cases}.$$

We suppose from now on that $d = F/m$ is even. Suppose first that $2 \parallel d$. We have $cx = -1 + an$ as above, and can take $c \equiv 0$ or $2 \pmod 4$ as we wish. Taking $c \equiv 2 \pmod 4$, we find that $(1+an)/nc = P(1/2xd, d/2)$, and taking $c \equiv 0 \pmod 4$, we get $P(1/4xd, d)$. Thus the neighbors of $P(x,1)$ are $P(1/2xd, d/2)$ and $P(1/4xd, d)$. Applying $w(y)$ and $w(2y)$, where $y \parallel d, y$ odd, we get in general

$$\ldots, P(1/4xd, d/y),\ P(x,y),\ P(1/2xd, d/2y), \ldots$$
$$\ldots, P(1/4xd, d/2y),\ P(x,2y),\ P(1/2xd, d/y), \ldots$$

and hence

$$(18) \ldots, P(x,y),\ P(1/2xd, d/2y),\ P(x/2, 2y),\ P(1/xd, d/y),\ P(x/4, y), \ldots.$$

This is in general of length $4k$, where k is the order of 4 in G. Exceptions can occur (with a cusp equal to one of its neighbors) only if $d = 2$, and $4x^2 = 1$ in G, so $P(x, 1)$ resp. $P(x, 2)$ is equal to its neighbor $P(1/4x, 1)$ resp. $P(1/4x, 2)$. Note that $P(x, 1)$ and $P(x, 2)$ are interchanged by $w(2)$. By Theorem 4, a component of $S(\mathbf{R})/(w(m))$ is a half-circle whose ends are fixed by $w(m)$ and interchanged by $w(2)$, and any E-point (present only if each prime dividing n is $\equiv 1 \pmod 4$) is fixed by $w(2)$. Thus, given $4x^2 = 1$ in G, and $d = 2$, so $P(x, 1)$ is connected to a fixed point of $w(m)$ or an E-point, this fixed point or E-point is in turn connected to the image $P(x, 2)$ of $P(x, 1)$ under $w(2)$. The component in question thus contains all $P(x \cdot 4^i, 1)$ and all $P(2x \cdot 4^i, 2)$, and is fixed by $w(2)$, so it contains all cusps $P(x \cdot 2^i, 1)$ and $P(x \cdot 2^i, 2)$, and no other cusps. Thus a special component has length $2l$, where l is the order of 2 in G. If $l = 2k$ is even, this is the same as the length $4k$ of ordinary components, while if $l = k$ is odd, then special components have length $2k$ while ordinary components have length $4k$. Thus:

Theorem 6.3. *Let $F = dm$, where $2 \parallel d$ and $m = n^2 (n \geq 3)$. Let k reps. l be the order of 4 reps. 2 in G.*

i) *If $d > 2$, then each new component of $S^{(m)}(\mathbf{R})$ has length $4k$, so*

$$\#(m) = \#(1)/2 + 2^{s-2}\varphi(n)/k$$

where s is the number of odd prime factors of d.

ii) *If $d = 2$ and $l = 2k$ is even, then all components have length $4k$, and*

$$\#(m) = \varphi(n)/4k.$$

iii) *If $d = 2$ and $l = k$ is odd, and A is the number of solutions of $x^2 = 1$ in G, then $S^{(m)}(\mathbf{R})$ has A special components of length $2k$, plus ordinary components of length $4k$, and*

$$\#(m) = \varphi(n)/4k + A/2.$$

Suppose next that $4 \parallel d$. By the usual argument, we find that the neighbors of $P(x, 1)$ are $P(1/2xd, d/2)$ and $P(1/4xd, d)$ and more generally $P(x, y)$ is betwen $P(1/2xd, d/2y)$ and $P(1/4xd, d/y)$, for an odd exact divisor y of d. Applying $w(4)$, we have $P(x, 4y)$ between $P(1/2xd, d/2y)$ and $P(1/4xd, d/4y)$, so a new component is

$$(19)\qquad \begin{aligned} &P(x,y), P(1/2xd, d/2y), , P(x,4y), P(1/4xd, d/4y),\\ &P(2x,2y), P(1/4xd, d/y), P(x,y), \end{aligned}$$

in general of length 6, but of length 3 if $d = 4$ and $16x^2 = 1$ in G, and only then. As to the old part, since $S(\mathbf{R})$ consists of 2^{r-2} circles, with $w(m)$ acting as a rotation on each component if $d = 4$ and carrying one circle onto another if $d > 4$, we have 2^{r-2} circles if $d = 4$ and 2^{r-3} if $d > 4$. Thus:

Theorem 6.4. *Let $F = dm$, where $4 \parallel d$ and $m = n^2 (n \geq 3)$.*

i) *If $d > 4$, then $\#(m) = 2^{r-3} + 2^{s-2}\varphi(n)$, where s is the number of odd prime factors of d.*

ii) *If $d = 4$, then $\#(m) = 2^{r-2} + \varphi(n)/4 + A/2$, where A is the number of solutions of $x^2 = 1$ in G.*

Finally, we have the case $8 \mid d$. By Theorem 4, $S(\mathbf{R})/(w(m))$ consists of 2^{r-2} circles. We find that the neighbors of $P(x,1)$ are $P(1/2xd, d/2)$ and $P(1/4xd, d)$; applying $w(d)$, we find $P(x,d)$ between $P(1/2xd, 2)$ and $P(1/4xd, 1)$. Similarly, solving $(c/2) \cdot x = -1 + an$, with $d \mid c$, we see that $P(x,2)$ lies between $P(1/2xd, d)$ and $P(1/xd, d/2)$ and $P(x, d/2)$ lies between $P(1/2xd, 1)$ and $P(1/xd, 2)$. Thus our component looks like

$$P(x,1),\ P(1/2xd, d/2),\ P(2x,2),\ P(1/4xd, d),\ P(x,1)$$

and has length 4 as do all new components (using the action of W). Thus:

Theorem 6.5. *Let $F = dm$, where $8 \mid d$ and $m = n^2 (n \geq 3)$. Then $\#(m) = 2^{r-2} + 2^{s-2}\varphi(n)$, where s is the number of odd prime factors of d.*

5. The Hyperelliptic Problem.

In this section our aim is to determine for which values of (D, F) the curve $S = S_{D,F}$ is hyperelliptic, and for each such value, whether S is hyperelliptic over $\mathbf{R}$ or $\mathbf{Q}$ or not, i.e., whether the curve $S/(v)$ of genus 0 obtained by dividing S by the hyperelliptic involution v has points raional over $\mathbf{R}$ or $\mathbf{Q}$ or not. If $D = 1$, then the first part was done in [12]-actually, this is the hardest case, and the second question does not arise, since S has rational cusps; we assume that $D > 1$ from now on.

The methods for determining whether S is hyperelliptic or not are well known (cf. [12] for the case $D = 1$ and [10] for the case $F = 1$), so we give here only an outline of the proof and a statement of the results. It turns out that the hyperelliptic involution v is always in W, so whether $S/(v)$ has real points or not is contained in the results of §3.

If S is hyperelliptic, and $F' \mid F$, then $S_{D,F'}$ is also "subhyperelliptic," i.e., either hyperelliptic or of genus ≤ 1 (since it has a function of degree 2 if S does). A hyperelliptic curve does not admit C_2^4 as a group of automorphisms (cf. [12], Proposition 1), so in our case at most three primes divide DF, i.e., $r \leq 3$. Thus, if S is hyperelliptic, then D is the product of only two primes, and F is 1 or a prime power; then hyperelliptic involution v is in W if $F > 1$.

Let l be any prime which does not divide DF. Then S has good reduction modulo l, a non-singular curve $\tilde{S}$ defined over $\mathbf{F}(l)$, of the same genus g. By the work of Ihara (cf.[7], esp. p. 293), the "supersingular points" on $\tilde{S}$ are all rational over $\mathbf{F}(l^2)$, and their number is

$$1 + \bar{g} - 2g, \tag{20}$$

where $\bar{g}$ is the genus of $\overline{S} = S_{D,lF}$. On the other hand, if S is hyperelliptic then so is $\tilde{S}$, and so $\tilde{S}$ (being a double cover of a curve of genus 0) has at most $2(l^2 + 1)$ points over $\mathbf{F}(l^2)$. Thus

$$1 + \bar{g} - 2g \leq 2(l^2 + 1) \qquad (l \nmid DF) \tag{21}$$

if S is hyperelliptic. The formula for the genus is

$$g = 1 + \varphi(D)\psi(F)/12 - e_2/4 - e_3/3 - \sigma/2, \tag{22}$$

where $\psi(F) = F\prod_{p|F}(1 + 1/p)$, e_k is the number of (inequivalent) fixed points of order k and σ is the number of cusps; $e_2 = \nu(\mathbf{Z}[i], \mathcal{O})$, $e_3 = \nu(\mathbf{Z}[\varsigma_3], \mathcal{O})$, and for us $\sigma = 0$ (since $D > 1$). In going from g to $\bar{g}$ we multiply $\psi(F)$ by $(l + 1)$, and e_k by 0, 1, or 2, depending on cases, so we have

$$1 + g - 2g \geq \varphi(D)\psi(F)(l - 1)/12. \tag{23}$$

In particular, if S is hyperelliptic we get (by (21) and (23))

$$\varphi(D)\psi(F) \leq 24(l^2+1)/(l-1) \qquad (l \not| DF). \tag{24}$$

In particular, if $F = 1$ we can take $l = 2$ or 3 (since S has genus 0 if $D = 6$), so $D = pq$ (p and q prime, $p < q$) with $(p-1)(q-1) \leq 120$. Then certainly $D < 300$ and we can use Table 5 (pp. 135-141) in [2], as follows. In this table, for $D = pq (p < q)$, the last column gives the dimensions of $S^{++}, S^{+-}, S^{-+}, S^{--}$, in order, where S is the space of new (à la Atkin-Lehner) cusp forms of weight 2 for $\Gamma_0(D)$ and, e.g., S^{+-} is the subspace on which $w(p) = 1$ and $w(q) = -1$. Now it has been known for a long time (since Eichler's work on the trace formula and its applications), that for any D the space S is essentially the same as the space of holomorphic differentials on S, i.e., the two spaces have the same dimension and the actions of the respective Hecke algebras are equivalent. This led Kazhdan and Mazur to surmise that the Jacobian $J(S)$ must be $\mathbf{Q}$-isogenous to the new part of the Jacobian $J_0(D)$ of $X_0(D)$, which was proved immediately by Ribet [13]. The actions of the two Atkin-Lehner groups on S are "opposite," i.e., if the Dirichlet series corresponding to a normalized newform is

$$\prod_{p|D}(1-a_p p^{-s})^{-1} \prod_{l \not| D}(1-a_l l^{-s}+l^{1-2s})^{-1},$$

then the operator $w(p)$ on new forms for $\Gamma_0(D)$ has eigenvalue $-a_p = \mp 1$, while the operator $w(p)$ on $J(S)$ has eigenvalue a_p; cf. Atkin-Lehner [1] and Eichler [6], respectively. Thus, for us, the last column of Table 5 should be read backwards; for example, when $D = pq$, the genus of S/W is the dimension of S^{--}. Checking through the table, we find twenty-four values $D = 26, \ldots, 206$ for which S is hyperelliptic, with hyperelliptic involution $v \in W$; these values are listed in Theorem 7 below. For example, for $D = 206 = 2 \cdot 103$, the last column in Table 5 reads $0, 5, 4, 0$, so S has genus 9 and $v = w(206)$, while for $D = 82$, it is $1, 0, 2, 0$, so $g = 3$ and $v = w(41)$. Instead of using the table, we could of course compute the genera in each case by formulas (22), (3), (4). To show that there are no further values of D for which S is hyperelliptic, i.e., to show that S is hyperelliptic but $v \not| inW$ is not possible, we need check only those S for which each $w(m)$, $m \neq 1$, has at most four fixed points, i.e., $g^{(m)} = [g/2]$ or $[(g+1)/2]$, by a result of Schoeneberg (cf. [12], p.452). Checking through Table 5, only $D = 133, 142, 177$ need a closer look. For $D = 142$, as we noted in the example at the end of §3, $w(2)$ has four fixed points, two over $\mathbf{Q}(i)$ and two over $\mathbf{Q}(\sqrt{-2})$, so it is not possible that C_2^3 acts as a group of rational automorphisms; this eliminates $D = 142$. Another argument is found in Michon [10], p. 224, and a proof that the cases $D = 133, 177$ do not arise; another way to eliminate these two values is to find a few more points on the curve besides the $(g-1) = 8$ supersingular points, so there are more that ten points over $\mathbf{F}(4)$. At any rate, we have the first part of:

Theorem 7. *There are exactly twenty-four values of D for which $S = S_{D,1}$ is hyperelliptic; in each case the hyperelliptic involution v is in W. These values, together with the genus g of S and the m for which $v = w(m,)$, are:*

D	g	m	D	g	m	D	g	m
26	2	26	62	3	62	95	7	95
35	3	35	69	3	69	111	7	111
38	2	38	74	4	74	119	9	119
39	3	39	82	3	41	134	6	134
51	3	51	86	4	86	146	7	146
55	3	55	87	5	87	159	9	159
57	3	19	93	5	31	194	9	194
58	2	29	94	3	94	206	9	206

For three of these values, namely $D = 57, 82, 93$, the curve S is not hyperelliptic over $\mathbf{R}$; for the other twenty-one values, S is hyperelliptic over $\mathbf{Q}$.

Proof of the last part. Since $\mathcal{O}$ is a maximal order, there are real points on $S^{(m)}$ if and only if $\sqrt{m} \in B$. This is always the case if $m = D$ and we find that $\sqrt{m} \notin B$ exactly in the three cases listed; for example $\sqrt{19} \notin B$ when $D = 57$ since the prime 3 splits in $\mathbf{Q}(\sqrt{19})$ but ramifies n B.

For the remaining twenty-one values, we can show that S is hyperelliptic over $\mathbf{Q}$ by finding in each case suitable "points of complex multiplication" which map to rational points on $S^{(m)}$. For any value of D, the construction is as follows. Let l be 1, 2, or a prime $\equiv 3 \pmod 4$, so $L = \mathbf{Q}(\sqrt{-l})$ has discriminant $-4, -8$, or $-l$ and odd class number h. We assume that each prime p dividing D does not split in L, so $L \hookrightarrow B$. To be specific, we can realize B as

$$B = \left\{ \begin{pmatrix} a & b \\ -lb' & a' \end{pmatrix} : a, b \in \mathbf{Q}(\sqrt{D}) \right\} \tag{25}$$

and take O to be a maximal order containing the ring of all such elements with a, b integral in $\mathbf{Q}(\sqrt{D})$. (Clearly B is indefinite, so it is enough to check that an odd prime ramifies in B if and only if it divides D). Then O contains anticommuting elements $\rho = \left(\begin{smallmatrix} 0 & 1 \\ -l & 0 \end{smallmatrix}\right)$ and $\beta = \left(\begin{smallmatrix} \sqrt{D} & 0 \\ 0 & -\sqrt{D} \end{smallmatrix}\right) \in U(m)$, so the point $P \in S$ defined by the fixed point of ρ in $\mathfrak{H}$ maps to a real point of $S^{(D)}$, by the results of §3. Shimura has shown (cf. [16], p.58) that the exact field of rationality of P is the class field $C(L)$ of the imaginary quadratic number-field L. Thus P is of degree h over L, and its image on $S^{(D)}$ is of degree h over $\mathbf{Q}$, so we have a rational point on $S^{(D)}$ if we can embed L in B for one of the nine values $l = 1, 2, 3, 7, 11, 19, 43, 67, 163$ for which $h = 1$. This can always be done if $D = pq < 2127$, as one checks; on particular, the twenty hyperelliptic curves S with $v = w(D)$ are hyperelliptic over $\mathbf{Q}$.

The only other case is $D = 58$. In this case S is of genus 2 and hence automatically hyperelliptic over its field of definition $\mathbf{Q}$, since there is a positive canonical divisor (of degree 2) defined over $\mathbf{Q}$. Also, we have already noted in the example at the end of §3 that $S^{(p)}$ has rational points if it has real points in the case $D = 2p$.

Finally, let $S = S_{D,F}$ be hyperelliptic, where $D > 1$ and $F > 1$. Then the hyperelliptic involution v must be in $W \simeq C_2^{\tau} = C_2^3$, so there are no difficult cases. The inequality (24) reduces the problem to checking out a finite number of cases; this number is reduced further by remembering that $S_{D,1}$ must by subhyperelliptic, and so D is among the values just determined (plus the values for which $g \leq 1$). For example, if $D = 6$ then $\psi(F) \leq 78$ if $5 \not| F$ and $\psi(F) \leq 100$ if $7 \not| F$, and we find exactly six hyperelliptic curves $S_{6,F}$, namely $F = 11, 19, 29, 31, 37$ with $v = w(6F)$ and $F = 17$ with $v = w(34)$. The result is:

Theorem 8. *There are exactly nineteen values of (D, F), with $D > 1$ and $F > 1$, such that $S = S_{D,F}$ is hyperelliptic. In each case the hyperelliptic involution v is in W. These values, together with the genus g and the m for which $v = w(m)$, are:*

D	F	g	m	D	F	g	m
6	11	3	66	14	3	3	14
	17	3	34		5	3	14
	19	3	114	15	2	3	15
	29	5	174		4	5	15
	31	5	186	21	2	3	7
	37	5	222	22	3	3	66
10	11	5	110		5	5	110
	13	3	65	26	3	5	26
	19	5	38	39	2	7	39
	23	9	230				

The curve S is not hyperelliptic over $\mathbf{R}$ *for exactly six of these values, namely* $(D,F) = (6,17),(10,13),(14,3),(15,4),(21,2),(26,3)$.

This leaves thirteen curves S which are hyperelliptic over $\mathbf{R}$. For the nine with $m = DF$, we find rational points on $S^{(m)}$ coming from complex multiplication by $\sqrt{-l}$, where l is a prime with $h(-l) = 1$, $l \equiv 3 \pmod 4$, and $\left(\frac{-l}{p}\right)$ is 1 if $p \mid F$ and is not 1 if $p \mid D$. We can take l as follows:

D	6					10		22	
F	11	19	29	31	37	11	23	3	5
l	19	67	67	43	67	43	43	11	11

Then $B = \left\{\begin{pmatrix} a & b \\ -lb' & a' \end{pmatrix} : a,b \in \mathbf{Q}(\sqrt{m})\right\}$ has discriminant D. We have anticommuting elements $\rho = \begin{pmatrix} 0 & 1 \\ -l & 0 \end{pmatrix}$ and $\beta = \begin{pmatrix} \sqrt{m} & 0 \\ 0 & -\sqrt{m} \end{pmatrix}$ with $\rho^2 = -l$ and $\beta^2 = m$. There is an obvious order $\overline{O}$ containing ρ and β. We construct an Eichler order O containing $\overline{O}$ by taking O_p to be a maximal order containing $\overline{O_p}$ if $p \nmid F$; if $p \mid F$, we map $\overline{O_p} = O_p$ onto the standard order by $\beta \mapsto \begin{pmatrix} 0 & 1 \\ m & 0 \end{pmatrix}$ and $\rho \mapsto \begin{pmatrix} \sqrt{-l} & 0 \\ 0 & -\sqrt{-l} \end{pmatrix}$; note that βO_p is the required two-sided ideal. Thus O is an Eichler order of level F, containing $\beta \in U(m)$ and ρ with $\rho^2 = -l$, with $\rho\beta = -\beta\rho$. The fixed points of ρ are in $\mathbf{Q}(\sqrt{-l})$ and map to real, i.e. rational, points on $S^{(m)}$. Thus S is hyperelliptic over $\mathbf{Q}$ in these nine cases.

This method will not work in two of the remaining four cases. The method depends on finding anticommuting ρ and β in $\mathcal{O}$ with $\rho^2 = -l$ and $\beta \in U(m)$. In the case $D = 10$, $F = 19$, $m = 38$, the discriminant of $\mathcal{O}$ would then divide $4 \cdot l \cdot 38$, so $l = 5$, which is not allowed (since $h(-l) = 1$). Thus we cannot prove that $S_{10,19}$ and similarly $S_{14,5}$ is hyperelliptic over $\mathbf{Q}$ by this method and must leave the problem until another occasion. For $S_{15,2}$ we use $l = 7$, $B = \left\{ \begin{pmatrix} a & b \\ -7b' & a' \end{pmatrix} : a, b \in \mathbf{Q}(\sqrt{15}) \right\}$, $\beta = \begin{pmatrix} \sqrt{15} & 0 \\ 0 & -\sqrt{15} \end{pmatrix}$ and $\rho = \begin{pmatrix} 0 & 1 \\ -7 & 0 \end{pmatrix}$. At the prime 2, we can map β onto $\begin{pmatrix} 1 & 1 \\ 14 & -1 \end{pmatrix}$ and ρ onto $\sqrt{-7}\begin{pmatrix} 1 & 0 \\ -2 & -1 \end{pmatrix}$ (note that $\sqrt{-7} \in \mathbf{Z}_2$). Thus $S_{15,2}$ is hyperelliptic over $\mathbf{Q}$, as is $S_{39,2}$, by the same proof.

Except for two unresolved cases, then, the curves S which are hyperelliptic over $\mathbf{R}$ are hyperelliptic over $\mathbf{Q}$.

References

[1] A. O. L. Atkin and J. Lehner, Hecke operators on $\Gamma_0(m)$. Math. Ann. 185 (1970), 134-160.

[2] B. Birch, W. Kuyk (ed.), Modular functins of one variable IV. Lecture Notes in Mathematics 476. Berlin, Heidelberg, New York: Springer, 1975.

[3] M. Eichler, Über die Einheiten der Divisionsalgebren. Math. Ann. 114 (1937), 635-654.

[4] M. Eichler, Über die Idealklassenzahl hyperkomplexer Systeme. Math. Z. 43 (1938), 481-494.

[5] M. Eichler, Zur Zahlentheorie der Quaternionen-Algebren. J. Reine Angew. Math. 195 (1955), 127-151.

[6] M. Eichler, Über die Darstellbarkeit von Modulformen durch Thetareihen. J. Reine Angew. Math. 195 (1955), 156-171.

[7] Y. Ihara, Congruence relations and Shimura curves. Proc. Symposia Pure Math. 33(1979), part 2, 291-311.

[8] B. Mazur, Rational isogenies of prime degree. Invent. Math. 44 (1978), 129-162.

[9] B. Mazur and H. P. F. Swinnerton-Dyer, Arithmetic of Weil curves. Invent. Math. 25 (1974), 1-61.

[10] J. -F. Michon, Courbes de Shimura hyperelliptiques. Bull. Soc. Math. France 109 (1981), 217-225.

[11] A. Ogg, Rational points on certain elliptic modular curves. Proc. Symposia Pure Math. 24 (1973), 221-231.

[12] A. Ogg, Hyperelliptic modular curves. Bull. Soc. Math. France 102 (1974), 449-462.

[13] K. Ribet, Sur les variétés abéliennes à multiplications réelles. C. R. Acad. Sci. Paris 291 (1980), Série A, 121-123.

[14] G. Shimura, On the zeta-functions of the algebraic curves uniformized by certain automorphic functions. J. Math. Soc. Japan 13 (1961), 275-331.

[15] G. Shimura, On the theory of automorphic functions. Ann. of Math. 70 (1959), 101-144.

[16] G. Shimura, Construction of class fields and zeta functions of algebraic curves. Ann. of Math. 85 (1967), 58-159.

[17] G. Shimura, On the real points of an arithmetic quotient of a bounded symmetric domain. Math. Ann. 215 (1975), 135-164.

[18] M. F. Vignéras, Arithmétique des Algèbres de Quaternions. Lecture Notes in Mathematics 800. Berlin, Heidelberg, New York: Springer, 1980.

Received February 26, 1982

Professor Andrew P. Ogg
Department of Mathematics
University of California
Berkeley, California 94720

Special Automorphic Forms on $PGSp_4$

I. I. Piatetski-Shapiro

To I.R. Shafarevich

In a classical situation special automorphic forms were studied by Maass. Let us recall their definition. Denote by H the Siegel half plane of genus 2. Consider Siegel's modular forms of a given weight with respect to the Siegel full modular group. It is known that they have the following Fourier decomposition:

$$f(Z) = \sum a_T \exp 2\pi i\, tr(TZ),$$

where T runs over the matrices of the form $\begin{pmatrix} n & r/2 \\ r/2 & m \end{pmatrix}$; $n, r, m \in Z$. Put $d_T = 4nm - r^2$, $e_T = (n, r, m)$. The Maass space (following Zagier) is the space of those $f(Z)$ such that the coefficients a_T depend only on d_T and e_T. The forms which lie in the Maass space do not satisfy the Ramanujan conjecture. That was one of the reasons why Maass studied these forms.

The aim of this paper is to study a similar space for the group $PGSp_4$ over an arbitrary global field k. Our special forms do not satisfy the Ramanujan conjectures either. We prove that all special forms are lifts from $\overline{SL}_2$. We also prove the local analog of this result.

Our results give an intrinsic characterization of the Weil lifting from $\overline{SL}_2$ to $PGSp_4 = SO_5$ in terms of Fourier coefficients. In §1 we introduce a local analog of the Maass space which are representations over a local field k having the U-property. A main result of §1 is that any preunitary representation with the U-property is a lift from $\overline{SL}_2$.

In §2 we introduce a notion of special automorphic forms (a global analog of Maass forms), and we prove that special cuspidal automorphic forms are lift from $\overline{SL}_2$. Our proof is completely different from that of Maass-Zagier and is based on representation theory and on R. Howe's general theory of dual reductive pairs. Howe's theory of the rank of representations of Sp_{2n} is also used here. In the appendix we give a simple proof of one corollary of Howe's theory.

I am very grateful to my student D. Soudry for his help in the preparation of this manuscript. I thank N. Wallach for his valuable remarks and linguistic corrections.

Notations and Basic Definitions

Let $J = \begin{pmatrix} 0 & I_2 \\ -I_2 & 0 \end{pmatrix}$.

$GSp_4 = \{g \in GL_4 \mid {}^t gJg = \lambda(g)J, \quad \lambda(g) \in k^*\}$

$G = PGSp_4 = GSp_4/\{\lambda I_4 \mid \lambda \in k^*\}$

$S = \left\{ S = \begin{pmatrix} I_2 & S \\ 0 & I_2 \end{pmatrix} \middle| S = {}^t S \right\}, \quad P = \left\{ \begin{pmatrix} A & * \\ 0 & x^t A^{-1} \end{pmatrix} \in GSp_4 \right\}$ and

$M = \left\{ \begin{pmatrix} A & 0 \\ 0 & x^t A^{-1} \end{pmatrix} \middle|_{x \in k^*}^{A \in GL_2} \right\}$. We have $P = M \cdot S$.

Denote by Q the parabolic subgroup of G which preserves a nontrivial vector. Let $Q = FN$ where F is the Levi subgroup and N the unipotent radical. Let Z be the center of N, and let D be the centralizer of Z in Q.

Let k be a global field with A its ring of adeles. Any character of $S_k \setminus S_A$ is obtained as $\psi_T(s) = \psi(tr\, TS)$, where ψ is a fixed nontrivial character of $k \setminus A$ and T is a symmetric matrix with entries in k. We call ψ_T nondegenerate if $\det T \neq 0$. Denote by O_T the stabilizer of ψ_T in M, and by O_T^c its connected component.

Let $\varphi(g)$ be an automorphic function on G. Introduce the Fourier coefficient

$$\varphi_T(g) = \int_{S_k \setminus S_A} \psi_T^{-1}(s)\varphi(sg)\, ds. \tag{0.1}$$

It is easy to see that

$$\varphi_T(\delta g) = \varphi_T(g), \qquad \forall \delta \in O_T(k). \tag{0.2}$$

We call $\varphi(g)$ *special* if it satisfies (0.2) for any $\delta \in O_T^c(A)$ and for any nondegenerate T.

Notations Related to the Weil Representation

The theory of the Weil representation is well known. We recall that the main facts are:

1) there exists a metaplectic group $\widetilde{Sp}_{2n}(k_p)$, where k_p is a local field. This group is a double covering of $Sp_{2n}(k_p)$. Similarly there exists a group of $\widetilde{Sp}_{2n}(A)$ where A is the adele ring of any global field k.

2) To any nontrivial character ψ of k_p we can attach the Weil representation ω_ψ of $\widetilde{Sp}_{2n}(k_p)$ which acts on $S(k_p^n)$, the space of Schwarz-Bruhat functions on an n-dimensional vector space over k_p. We will view $\widetilde{Sp}_{2n}(k_p)$ as the collection of pairs (g,ϵ) where $g \in \widetilde{Sp}_{2n}(k_p)$ and $\epsilon = \pm 1$. Then we have that the Weil representation ω_ψ is the unique representation satisfying

$$\begin{aligned} (i)\quad & \omega_\psi\left(\begin{pmatrix} \alpha & \beta \\ 0 & t_\alpha - 1 \end{pmatrix}, \epsilon\right)\varphi(X) \\ & = \epsilon\,\frac{\gamma(1)}{\gamma(\det\alpha)}|\det\alpha|^{1/2}\psi\left(\frac{1}{2}\langle X_\alpha, X_\beta\rangle\right)\varphi(X_\alpha) \\ (ii)\quad & \omega_\psi\left(\begin{pmatrix} 0 & I \\ -I & 0 \end{pmatrix}, \epsilon\right)\varphi(X) \\ & = \epsilon\left(\frac{1}{\gamma(1)}\right)^n \int_{k_p^n}\psi\left(\langle y, x\begin{pmatrix} 0 & I \\ -I & 0 \end{pmatrix}\rangle\right)\varphi(y)dy \end{aligned} \tag{0.3}$$

where $\gamma(t)$ is the well-known Weil constant. The similar formulas will be true for the adelic case.

3) We now introduce the notion of a dual reductive pair due to R. Howe [8] in a slightlly modified form. Let k be any field. Let G_1, G_2 be two reductive subgroups of $Sp_{2n}(k)$. We will call the pair (G_1, G_2) a dual reductive pair if

(a) $g_1 g_2 = g_2 g_1; \quad \forall g_1 \in G_1, \forall g_2 \in G_2$.
  (Hence G_1G_2 is a subgroup of $Sp_{2n}(k)$.)
(b) The full centralizer of G_i in $Sp_{2n}(k)$ lies in G_1G_2 $(i = 1,2)$.

The typical example of a dual reductive pair will be the following: $G_1 = SO_m$, $G_2 = Sp_{2l}$. In our paper we consider the case $m = 5$ and $l = 1$.

4) Let k be a global field and let (G_1, G_2) be a dual reductive pair lying in $Sp_{2n}(k)$. Denote by Z the $2n$ dimensional space in which $Sp_{2n}(k)$ acts. Put $Z = Z_1 \otimes Z_2$ where Z_1 and Z_2 are Lagrangian (isotropic) subspaces. We know that the Weil representation of $\widetilde{Sp}_{2n}(A)$ acts on the space $S(Z_1(A))$ — the space of Schwarz-Bruhat functions on $Z_1(A)$. We define

$$\theta_\psi^\phi(g_1, g_2) = \sum_{z_1 \in Z_1(k)} \omega_\psi(g_1 \cdot g_2)\phi(z_1), \qquad \phi \in (S(Z_1)(A)). \tag{0.4}$$

It is known that this function has moderate growth and hence the integral

$$\int \theta_{\psi}^{\phi}(g_1, g_2) f_2(g_r) dg_r \tag{0.5}$$

makes sense, where f_2 is any cuspidal automorphic form.

Assume that $f_2 \in \pi_2$, where π_2 is a cuspidal automorphic representation; then the set of functions of the form (0.5) is an automorphic representation [3], [10].

Now we describe a dual reductive pair, which is important for our purposes.

Consider the space

$$X = \{T \in M_4(k) \mid TJ \text{ is skew symmetric and } tr(T) = 0\}.$$

X is a 5-dimensional space over k. We get a representation of GSp_4 on X by

$$g \colon T \longrightarrow g^{-1}T_g.$$

It is easy to check that this action preserves X. We put a symmetric non-degenerate from (,) on X by defining

$$(T_1, T_2) = tr(T_1 T_2).$$

Put $G = PGSp_4$. Then the above action of G on X imbeds G as the connected component of the orthogonal group O_5 preserving this form. Denote it by O_5^c.

There exists a basis of X such that the symmetric form is represented by

$$\begin{pmatrix} 0&1&0&0&0\\ 1&0&0&0&0\\ 0&0&0&1&0\\ 0&0&1&0&0\\ 0&0&0&0&1 \end{pmatrix}$$

For this realization of G, P is a parabolic subgroup preserving an isotropic line and Q a parabolic subgroup preserving a two-dimensional isotropic subspace.

We will denote (,) the symmetric form on X which is preserved by $PGSp_4$. We will denote by $\langle\ ,\ \rangle$ the skew symmetric form on a two-dimensional space Y preserved by SL_2. This way we define a dual reductive pair G, SL_2. In fact, it defines a lifting from $\overline{SL}_2$ to G.

Assume that the automorphic form $\varphi(g)$ is a lift, i.e., it can be written in the form

$$\varphi(g) = \int_{SL_2(k)\backslash SL_2(A)} \theta_{\psi}^{\phi}(g,h)\alpha(h)\,dh \tag{0.6}$$

where $\alpha(h)$ is a cuspidal automorphic function on $\overline{SL}_2(A)$. Then [3]

$$\varphi_T(g) = \int_{N_A\backslash SL_2(A)} (\omega_\psi(g,h)\phi)(z_T)W(h;\psi_{\det T},\alpha)\,dh \tag{0.7}$$

where $W(h,\psi_\beta,\alpha) = \int_{K\backslash A}\psi(-\beta z)\alpha(\left(\begin{smallmatrix}1&z\\0&1\end{smallmatrix}\right)h)\,dz$ and $z_T = (y_2,-T)$ is the special point in Z_1 which depends only on T. N is the unipotent subgroup of SL_2, which preserves y_2. Since any $\delta \in O_T^c(A)$ preserves z_T, we have

$$\varphi_T(\delta g) = \varphi_T(g), \quad \delta \in O_T^c(A). \tag{0.8}$$

This shows that any form which comes as a lifting is special.

§1 Representations over a Local Field with U-Property

Let (π, V) be an irreducible smooth representation of G_k where k is a local field. Let ψ_T be a nondegenerate character of S_k. We say that (π, V) has the U-property (uniqueness) if, whenver a linear functional l_T satisfies $l_T(\pi(s)v) = \psi_T(s)l_T(v)$ ($s \in S_k, v \in V, T$ nondegenerate), then $l(\pi(\delta)v) = l_T(v)$ for any $\delta \in O_T^c$. Assume that (π, V) is a preunitary irreducible smooth admissible representation with the U-property. We will prove that $\pi|_Q$ is irreducible when we consider π as a unitary representation. We also prove the following rigidity property. If (π_1, V_1) and (π_2, V_2) are two such representations and $\pi_1|_Q \simeq \pi_2|_Q$, then $\pi_1 \simeq \pi_2$.[1] It is easy to deduce from [3] that the Weil lifting from $\overline{SL}_2$ has the U-property. In this paper we prove the converse, that is, that any representation satisfying the U-property comes as a lift from $\overline{SL}_2$.

---

[1]This rigidity property is false without the U-property, even if $\pi_i|_Q$ is irreducible.

Lemma 1.1. *Let (π, V) have the U-property; the space of the linear functionals on V satisfying*

$$l_T(\pi(s)v) = \psi_T(s)\, l_T(v); \quad s \in S, \quad v \in V,$$

is at most one-dimensional.

Proof. Consider the group $R = O_T^c.S$. Let μ be any character O_T^c. Let l be a linear functional satisfying

$$l(\pi(\delta s)v) = \mu(\delta)\psi_T(s)\, l\,(v). \tag{1.1}$$

It is known that the space of linear functionals satifying (1.1) is at most one dimensional [4]. In our case $\mu = 1$.

We describe a generalized Kirillov model. Let T be a non-degenerate symmetric matrix of order two. Let (π, V) be an irreducible smooth representation of G (k a local nonarchimedean field). We say that the integral

$$\int_S \psi_T^{-1}(s)\pi(s)v\, ds$$

exists if it stabilizes for large open compact subgroups of S. Let $T_1, \ldots, T_2$ be representatives of classes M-equivalent symmetric matrices. Put

$$V_i = \{v \in V \mid \textstyle\int_S \psi_{T_i}^{-1}(s)\pi(s)v\, ds = 0\}.$$

Denote $\mathbf{L}_i = V/V_i$ and $\varphi^i\colon V \longrightarrow \mathbf{L}_i$ the natural projection. Introduce the functions

$$\varphi_v^i(m) = \varphi^i(\pi(m)v). \tag{1.2}$$

We have

$$\varphi_v^i(sm) = \psi_{T_i}(s)\varphi_v^i(m). \tag{1.3}$$

Lemma 1.2 *Let $\lambda(m)\colon M \longrightarrow \mathbf{L}_{i_0}$ be a smooth function with a compact support. Then there exists $v_0 \in V$ such that*

$$\varphi_{v_0}^{i_0}(m) = \lambda(m) \quad \textit{and} \quad \varphi_{v_0}^i(m) = 0, \quad i \neq i_0. \tag{1.4}$$

Proof. It suffices to prove the lemma for $\lambda(m) = l_0 \cdot e_m(m)$ where $e_0(m)$ is a characteristic function of a sufficiently small open compact neighborhood of a point $m_0 \in M$ and $l_0 \in \mathbf{L}_{i_0}$. First we construct $v_1 \in V$ such that

$\varphi_{v_1}^{i_0}(m_0) = l_0$. Now define

$$v_0 = \frac{\int_{s_0} \psi_{T_{i_0}}^{-1}(s)\pi(s)v_1\,ds}{\int_{s_0} ds} \tag{1.5}$$

where s_0 is a sufficiently large open compact subgroup of S. This v_0 satisfies the properties of the lemma.

Remark. If $v_0 \in V$ satisfies (1.4), then for any T with $\det T = 0$ we have $l_T(v_0) = 0$. In order to prove this, if suffices to verify that a set of T such that $l_T(v_0) \neq 0$ consists of T of the form $mT_{i_0}m^{-1}$, $m \in \sup\lambda(m)$. This last set is closed and does not contain any T such that $\det T = 0$.

Let l_T be a linear functional such that (T-nondegenerate)

$$l_T(\pi(s)v) = \psi_T(s)\,l(v), \quad \forall v \in V, \quad s \in S. \tag{1.6}$$

We denote by l_T^{inv} (the unique up to a scalar multiple) functional l_T such that

$$l_T(\pi(\delta)v) = l_T(v), \quad \forall v \in V, \quad \forall s \in O_T^c. \tag{1.7}$$

Let l^{inv}: $\mathbf{L}_i \longrightarrow C$ be any functional such that $l^{\mathrm{inv}}(\pi(\delta)x) = l^{\mathrm{inv}}(x)$, $\forall x \in \mathbf{L}_i$, $\forall \delta \in O_{T_i}^c$. We denote by π the natural action of O_{T_i} on $\mathbf{L}_i$. Denote by $\mathbf{L}_l^0 = \{l \in \mathbf{L}_i \mid l^{\mathrm{inv}}(l) = 0\}$.

$$l(v) = l^{\mathrm{inv}}(\varphi_v^i(m_1)) \tag{1.8}$$

will be an invariant functional for some T. It is easy to check that any invariant functional has the form (1.8).

Lemma 1.3. *If (π, V) is an irreducible smooth representaion, which does not satisfy the U-property, then there exists a vector $v_0 \in V$, $v_0 \neq 0$ with $l_T^{\mathrm{inv}}(v_0) = 0$, $\forall T$.*

Proof. If (π, V) does not possess the U-property then there exists i_0 such that $\mathbf{L}_{i_0}^0$ is nontrivial. Using Lemma 1.1 we can construct $v_0 \neq 0$ such that $\varphi_{v_0}(m) \in \mathbf{L}_i^0$, $\forall m$, $\forall i$. Thus v_0 satisfies the assumptions of our lemma.

If (π, V) has the U-property, then according to Lemma 1, the spaces $\mathbf{L}_i$ are one-dimensional and the group $O_{T_i}^c$ acts on $\mathbf{L}_i$ trivially. Hence we can view $\varphi^i(m)$ as functions with values in $\mathbf{C}$.

Assume now that (π, V) is an irreducible preunitary representation satisfying the U-property.

Lemma 1.4. *Under the appropriate choice of the measure dm on $O^c_{T_i} \setminus M$, we have*

$$(v,v) = \sum_{i=1}^{r} \int_{O^c_{T_i} \setminus M} | \varphi^i_v(m) |^2 \, dm. \tag{1.9}$$

Moreover,

$$\overline{V} \simeq \oplus L^2(O^c_{T_i} \setminus M), \tag{1.10}$$

where the summation is with respect to those i such that φ^i is nontrivial.

Proof. Consider the completion $\overline{V}$ of V. According to the result of the appendix, we have the following spectral decomposition with respect to the unitary operators $\pi(s)$, $s \in S$

$$\overline{V} = \int \overline{V}_T \, dT.$$

The integration is with respect to all nondegenerate symmetric matrices. (Recall that any character of S has the form $\psi_T(s) = \psi(tr\, TS)$ where T is a symmetric matrix of order 2.) The meaning of (1.11) is that there exists a map that attaches to each $v \in \overline{V}$ a measurable function $f_v(T) \in \overline{V}_T$ such that

$$(v,v) = \int |f_v(T)|^2 dT,$$

where $dT = d\mu_\pi(T)$ is a spectral measure, corresponding to π/S.

Let $v \in V$ be a smooth vector. Then there exists an open compact subgroup $M_0 \subset M$ which stabilizes v. $f_v(T)$ is preserved by M_0 as well.

Let T_1 be a nondegenerate symmetric matrix of order 2. Then its orbit under M_0 is an open neighborhood of T_1. This implies that $f_v(T)$ is locally constant on the set of the nondegenerate T's. According to our construction, $f_v(T)$ defines a map $V \to \overline{V}_T$ which commutes with the natural action of S. (On $\overline{V}_T$, S acts according to the character ψ_T.) Since π has the U-property, all the spaces $\overline{V}_T$ should be one-dimensional, and O^c_T acts trivially on O^c_T. Thus we can identify $f_v(T)$ with one of the function $\varphi^i_v(m)$.

Lemma 1.5. *Assume that $P = MS$ and $Q = FN$ contain the same Borel subgroup B of G. Then for any nondegenerate symmetric T except for a subset of a smaller dimension,*

$$\overline{O_T^c B_1} = M,$$

where $B_1 = B \cap M$.

Proof. It is well known that $B_1 \setminus M$ is a complete algebraic manifold of dimension 1. O_T^c is a connected algebraic group acting on this manifold. Hence either its image is one point or dense in the manifold. It is easy to determine when its image is a point. In this case $O_T^c \subset B_1$. Such T lies in a subset of positive codimension.

Theorem 1.1.

(1) *If (π, V) is a preunitary smooth irreducible representation with the U-property, then for any Q-invariant subspace V_1, we have*

$$\overline{V}_1 = \overline{V}.$$

(2) *If (π_1, V_1), (π_2, V_2) are two such representations, satisfying $\pi_1|_Q \cong \pi_2|_Q$ as unitary representations, then $\pi_1 \cong \pi_2$.*

Proof. Let V_1 be a Q-invariant subspace of V. Denote by

$$E_i = \{m \in O_{T_i}^c \setminus M \mid \exists v \in V_1, \quad \varphi_v^t(m) \neq 0\}.$$

First we show that $\overline{E}_i$ is an M invariant set. Indeed, let $m_0 \in \overline{E}_i$ and $m_1 \in M$. We show that $m_0 m_1 \in \overline{E}_1$. There exists $\tilde{m}_0$ such that $\varphi_v^{T_i}(\tilde{m}_0) \neq 0$, and there exists $\tilde{m}_1$, close to m_1, such that $\tilde{m}_1 = \tilde{m}_0^{-1} \delta \tilde{m}_0 b$, $b \in B_1$, $\delta \in O_{T_i}^c$. (Lemma 1.5). We get that $\tilde{m}_0 \tilde{m}_1 \in E_i$, which implies $m_0 m_1 \in \overline{E}_i$. This shows that $\overline{E}_i = O_{T_i}^c \setminus M$. Using the same argument as in the proof of Lemma 1.2, one can show that for any $m_0 \in E_{i_0}$, $\exists v_0 \in V_1$ such that $\varphi_{v_0}^{i_0}(m) = 1$, $m \in U(m_0)$ and $\varphi_{v_0}^i(m) = 0$, if $i \neq i_0$ or $m \notin U(m_0)$. Since $\overline{E}_i = O_{T_i}^c \setminus M$, $\forall i$, such that φ^i is nontrivial, we have (see 1.10) that the set of such functions is dense in $\overline{V}$.

Now we prove the second statement of our theorem. Let α: $\overline{V}_1 \longrightarrow \overline{V}_2$ be a Q-isomorphism, which exists according to our assumption. Put $B_1 = M \cap Q$. It is clear that if $v \in V_1$ is invariant with respect to a compact open subgroup U of B_1, then $\varphi_v^i(m)$ is locally constant. We show that if $\varphi_v^i(m_1) = 0$, then $\varphi_{\alpha_v}^i(m_1) = 0$ (for v as above). Indeed, $\varphi_v^i(m_1)$ is a linear functional which is uniquely defined by its transformation law with respect to S.

It is obvious that φ^i is nontrivial for π_1 iff the corresponding map for π_2 is nontrivial. Denote by S^{i_0} the subset

$$\{v \in \overline{V}_1 \mid \varphi_v^{i_0}(m) \text{ is locally constant with compact support and } \varphi_v^i(m) = 0 \text{ for } i \neq i_0\}.$$

Similarly we introduce $S_2^{i_0}$. It is clear that α sends $S_1^{i_0}$ to $S_2^{i_0}$. Since α is a linear map preserving zero and commuting with right translations by elements of B_1, α is a multiplication by a constant on $S_1^{i_0}$. Hence, on $S_1^{i_0}$, α commutes with the action of M. Since $\bigoplus S_1^i$ is dense in $\overline{V}_1$ and since M and Q generate G, λ commutes with the action of G. This proves our theorem.

Lemma 1.6. *If $Z \subset G$ is any unipotent subgroup and (π, V) is an irreducible infinite dimensional unitary representation of G, then it has no vector in V which is invariant with respect to Z.*

Proof. It follows from the Howe-Moore theorem that any matrix coefficient of an infinite dimensional unitary irreducible representation tends to zero [7].

Put $Q = FN$ where F is the reductive part and N the unipotent radical of Q. Denote by Z the center of N. Let D be the centralizer of Z in Q. It is well known that any representation τ of D on which Z acts according to a given nontrivial character ψ is of the form $\sigma \otimes \omega_\psi$, where ω_ψ is the Weil representation of $\overline{D} = \overline{SL_2}$. N which corresponds to the character ψ and σ is an irreducible unitary representation of $\overline{SL_2}$. τ is irreducible (unitary) iff σ is irreducible (unitary).

Theorem 1.2. *Any irreducible smooth preunitary representation (π, V) with the U property can be obtained as a Weil lifting from some irreducible unitary representation of $\overline{SL_2}$. (We assume here that (π, V) is not a trivial representation.)*

Proof. Let (π, V) be as above. It has to be infinite-dimensional. According to Lemma 1.6, it has no vector which is invariant with respect to Z.

This implies (see Theorem 1.1) that (π, V), viewed as a representation of Q, can be written as $\mathrm{ind}_D^Q \tau$, where τ is an irreducible unitary representation of D, on which Z acts according to a nontrivial character ψ. Hence, τ is of the form $\sigma \otimes \omega_\psi$, where Q is a unitary representation of $\overline{SL}_2$. Consider the Weil lifting $\theta(\sigma, \psi)$. $\theta(\sigma, \psi)$ has the U-property, and its restriction to Q is isomorphic to $\sigma \otimes \omega_\psi$ [3]. The rigidity theorem (Theorem 1.1) can be proved in the case when we assume *a priori* that only one representation is unitary. This implies that $\pi = \theta(\sigma, \psi)$.

We consider now the archimedean case. The idea of the proof is similar to the nonarchimedean case. It is easy to see that if $k = \mathbf{R}$, then there are two equivalence classes of nondegenerate symmetric matrices of order 2. The first class consists of definite matrices and the second consists of indefinite matrices. In the case $k = \mathbf{C}$, there is only one equivalence class of nondegenerate symmetric matrices.

Let (π, V) be a smooth, preunitary, irreducible admissible infinite dimensional representation of G. Let us recall that this means that there exists in V an invariant hermitian form. We let H be the completion of V with respect to this form. Then V consists of all the vectors in H lying in the domain of definition for all the operators of the enveloping algebra of G. It is easy to see that V is a nuclear space. Denote by V' the space of continuous linear functionals on V. Then we have a rigged (equipped) space:

$$V \subset H \subset V'. \tag{1.12}$$

For the definition of a rigged space see [6].

Let T be a symmetric matrix of order 2. Denote by $\mathbf{L}_{T'}$ the space of all continuous linear functionals on V satisfying

$$l_T(\pi(s)v) = \psi_T(s)\, l_T(v) \tag{1.13}$$

and denote by V_T^0 the subspace $\{v \in V \mid l_T(v) = 0, \forall l_T \in \mathbf{L}_T\}$.

Let φ^T be the natural projection from V onto $V_T = V_T^0 \setminus V$. Put

$$\varphi_v^i(m) = \varphi^{T_i}(\pi(m)v).$$

We now consider the equivalence classes of *all* symmetric matrices. According to the spectral theorem which is proved in [6], we have

$$(v, v) = \sum_i \int_{M_i \setminus M} (\varphi_v^i(m), \varphi_v^i(m))\, d\sigma^i(m), \tag{1.14}$$

where M_i is the stabilizer of T_i.

In the appendix it is proved that except for the case $k = \mathbb{C}$ and π the Weil representation — the contribution of the degenerate terms is zero.

We now explain how to prove a lemma similar to Lemma 1.2. In order to prove it, we have to use the same argument as in Lemma 1.2 and also the fact[2] that the Lie operators act locally on the functions $\varphi_v^i(m)$. This implies that any smooth function with compact support on $M_i \setminus M$ defines some vector in V.

Now theorems 1.1, 1.2 can be proved exactly as in the nonarchimedean case.

§2 Special Automorphic Forms

In this section we will prove that any cuspidal special automorphic form on G is a lifting from $\overline{SL}_2$ (Theorem 2.2). It seems that the assumption that the automorphic form should be cuspidal is not necessary. We need it since the lifting of noncuspidal automorphic forms from $\overline{SL}_2$ was not yet investigated.

Theorem 2.1. *Let $\pi = \bigotimes \pi_p$ be an irreducible automorphic special representation; then each component π_p has the U-property.*

Proof. Let $f(g) \in \pi$. It is easy to see that

$$f(g) = \sum_{T \in M_2^S(k)} f_T(g). \tag{2.1}$$

where

$$f_T(g) = \int_{S_k \setminus S_A} \psi_T^{-1}(s) f(sg)\, ds. \tag{2.2}$$

Assume that π_{p_0} does not have the U-property. Then, according to lemma 1.3, there exists a non-zero vector $u_0 \in \pi_{p_0}$ such that for any linear functional of type l_T^{inv}, we have

$$l_T^{\text{inv}}(u_0) = 0. \tag{2.3}$$

[2]J. N. Bernstein told me a very simple proof of this statement based on an embedding into a representation which is induced from the parabolic subgroup.

Take a finite place $p_1 \neq p_0$. Then take the non-zero vector $u_1 \in \pi_{p_1}$ which was defined in Lemma 1.2 (see the remark after the lemma). Let $u = \bigotimes u_p \in \pi$ be a non-zero vector such that $u_{p_0} = u_0$, $u_{p_1} = u_1$. Let $f(g)$ be the automorphic function in π which corresponds to u. We will show that under our assumptions

$$f\,|_{P_A} \equiv 0. \tag{2.4}$$

From the definition of u_1, it follows that

$$f_T(p) = 0 \quad \forall p \in P_A \quad \forall T,\ \det T = 0.$$

If $\det T \neq 0$, then since $f(g)$ is a special automorphic form, $f_T(P)$ defines an invariant linear functional on π. Using the property of u_0 we obtain that $f_T(p) = 0$ for $p \in P_A$ and invertible T. It is well known that $P_k \setminus P_A$ is a dense subset of $G_k \setminus G_A$. Hence, from (2.1) it follows that: $f \equiv 0$, a contradiction.

Corollary. *Let π be a preunitary irreducible special automorphic representation. Then $\pi|_{Q_A}$ is irreducible as a unitary representation.*

Proof. This follows from Theorem 2.1 and Theorem 1.1.

Let π be a preunitary irreducible special automorphic representation. Let $f \in \pi$. Put

$$f_\psi(g) = \int_{Z_k \setminus Z_A} \psi^{-1}(zg) f(zg) dz.$$

Then $f_\psi(\delta g) = f_\psi(g)$, $\forall \delta \in D_k$. Hence the set of $f_\psi|_{D_A}$ is an automorphic representation of D_A. Consider the map $f \to f_\psi|_{D_A}$. Denote it by $\overline{Q}$. It defines a homomorphism $\pi \to \operatorname{Ind}_{D_A}^{Q_A}\tau$. We know that the image of π is nontrivial. From the above corollary we know that $\pi|_{Q_A}$ is irreducible as a unitary representation. This implies that $\pi|_{Q_A} = \operatorname{Ind}_{D_A}^{Q_A}\tau$ as unitary representations and τ is irreducible.

Theorem 2.2. *Let π be an irreducible cuspidal special automorphic representation of G. Then there exists an irreducible cuspidal automorphic representation σ of $\overline{SL}_2(A)$ with a missing character ψ such that $\pi = \theta(\sigma, \psi)$.*

Proof. We have already proved that $\pi|_{Q_A} = \operatorname{Ind}_{D_A}^{Q_A}\tau$ where τ is the unitary irreducible automorphic representation of D_A. It is known that any irreducible representation of D_A with a given restriction to the group Z_A (Z is the center of D) can be uniquely written in the form $\sigma \otimes \omega_\psi$, where σ is an irreducible representation of $\overline{SL}_2(A)$ and ω_ψ is the Weil representation of D_A which corresponds to the character ψ. It is easy to see that if τ is automorphic, then so is σ. It remains to prove that σ is cuspidal with a missing character ψ. We first see that

$$\int_{Z_k \backslash Z_A} f(z)\,dz = 0 \quad \forall f \in \pi, \tag{2.5}$$

since otherwise π possesses a Whittaker model [3]. However, any local component of π cannot have a Whittaker model because it represents a lifting from $\overline{SL}_2$[3]. Since the unipotent subgroup of $\overline{SL}_2 \subset D$ and Z are conjugate in G, we get that the restriction of any automorphic function $\varphi \in \tau$ to $\overline{SL}_2$, is a cuspidal function, and hence, using the results of [3] we have that ψ is a missing character of σ.

We now prove that $\pi = \theta(\sigma, \psi)$. We know that π and $\theta(\sigma, \psi)$ are special automorphic representations and we also know that their restrictions to Q_A are of the form $\mathrm{Ind}_{D_A}^{Q_A}$, where $\tau = \sigma \otimes \omega_\psi$ [3]. Using Theorem 1.1 we obtain that $\pi \cong \theta(\sigma, \psi)$. Since the multiplicity one theorem is true for such representations, we obtain that $\pi = \theta(\sigma, \psi)$ [3].

Appendix

The aim of this appendix is to prove the following theorem, which is due to R. Howe.

Theorem. *Let (π, H) be an infinite dimensional irreducible unitary representation of $G = PGSp_4(k)$, where k is any local field. In the case $k = \mathbf{C}$ we exclude the Weil representation. Consider the spectral decomposition with respect to S:*

$$H = \int H_T \, d\mu_T, \tag{1}$$

where T is the symmetric 2×2 matrix which defines the character $\psi_T(S) = \psi(tr\, TS)$. Then the spectral measure of the set of all degenerate matrices equals zero.

Proof. We first prove the following lemma. Let $k \neq \mathbf{C}$.

Lemma 1. *Let (τ, H) be an irreducible unitary representation of D such that Z acts nontrivially. Consider the spectral decomposition of H with respect to S. Then the spectral measure of the set of degenerate characters of S is zero.*

Proof. It is known that any irreducible unitary representation of D such that Z acts according to a nontrivial character ψ is of the form $\sigma \otimes \omega_\psi$ where ω_ψ is the Weil representation corresponding to ψ and σ is an irreducible unitary representation of $\overline{SL}_2$ and is hence infinite dimensional. In the case $k = \mathbf{C}$, it is not true because ω_ψ is a representation of D itself and hence we can take σ to be any representation of $\overline{SL}_2(\mathbf{C})$; for instance, the trivial representation. Write $S = XY$ where X is the unipotent subgroup of SL_2 and Y is the maximal abelian subgroup of N.

Taking $S = \left\{ \begin{pmatrix} I_2 & S \\ 0 & I_2 \end{pmatrix} \mid S\text{-symmetric} \right\}$, X corresponds to $\begin{pmatrix} I_2 & \left(\begin{smallmatrix} 0 & 0 \\ 0 & x \end{smallmatrix}\right) \\ 0 & I_2 \end{pmatrix}$ and Y corresponds to $\begin{pmatrix} I_2 & \left(\begin{smallmatrix} * & * \\ * & 0 \end{smallmatrix}\right) \\ 0 & I_2 \end{pmatrix}$. Consider the spectral decomposition of ω_ψ with respect to S. The spectral measure is concentrated on matrices of the form $T = \begin{pmatrix} 1 & \beta \\ \beta & \beta^2 \end{pmatrix}$ and we can take as a spectral measure just the standard Lebesgue measure $d\beta$. For the representation σ the spectral measure is concentrated on matrices of th form $T' = \begin{pmatrix} 0 & 0 \\ 0 & \alpha \end{pmatrix}$ where α is in a union of cosets module $(K^*)^2$, and we can take the spectral measure to be $d\alpha$. Hence the spectral decomposition of τ with respect to S is concentrated on matrices of the form $\begin{pmatrix} 0 & 0 \\ 0 & \alpha \end{pmatrix} + \begin{pmatrix} 1 & \beta \\ \beta & \beta^2 \end{pmatrix} = \begin{pmatrix} 1 & \beta \\ \beta & \beta^2 + \alpha \end{pmatrix}$ with measure $d_\beta d_\alpha$. The degenerate matrices are those with $\alpha = 0$. Clearly, the spectral measure of the set of such matrices equals zero.

Lemma 2. *Let $k \neq \mathbf{C}$ and (τ, H) be a unitary representation of D on which Z acts according to a nontrivial character; then the spectral measure of the set of degenerate characters of S is zero.*

Proof. This follows from the fact that H is a direct sum (maybe continuous) of irreducible unitary representations satisfying the assumptions of Lemma 1. We now prove our theorem for the case $k \neq \mathbf{C}$. Consider the spectral decomposition of H with respect to Z. We have

$$H = \int_k H_\gamma \, d_\mu(\gamma), \tag{2}$$

where γ corresponds to the character $\psi(\gamma z) = \psi_\gamma$. The spectral measure of $\{\gamma = 0\}$ equals zero, since otherwise there exists a vector $v \in H$ which is invariant with respect to $\pi(Z)$, which contradicts a well-known theorem of Howe and Moore about the vanishing of matrix coefficients at infinity [7]. For any $\gamma \neq 0$, Z acts according to a nontrivial character on H, and hence we can apply Lemma 2. This proves the theorem in the case $k \neq \mathbf{C}$.

We now consider the case $k = \mathbf{C}$.

We consider the spectral decomposition (2). For the same reasons the spectral measure of $\{\gamma = 0\}$ equals zero. Each of the spaces H_γ can have a subspace H_γ^0 which is a multiple of ω_{ψ_γ}. H_γ^0 is the sub-space of H_γ such that the spectral measure for the spectral decomposition of H_γ with respect to S is concentrated on degenerate characters of S. The reason for this is the following: any irreducible unitary representation of D, such that Z acts according to a nontrivial character, has the form $\sigma \otimes \omega_{\psi_\gamma}$, where σ is an irreducible unitary representation of SL_2. Hence σ can be either infinite dimensional or trivial. In the first case the spectral measure concentrated on degenerate characters is zero. The proof of this fact is the same as that of Lemma 1. In the second case the spectral measure is concentrated on degenerate characters.

Now consider the space

$$H^0 = \int H_\gamma^0 \, d\gamma.$$

By construction H^0 is invariant with respect to Q. We have that H^0 is invariant with respect to the Levi subgroup M of P, since M transforms degenerate characters of S to degenerate characters of S. Thus H^0 is G-invariant and hence $H = H^0$. R. Howe proved that if the spectrum of an infinite dimensional irreducible unitary representation of G is degenerate, then H is the Weil representation [9].

References

[1] D. Zagier, Sur la conjecture de Saito-Kurokawa (D'apres H. Maass), Seminaire de Théorie des Nombres, Paris 1979–80. Séminaire Délange-Pisot-Poitou. Birkhäuser.

[2] H. Maass, Über eine Spezialschar von Modulformen zweiten Grades. Invent. Math. 52 (1979), 95–104.

[3] I. I. Piatetski-Shapiro, On the Saito-Kurokawa lifting. Preprint. February 1981.

[4] M. E. Novodvorsky and I. I. Piatetski-Shapiro, Generalized Bessel models for the symplectic group of rank 2, Mat. Sb. 90(2) (1973), 246–256.

[5] I. I. Piatetski-Shapiro, L-functions for $GSp_{(4)}$. Preprint.

[6] I. M. Gelfand and N. Ya. Vilenkin, Generalized functions, Vol. 4, Academic Press, 1964.

[7] R. Howe and C. Moore, Asymptotic properties of unitary representations. J. Fun. Anal. 32 (1979), 72–96.

[8] R. Howe, θ series and invariant theory, Proc. Symp. Pure Math., XXXIII, Part I, 275–286.

[9] R. Howe, The notion of rank for representations. Preprint.

[10] R. Howe and I. I. Piatetski-Shapiro, Some examples of automorphic forms on GSp_4. Preprint.

Received June 22, 1982

Professor I. I. Piatetski-Shapiro
Department of Mathematics
Yale University
Box 2155 Yale Station
New Haven, Connecticut 06520

Sous-variétés d'une variété abélienne et points de torsion

M. Raynaud

To I.R. Shafarevich

Soient A une variété abélienne définie sur le corps des nombres complexes, T le sous-groupe de torsion de A et X un sous-schéma fermé intègre de A.

Théorème. *Si $T \cap X$ est dense dans X pour la topologie de Zariski, alors X est le translaté, par un point de torsion, d'une sous-variété abélienne de A.*

Ce résultat généralise d'une part [9], qui traite du cas où X est une courbe et d'autre part [2] où T est remplacé par sa composante n primaire pour n entier > 1.

Indiquons le principe de la démonstration dans le cas où X ne contient pas de translaté d'une sous-variété abélienne non nulle de A. Pour établir la finitude de $X \cap T$, on choisit un bon nombre premier p et on démontre d'une part que la torsion p-primaire contenue dans $X + a$ est finie et bornée indépendamment de $a \in A$, ce qui se fait par un raffinement facile du résultat de Bogomolov; d'autre part, on prouve la finitude de la torsion première à p, par une méthode p-adique.

Pour cela on se ramène au cas où A est un R-schéma abélien, où R est un anneau de valuation discrète complet, de corps des fractions K de caractéristique 0, de corps résiduel k algébriquement clos de caractéristique p. On suppose de plus que R est non ramifié, donc isomorphe à l'anneau $W(k)$ des vecteurs de Witt à coefficients dans k. Le schéma X est maintenant un sous-schéma de la fibre générique A_K de A.

Examinons d'abord le cas où A est un relèvement canonique (au sens de Serre-Tate), de sa fibre spéciale A_o. Dans ce cas, le Frobenius absolu F de A_o se relève en un endomorphisme ϕ de A, σ_R-linéaire, où σ_R est le Frobenius de $R = W(k)$. Alors ϕ laisse fixe les points de torsion de $A(R)$ d'ordre premier à p, et donc aussi X si on suppose que ces derniers sont denses dans X, pour la topologie de Zariski. Une étude analytique locale (cf. 5.3.1) permet alors d'en déduire que X est le translaté d'une sous-variété abélienne de A.

Dans le cas général, le Frobenius F de A_0 ne se relève pas en un endomorphisme de A, et on doit remplacer A par son extension vectorielle universelle formelle $\mathcal{E}$. Cette modification crée quelques complications plus techniques que théoriques. Tout d'abord $\mathcal{E}$ n'est plus un schéma en groupes mais un schéma formel en groupes et sa fibre générique $\mathcal{E}_K$ est un groupe analytique rigide. Ensuite l'endomorphisme ϕ agissant sur l'algèbre de Lie de $\mathcal{E}_K$ fait intervenir plusieurs pentes; on doit séparer la contribution de chacune d'elles et traiter spécialement la pente 0.

Comme application, on montre que la torsion située sur une courbe non elliptique de A, est finie, bornée uniformément après translation, du moins si X n'est pas contenue dans une surface abélienne de A.

1. Extensions vectorielles universelles.

Les résultats énoncés dans ce numéro sont extraits de [7].

1.1. Soient S un schéma, A un S-schéma abélien, A' le S-schéma abélien dual de A. Il existe une extension universelle de S-schémas en groupes:

$$0 \to V(A) \to E(A) \to A \to 0, \tag{1}$$

de A par un S-groupe vectoriel $V(A)$. Vu son caractère universel, l'extension (1) est unique à isomorphisme unique près. De plus, on montre que sa formation commute aux changements de base $S' \to S$ et que, si on suppose S affine pour simplifier, on a un isomorphisme canonique de suites exactes de $H^0(S, \mathcal{O}_S)$-modules localement libres:

$$\begin{array}{ccccccccc} 0 \to & V(A) & \to & \operatorname{Lie} E(A) & \to & \operatorname{Lie}(A) & \to 0 \\ & \downarrow\wr & & \downarrow\wr & & \downarrow\wr & \\ 0 \to & H^0(A', \Omega^1_{A'}) & \to & H^1_{DR}(A') & \to & H^1(A', \mathcal{O}_{A'}) & \to 0, \end{array} \tag{2}$$

où la première ligne est la suite exacte d'algèbres de Lie déduite de (1) et la seconde est la suite exacte courte qui décrit la cohomologie de De Rham relative de A' sur S, en degré 1.

1.2. Soit R un anneau de valuation discrète complet, de corps des fractions K de caractéristique zéro et de corps résiduel k algébriquement clos de caractéristique $p > 0$. On suppose que pR est l'idéal maximal de R (i.e., R est abolument non ramifié) et pour $n \geq 0$, on pose $R_n = R/p^{n+1}R$.

Pour tout $n \geq 0$, soit A_n un R_n-schéma abélien tel que

$$A_{n+1} \times_{Rn+1} R_n = A_n$$

et soit $\mathcal{A} = \varinjlim A_n$ le R-schéma abélien formel correspondant. Si A'_n est le R_n-schéma abélien dual de A_n, nous dirons que $\mathcal{A}' = \varinjlim A'_n$ est le schéma abélien formel dual de $\mathcal{A}$. Par exemple, dans le cas algébrisable, $\mathcal{A}$ et $\mathcal{A}'$ proviennent par complétion le long de leurs fibres fermées d'un R-schéma abélien A et de son dual A'.

Pour tout $n \geq 0$, l'analogue de (1) fournit une suite exacte de R_n-schémas en groupes lisses:

(3) $$0 \to V(A_n) \to E(A_n) \to A_n \to 0,$$

qui fait de $E(A_n)$ une extension vectorielle universelle de A_n. Passant à la limite inductive sur n, on obtient une suite exacte de R-schémas formels en groupes lisses:

(4) $$0 \to \mathcal{V}(\mathcal{A}) \to \mathcal{E}(\mathcal{A}) \to \mathcal{A} \to 0$$

que nous appellerons l'extension vectorielle formelle universelle de $\mathcal{A}$. Donc, par construction, si l'on part d'un R-schéma abélien A, de complétion formelle $\mathcal{A}$, (4) se déduit de (1) par complétion formelle le long des fibres fermées et redonne (3) par réduction mod. p^{n+1}.

1.3. Il nous faut maintenant rappeller le caractère cristallin de (4). Notons d'abord que R n'étant pas ramifié, l'idéal maximal pR_n de R_n a une structure naturelle d'idéal à puissances divisées. Par passage à la limite sur n, on en déduit un isomorphisme canonique de R-modules:

(5) $$H^1_{DR}(\mathcal{A}') \xrightarrow{\sim} H^1_{cris}(A'_o),$$

où le second membre désigne la cohomologie cristalline de A_o à coefficients dans $W(k) = R$, anneau des vecteurs de Witt de longueur infinie à coefficients dans k. L'analogue de (2) fournit un isomorphisme:

$$\text{Lie}\,\mathcal{E}(\mathcal{A}) \xrightarrow{\sim} H^1_{DR}(\mathcal{A}'), \tag{6}$$

d'où l'on déduit un isomorphisme canonique:

$$\tau_{\mathcal{A}} : \text{Lie}\,\mathcal{E}(\mathcal{A}) \xrightarrow{\sim} H^1_{DR}(\mathcal{A}') \xrightarrow{\sim} H^1_{cris}(A'_o). \tag{7}$$

Soit maintenant $\mathcal{B} = \varinjlim_n B_n$ un second R-schéma abélien formel, $\mathcal{B}'$ son dual et $\mathcal{E}(\mathcal{B})$ son extension vectorielle formelle universelle.

Soit u_o: $A_o \longrightarrow B_o$ un k-morphisme de schémas abéliens. Alors, d'après ([7] chap. II. §1) pour tout $n \geq 0$, il existe un morphisme de R_n-schémas en groupes:

$$E_n(u_o)\colon E(A_n) \longrightarrow E(B_n),$$

caractérisé par les propriétés suivantes:

i) $E_o(u_o)$ est le k-morphisme canonique provenant du caractère universel des extensions $E(A_o)$ et $E(B_o)$.

ii) $E_{n+1}(u_o)$ se réduit en $E_n(u_o)$ modulo p^{n+1}, et donc les $E_n(u_o)$ définissent un morphisme $\mathcal{E}(u_o)$: $\mathcal{E}(\mathcal{A}) \longrightarrow \mathcal{E}(\mathcal{B})$ de R-schémas formels en groupes.

iii) Le diagramme suivant est commutatif:

$$\begin{array}{ccc} \text{Lie}\,\mathcal{E}(\mathcal{A}) & \xrightarrow{\text{lie}\,\mathcal{E}(u_o)} & \text{Lie}\,\mathcal{E}(\mathcal{B}) \\ \tau_{\mathcal{A}} \downarrow \wr & & \wr \downarrow \tau_{\mathcal{B}} \\ H^1_{cris}(A'_o) & \xrightarrow[H^1_{cris}(u'_o)]{} & H^1_{cris}(B'_o), \end{array} \tag{8}$$

où u'_o: $B'_o \longrightarrow A'_o$ est le morphisme dual de u_o et $H^1_{cris}(u'_o)$ est l'application déduite de u'_o par passage à la cohomologie cristalline.

1.4. Notons σ_k: Spec$(k) \longrightarrow$ Spec(k), l'isomorphisme défini par le Frobenius de k qui envoie $a \in k$ sur a^p. On note σ_R l'automorphisme de Spec(R) = Spec$(W(k))$ qui relève σ_k et σ_K la restriction de σ_R à Spec(K). Pour tout R-schéma (ou schéma formel) X on note $X^{(1)}$ le R-schéma (ou schéma formel) déduit de X par le changement de base σ_R. On a donc un isomorphisme absolu $X^{(1)} \rightarrow X$ noté encore σ_R.

Considérons sur le k-schéma abélien A_o le morphisme de Frobenius absolu F: $A_o \longrightarrow A_o$ et le morphisme de Frobenius relatif $F_{/k}$: $A_o \longrightarrow A_o^{(1)}$, de sorte que l'on a $F = \sigma_k \circ F_{/k}$. Par le changement de base σ_R, on déduit de $\mathcal{A}$ (resp. $\mathcal{E}(\mathcal{A})$) des R-schémas formels en groupes $\mathcal{A}^{(1)}$ et $\mathcal{E}(\mathcal{A})^{(1)}$ et par transport de structure par l'automorphisme σ_R, on a $\mathcal{E}(\mathcal{A})^{(1)} = \mathcal{E}(\mathcal{A}^{(1)})$.

D'après 1.3, au k-morphisme $F_{/k}$: $A_o \longrightarrow A_o^{(1)}$, correspond un morphisme de R-schémas formels en groupes $\mathcal{E}(F_{/k})$: $\mathcal{E}(\mathcal{A}) \longrightarrow \mathcal{E}(\mathcal{A})^{(1)}$, que nous noterons $\phi_{/R}$. En composant $\phi_{/R}$ avec l'isomorphisme absolu σ_R: $\mathcal{E}(\mathcal{A})^{(1)} \longrightarrow \mathcal{E}(\mathcal{A})$, on obtient un endomorphisme absolu, σ_R-linéaire ϕ: $\mathcal{E}(\mathcal{A}) \longrightarrow \mathcal{E}(\mathcal{A})$. Au morphisme de Frobenius relatif F/k: $A_o \longrightarrow A_o^{(1)}$, correspond par dualité le morphisme du Verschiebung V/k: $A_o'^{(1)} \longrightarrow A_o'$. Il en résulte que si on identifie Lie $\mathcal{E}(\mathcal{A})$ à $H^1_{cris}(A_o')$ par τ, Lie ϕ s'identifie au Verschiebung V, σ_R^{-1}-linéaire, de la cohomologie cristalline.

1.5. Dans la suite, le schéma abélien formel $\mathcal{A}$ étant fixé, on pose $\mathcal{E} = \mathcal{E}(\mathcal{A}), \mathcal{V} = \mathcal{V}(\mathcal{A})$ et on note π: $\mathcal{E} \longrightarrow \mathcal{A}$ l'application de passage au quotient qui intervient dans (4). On a donc une suite exacte de R-schémas formels en groupes:

$$0 \to \mathcal{V} \to \mathcal{E} \xrightarrow{\pi} \mathcal{A} \to 0, \tag{9}$$

qui par réduction modulo k, fournit une suite exacte de k-groupes algébriques

$$0 \to V_o \to E_o \to A_o \to 0 \tag{10}$$

qui fait de E_o l'extension vectorielle universelle de A_o. Aux suites exactes (9) et (10) correspondent des suites exactes d'algèbres de Lie:

$$0 \to L \to M \to A \to 0 \tag{9'}$$

$$0 \to L/pL \to M/pM \to A/pA \to 0. \tag{10'}$$

1.6. Il résulte de 1.3 que $\mathcal{E}$ ne dépend pas du choix du R-schéma abélien formel $\mathcal{A}$ qui relève A_o. Plus précisément, si $\overline{\mathcal{A}}$ est un autre R-schéma abélien formel qui relève A_o, on a une suite exacte de R-schéma formels en groupes, analogue de (9):

$$(11) \qquad 0 \to \overline{\mathcal{V}} \to \mathcal{E} \xrightarrow{\pi} \overline{\mathcal{A}} \to 0,$$

qui s'identifie à (10) par réduction modulo p. Le groupe formel vectoriel $\overline{\mathcal{V}}$ s'identifie par l'application logarithme à un sous-R-module $\overline{L}$ de $M = \operatorname{Lie}\mathcal{E}$.

De plus, la correspondance:

$$\overline{L} \mapsto \overline{\mathcal{V}} \mapsto \overline{\mathcal{A}} = \mathcal{E}/\overline{\mathcal{V}}$$

établit une bijection entre:

– d'une part les R-modules $\overline{L}$, facteur direct de M, tels que $\overline{L}/p\overline{L} = L/pL$ (comme sous-k-vectoriels de M/pM)

– d'autre part, les R-schémas abéliens formels $\overline{\mathcal{A}}$ qui relèvent A_o.

2. Géométrie formelle et géométrie rigide.

2.1. Soit $\mathcal{X}$ un R-schéma formel de type fini. Rappelons [8] qu'il lui correspond canoniquement un K-espace analytique rigide X_K, la "fibre générique" de $\mathcal{X}$. Décrivons cette correspondance dans le cas oú $\mathcal{X}$ est affine. Soit A l'algèbre de définition de $\mathcal{X}$. Donc A est une R-algèbre, complète pour la topologie p-adique, topologiquement de type fini. Alors la fibre générique $\mathcal{X}_K$ de $\mathcal{X}$ est l'espace rigide analytique, affinoïde, associé à l'algèbre de Tate $A_K = A \otimes_R K$.

2.2. Soit $\mathcal{Y}$ un sous-schéma formel fermé de $\mathcal{X}$, alors $\mathcal{Y}_K$ est un sous-espace analytique rigide fermé de $\mathcal{X}_K$ · Avec les notations précédentes si $\mathcal{Y}$ est défini par l'idéal I de A, $\mathcal{Y}_K$ est défini par l'idéal $I_K = I \otimes_R K$ de A_K.

Réciproquement, si $\mathcal{Y}_K$ est un sous-espace anlytique rigide fermé de $\mathcal{X}_K$, il existe un plus petit sous-schéma formel fermé $\mathcal{Y}$ de $\mathcal{X}$, de fibre générique $\mathcal{Y}_K$. De plus, $\mathcal{Y}$ est R-plat: si $\mathcal{Y}_K$ est défini par l'idéal J de A_K, $\mathcal{Y}$ est défini par le plus grand idéal I de A tel que $I \otimes_A K = J$.

La correspondance $\mathcal{Y} \mapsto \mathcal{Y}_K$ établit une bijection entre sous-schémas formels fermés R-plats de $\mathcal{X}$ et sous-espaces analytiques rigides fermés de $\mathcal{X}_K$. Par analogie avec l'opération d'adhérence schématique dans le cas algébrique, nous dirons que $\mathcal{Y}$ est l'adhérence formelle de $\mathcal{Y}_K$ dans $\mathcal{X}$.

2.3. Reprenons les notations du numéro précédent. A la suite exacte (9) de R-schémas formels en groupes lisses correspond, par passage à la fibre générique, une suite exacte de K-groupes rigides analytiques lisses:

$$0 \to \mathcal{V}_K \to \mathcal{E}_K \to \mathcal{A}_K \to 0. \tag{12}$$

La suite exacte d'algèbres de Lie, associée à (12) se déduit de (9′) par tensorisation avec K. En particulier, nous noterons H l'algèbre de Lie de $\mathcal{E}_K$ identifiée à $H^1_{cris}(A'_o) \otimes_R K$ par $\tau_{\mathcal{A}} \otimes_R K$, de sorte que l'on a la suite exacte:

$$0 \to L \otimes_R K \to H \to N \otimes_R K \to 0. \tag{12'}$$

L'endomorphisme absolu, σ_R-linéaire ϕ de $\mathcal{E}$ induit un endomorphisme σ_K-linéaire de l'espace rigide $\mathcal{E}_K$ et noté encore ϕ; Lie(ϕ) induit le Verschiebung V sur H.

Remarque 2.3.1. Supposons que le R-schéma abélien formel $\mathcal{A}$ soit la complétion formelle d'un R-schéma abélien A et soit A_K la K-variété abélienne fibre générique de A. Par restriction de (1) à la fibre générique, on obtient un K-groupe algébrique E_K extension vectorielle universelle de A_K. Si l'on applique à E_K le foncteur "Gaga" analytique rigide, on obtient un groupe analytique rigide $(E_K)_{rig}$ (qui n'est pas réunion d'un nombre fini d'ouverts affinoïdes si $A \neq 0$) et $\mathcal{E}_K$ est un sous-groupe ouvert rigide de $(E_K)_{rig}$ (plus précisément, on déduit $(E_K)_{rig}$ de $\mathcal{E}_K$ à partir de (12), en étendant $\mathcal{V}_K$ en un K-vectoriel). Mais en général, l'endomorphisme ϕ de $\mathcal{E}_K$ ne s'étend pas à $(E_K)_{rig}$ et, a fortiori, ne provient pas d'un endomorphisme algébrique, σ_K-linéaire de E_K. C'est pour cette raison que nous allons devoir travailler avec les structures formelles et rigides analytiques.

3. Eléments indéfiniment p-divisibles.

3.1. Considérons un groupe commutatif G, un nombre premier p et le système projectif:

$$\cdots \to G \xrightarrow{p} G \xrightarrow{p} G, \tag{13}$$

dans lequel l'application $G \xrightarrow{p} G$ est la multiplication par p. Soit $\varprojlim_{\text{"}p\text{"}} G$ la limite projective de ce système. Un élément de $\varprojlim_{\text{"}p\text{"}} G$ s'identifie à une suite (x_n), $n \geq 0$, d'éléments de G tels que $px_{n+1} = x_n$. Soit ${}_{p^n}G$ le noyau de la multiplication par p^n dans G et soit $\varprojlim\limits_{\text{“}p,n\text{”}} {}_{p^n}G$ la limite projective du système projectif:

$$\cdots \to {}_{p^2}G \xrightarrow{p} {}_pG \to 0. \tag{14}$$

Par passage à la limite sur les suites exactes:

$$0 \to {}_{p^n}G \to G \xrightarrow{p^n} p^nG \to 0,$$

on obtient la suite exacte:

$$0 \to \varprojlim_{\text{“}p,n\text{”}} {}_{p^n}G \to \varprojlim_{\text{“}p\text{”}} G \xrightarrow{\alpha_o} \bigcap_n p^nG$$

où $\alpha_o((x_n)) = x_o$. Si de plus le système projectif (14) satisfait à la condition de Mittag-Leffler, α_o est surjectif.

3.2. Soit R comme dans 1.2 et soit $\mathcal{X}$ un R-schéma formel en groupes commutatif et lisse, à fibre fermée $\mathcal{X}_o = \mathcal{X}_R \otimes k$ connexe

Lemme 3.2.1. *Les applications:*

$$\varprojlim_{\text{“}p\text{”}} \mathcal{X}(R) \to \varprojlim_{\text{“}p\text{”}} \mathcal{X}(R_n) \to \varprojlim_{\text{“}p\text{”}} \mathcal{X}(k)$$

déduites des applications canoniques $\mathcal{X}(R) \to \mathcal{X}(R_n) \to \mathcal{X}(k)$, *sont bijectives.*

En effet, on a $\mathcal{X}(R) = \varprojlim\limits_{n} \mathcal{X}(R_n)$ et donc, par associativité des limites projectives, $\varprojlim\limits_{\text{“}p\text{”}} \mathcal{X}(R) = \varprojlim\limits_{n} \varprojlim\limits_{\text{“}p\text{”}} \mathcal{X}(R_n)$. Par suite, il suffit de montrer que pour tout entier $n \geq 0$, l'application $\varprojlim\limits_{\text{“}p\text{”}} \mathcal{X}(R_{n+1}) \to \varprojlim\limits_{\text{“}p\text{”}} \mathcal{X}(R_n)$ est bijective. Or comme $\mathcal{X}$ est lisse, l'application $\mathcal{X}(R_{n+1}) \to \mathcal{X}(R_n)$ est surjective; par ailleurs son noyau est annulé par p, étant

Notons que la multiplication par p^n dans $\mathcal{X}$ est étale sur la fibre générique $\mathcal{X}_K$, en particulier ${}_{p^n}\mathcal{X}(R)$ est fini et donc on a la suite exacte:

$$(15) \qquad 0 \to \varprojlim_{\text{“}p,n\text{”}} {}_{p^n}\mathcal{X}(R) \to \varprojlim_{\text{“}p\text{”}} \mathcal{X}(R) \overset{a_o}{\to} \bigcap_n p^n\mathcal{X}(R) \to 0.$$

Nous dirons que les éléments de $\bigcap_n p^n\mathcal{X}(R)$ sont les *éléments indéfiniment p-divisibles de* $\mathcal{X}(R)$.

Notons $T_{p^\infty}\mathcal{X}(R)$ (resp. $T_{p^\infty}\mathcal{X}(k)$) la torsion p-primaire p-divisible de $\mathcal{X}(R)$ (resp. $\mathcal{X}(k)$).

Lemme 3.2.2. *L'application* $\bigcap_n p^n\mathcal{X}(R) \to \mathcal{X}(k)$ *a un noyau fini (resp. est injective) si et seulement si l'application* $T_{p^\infty}\mathcal{X}(R) \to T_{p^\infty}\mathcal{X}(k)$ *est surjective (resp. bijective).*

En effet, le diagramme commutatif à lignes exactes:

$$\begin{array}{ccccccc}
0 \to & \varprojlim_{\text{“}p,n\text{”}} {}_{p^n}\mathcal{X}(R) \to & \varprojlim_{\text{“}p\text{”}} \mathcal{X}(R) \to & \bigcap_n p^n\mathcal{X}(R) \to 0 \\
& \downarrow & \downarrow\wr & \downarrow \\
0 \to & \varprojlim_{\text{“}p,n\text{”}} {}_{p^n}\mathcal{X}(k) \to & \varprojlim_{\text{“}p\text{”}} \mathcal{X}(k) \to & \mathcal{X}(k)
\end{array}$$

montre que le noyau de l'application $\bigcap_n p^n\mathcal{X}(R) \to \mathcal{X}(k)$ est isomorphe au conoyau de la flèche $\varprojlim_{\text{“}p,n\text{”}} {}_{p^n}\mathcal{X}(R) \to \varprojlim_{\text{“}p,n\text{”}} {}_{p^n}\mathcal{X}(k)$. Cette dernière application se déduit de l'application $T_{p^\infty}\mathcal{X}(R) \to T_{p^\infty}\mathcal{X}(k)$ par application du foncteur $\mathrm{Hom}(\mathbb{Q}_p/\mathbb{Z}_p, .)$, d'où le lemme.

Exemples 3.2.3. Supposons que $\mathcal{X}$ soit un schéma abélien formel. Alors $\mathcal{X}_o(k)$ est p-divisible, donc $\bigcap_n p^n\mathcal{X}(k) = \mathcal{X}(k)$.

Notons $\mathcal{X}_{p^\infty}$ le R-groupe p-divisible construit sur les noyaux des multiplications par p^n dans $\mathcal{X}$. On a une suite exacte de R-groupes p-divisibles:

$$(16) \qquad 0 \to (\mathcal{X}_{p^\infty})_{inf} \to \mathcal{X}_{p^\infty} \to (\mathcal{X}_{p^\infty})_{et} \to 0$$

où $(\mathcal{X}_{p^\infty})_{inf}$ est connexe et $(\mathcal{X}_{p^\infty})_{et}$ est étale isomorphe à $(\mathbb{Q}_p/\mathbb{Z}_p)^h$, où h est le p-rang de $\mathcal{X}_o$. La suite exacte (16) est scindée en réduction modulo p et il résulte de 3.2.2. que l'application $\bigcap_n p^n\mathcal{X}(R) \to \mathcal{X}(k)$ est bijective si et seulement si (16) est scindée. Ce sera le cas si $h = 0$, ou bien si $\mathcal{X}_o$ est ordinaire et si $\mathcal{X}$ est le relèvement canonique de Serre-Tate.

3.3. Reprenons les notations de 1.5.

Lemme 3.3.1. *L'application* $\varprojlim_{\text{“}p\text{”}} \mathcal{E}(R) \to \varprojlim_{\text{“}p\text{”}} \mathcal{A}(R)$, *déduite de* π *(9), est bijective; en particulier, tout élément indéfiniment p-divisible de* $\mathcal{A}(R)$ *se relève en un élément indéfiniment p-divisible de* $\mathcal{E}(R)$.

D'après 3.2.1, il suffit de voir que l'application $\varprojlim_{\text{“}p\text{”}} \mathcal{E}(k) \to \varprojlim_{\text{“}p\text{”}} \mathcal{A}(k)$ est bijective, ce qui résulte de la suite exacte:

$$(12') \qquad 0 \to \mathcal{V}(k) \to \mathcal{E}(k) \to \mathcal{A}(k) \to 0$$

déduite de (10) et du fait que $\mathcal{V}(k)$ est annulé par p.

Remarques 3.3.2. i) Pour la démonstration du théorème annoncé dans l'introduction, les seuls points indéfiniment p-divisibles de $\mathcal{A}(R)$ que nous aurons à considérer sont ceux de torsion.

ii) On peut montrer que *la torsion p-primaire* indéfiniment p-divisible de $\mathcal{E}(R)$ est nulle, de sorte que l'application α_o: $\varprojlim_{\text{“}p\text{”}} \mathcal{E}(R) \to \mathcal{E}(R)$ est injective.

3.4. Considérons l'endomorphisme absolu ϕ de $\mathcal{E}$ et rappelons que $\phi = \sigma_R \circ (\phi_{/R})$ (1.4). Notons que ϕ induit un endomorphisme du groupe $\mathcal{E}(R)$. En effet, soit $x \in \mathcal{E}(R)$, vu comme une section de $\mathcal{E}$ au-dessus de $Spec(R)$. Composant x avec $\phi_{/R}$ on obtient une section de $\mathcal{E}^{(1)}$ au-dessus de $Spec(R)$, puis on transporte cette section en une section de $\mathcal{E}$ au-dessus de $Spec(R)$ par le changement de base $(\sigma_R)^{-1}$. On a donc $\phi \circ x = \phi(x) \circ \sigma_R$.

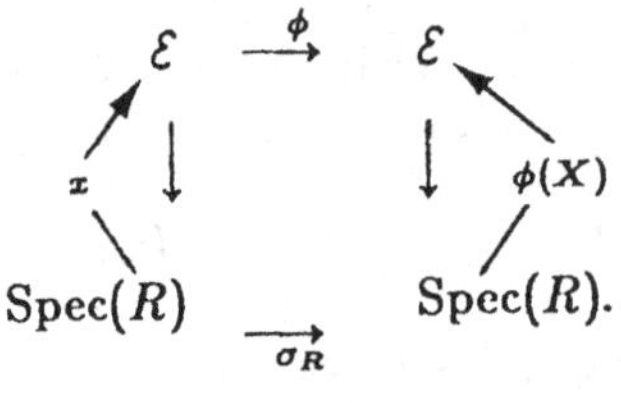

Proposition 3.4.1. *Les éléments indéfiniment p-divisibles de* $\mathcal{E}(R)$ *sont fixés par* ϕ.

Il suffit de montrer que ϕ induit l'identité sur $\varprojlim_{"p"} \mathcal{E}(R)$ et d'après 3.2.1. il suffit de voir que ϕ_o induit l'identité sur $\varprojlim_{p} \mathcal{E}_o(k)$, où ϕ_o est l'endomorphisme absolu de $\mathcal{E} \otimes_R k = \mathcal{E}_o$ induit par ϕ. Par construction, ϕ_o provient, par fonctorialité de l'extension vectorielle universelle, du Frobenius absolu F de A_o, de sorte que l'on a un diagramme commutatif:

$$\begin{array}{ccc} \mathcal{E}_o(k) & \xrightarrow{\phi_o} & \mathcal{E}_o(k) \\ {\scriptstyle \pi_o}\downarrow & & \downarrow{\scriptstyle \pi_o} \\ A_o(k) & \xrightarrow{F} & A_o(k). \end{array}$$

Or F induit l'identité sur $A_o(k)$ et on a déjà remarqué que l'application $\varprojlim_{"p"} E_o(k) \to \varprojlim_{"p"} A_o(k)$ était bijective.

3.5. Soient R comme dans 1.2 et A un R-schéma abélien de fibre générique A_k.

Théorème 3.5.1. *Soit X_K un sous-schéma fermé intègre de A_K contenant un ensemble Zariski dense $\sum$ de points indéfiniment p-divisibles de $A(K) = A(R)$. On suppose que X n'est stable par aucune translation non nulle de A_K. Alors il existe une partie $\sum_1$ de $\sum$, Zariski dense dans X_K, dont l'image dans $A(k)$ est un seul point.*

Corollaire 3.5.2. *Soit X_K un sous-schéma fermé intègre de A_K contenant un ensemble Zariski dense $\sum$ de points indéfiniment p-divisibles de $A(k)$. On suppose de plus que l'une des conditions suivantes est réalisée:*

i) *$\sum$ est formé de points de torsion.*
ii) *L'application $\bigcap_n p^n A(R) \to A(K)$ a un noyau fini (cf. 3.2.2).*

Alors X_K est le translaté, par un point indéfiniment p-divisible, d'une sous-variété abélienne de A_K.

La démonstration du théorème fera l'objet des trois numéros suivants. Montrons comment le théorème entraîne le corollaire.

Soit B_K le sous-schéma en groupes de A_K formé des translations de A_K qui laissent stable X_K. Soit B le sous-schéma en groupes de A, adhérence schématique de B_K dans A et soit C le R-schéma abélien quotient A/B. Notons $\sum''$ (resp. X''_K) l'image de $\sum$ (resp. X_K) dans C_K. Alors $\sum''$ est Zariski dense dans X''_K et X''_K n'est stable par aucune translation non nulle de C_K. Il reste à voir que X''_K est réduit à un point rationnel. D'après le théorème 3.5.1, il existe une partie $\sum''_1$ de $\sum''$, Zariski dense dans X''_K dont l'image dans $C(k)$ est réduite à un point. Les hypothèses i) et ii) se conservent quand on remplace $\sum$ et Λ par $\sum''$ et C (dans le cas ii) cela résulte par exemple de 3.2.2) et assurent que les fibres de l'application $\sum'' \to C(k)$ sont finies, donc X''_K est fini et par suite réduit à un point rationnel.

4. Etude rigide analytique.

Dans les numéros 4 et 5 on reprend les hypothèses et notations de 1.5 et 1.6.

4.1. Soit Λ une partie de $\mathcal{E}(R) = \mathcal{E}_K(K)$.

Lemme 4.1.1. *Il existe un plus petit sous-schéma formel fermé $\mathcal{Y}$ de $\mathcal{E}$ tel que $\Lambda \subset \mathcal{E}(R)$. De plus $\mathcal{Y}$ est R-plat et sa fibre générique $\mathcal{Y}_K$ (2.1) est le plus petit sous-espace analytique rigide fermé de $\mathcal{E}_K$ contenant Λ.*

En effet, soit $\mathcal{U}_i$ un recouvrement ouvert affine du R-schéma formel $\mathcal{E}$ et soit A_i l'anneau de $\mathcal{U}_i$. Considérons la famille $I_j, j \in J$, des faisceaux cohérents d'idéaux de $\mathcal{O}_\mathcal{X}$ qui s'annulent sur Λ. Cette famille est non vide, car elle contient 0; elle est filtrante croissante; enfin comme les anneaux A_i sont noethériens, elle contient un élément maximal I. Alors I définit le plus petit sous-schéma formel fermé $\mathcal{Y}$ de $\mathcal{E}$ tel que $\Lambda \subset Y(R)$. Si $\overline{\mathcal{Y}}$ désigne le sous-schéma formel fermé de $\mathcal{Y}$, déduit de Y en divisant $\mathcal{O}_\mathcal{Y}$ par son faisceau d'idéaux de R-torsion, alors $\overline{\mathcal{Y}}(R)$ contient aussi Λ, donc $\mathcal{Y} = \overline{\mathcal{Y}}$ et par suite $\mathcal{Y}$ est R-plat. La dernière assertion résulte de 2.2.

Remarques 4.1.2.

i) La définition de $\mathcal{Y}$ et de $\mathcal{Y}_K$ est globale; en particulier la formation de $\mathcal{Y}$ ne commute pas à la restriction à un sous-schéma formel ouvert de $\mathcal{E}$.

ii) L'espace rigide $\mathcal{Y}_K$ est réduit (c'est-à-dire ses anneaux locaux sont réduits) comme il résulte de l'existence d'un espace rigide réduit, sous-jacent à un espace rigide donné (conséquence de l'excellence des anneaux de la géométrie rigide [1] (3.3)).

Nous dirons que $\mathcal{Y}$ (resp. $\mathcal{Y}_K$) est *l'adhérence formelle* (resp. *rigide*) de Λ dans $\mathcal{E}$ (resp. $\mathcal{E}_K$).

4.2. Soit Y l'adhérence rigide dans $\mathcal{E}_K$ d'une famille Λ de points indéfiniment p-divisibles de $\mathcal{E}(R)$. Les points de Λ sont fixés par ϕ (3.4.1), donc Y est stable par ϕ. On suppose de plus que Y est un espace rigide irréductible. Soit d sa dimension. Comme Y est réduit et K de caractéristique 0, Y est alors lisse sur K en dehors d'un fermé rigide $Z \neq Y$; l'ouvert complémentaire U est le plus grand ouvert de Y où le faisceau des formes différentielles de degré 1, Ω_Y, est localement libre de rang d.

Rappelons (2.3) que l'on a noté H l'algèbre de Lie de $\mathcal{E}_K$ et que ϕ induit sur H le Verschiebung V issu de la cohomologie cristalline de A'_o. Soit

$$H = \bigoplus_{\lambda \in \mathbb{Q}} H_\lambda \tag{17}$$

la décomposition de H suivant les pentes de V. Par translation, on étend la décomposition (17) en une décomposition du fibré tangent T à $\mathcal{E}_K$:

$$T = \bigoplus_{\lambda \in \mathbb{Q}} T_\lambda. \tag{18}$$

Soit $y \in \Lambda$. L'espace tangent de Zariski à Y en y, $T_Y(y)$ est un sous-K-vectoriel de la fibre $T(y)$ de T en y. Comme y est un point fixe sous ϕ, l'application tangente à ϕ en y, $T_\phi(y)$ induit une application σ_K-linéaire $T(y) \to T(y)$ qui coïncide avec V: $H \longrightarrow H$ quand on identifie H à $T(y)$ par translation. Comme Y est stable par ϕ, $T_Y(y)$ est un sous-K-espace vectoriel de $T(y)$ stable par $T_\phi(y)$ et par suite:

$$T_Y(y) = \bigoplus_{\lambda \in \mathbb{Q}} T_Y(y) \bigcap T_\lambda(y). \tag{19}$$

Proposition 4.2.1. *Sur l'ouvert de lissité U de Y, le fibré tangent T_U á U est localement libre et l'on a la décomposition:*

$$T_U = \bigoplus_{\lambda \in \mathbb{Q}} T_U \bigcap T_\lambda \mid U \tag{20}$$

où les faisceaux $T_U \cap T_\lambda \mid U$ sont localement libres de rang constant.

Nous allons travailler avec le faisceau des formes différentielles sur Y, Ω_Y, plutôt qu'avec le fibré tangent T_U qui n'existe que sur U. Soit Ω le faisceau des formes différentielles sur $\mathcal{E}_K$. A la décomposition (18) correspond une décomposition duale

$$\Omega = \bigoplus_{\lambda \in \mathbb{Q}} \Omega_\lambda .$$

On a une surjection canonique u: $\Omega \mid Y \longrightarrow \Omega_Y$ dont la fibre en un point $y \in Y(K)$ est la surjection $\Omega(y) \to \Omega_Y(y)$, duale de l'injection sur les espaces tangents $T_Y(y) \hookrightarrow T(y)$. Pour tout $\lambda \in \mathbb{Q}$, considérons le faisceau cohérent N_λ sur Y, conoyau de l'application composée:

$$\bigoplus_{\mu \neq \lambda} \Omega_\mu \mid Y \hookrightarrow \Omega \mid Y \xrightarrow{u} \Omega_Y .$$

Soit $N = \bigoplus_{\lambda \in \mathbb{Q}} N_\lambda$ et soit v: $\Omega_Y \longrightarrow N$, l'application somme des surjections canoniques $\Omega_Y \to N_\lambda$. Alors v est encore surjective et sa fibre en $y \in Y(K)$ est bijective si et seulement si la relation (19) est vérifiée.

Soit Y' le sous-espace rigide analytique fermé de Y formé des points où le rang de N est $\geq d$ (cf. [6], lemme 3.6). Il résulte des considérations précédentes et de (19) que Y' contient Λ, et par suite $Y' = Y$. En particulier $N \mid U$ est partout de rang $\geq d$. Comme U est réduit et Ω_U localement libre de rang d, la flèche surjective $v \mid U$: $\Omega_U \longrightarrow N \mid U$ est nécessairement un isomorphisme. Par suite $N_\lambda \mid U$ est localement libre, de rang constant sur U, puisque Y est irréductible. Par dualité entre T_U et Ω_U, on en déduit 4.2.1.

5. Etude analytique "molle."

5.1. Reprenons les décompositions (17) et (18):

$$H = \bigoplus_{\lambda \in \mathbb{Q}} H_\lambda, \ T = \bigoplus_{\lambda \in \mathbb{Q}} T_\lambda .$$

La décomposition de T ne provient pas d'une décomposition en produit du groupe rigide $\mathcal{E}_K$, mais elle en provient localement au sens analytique "mou".

De façon précise appelons réseau P de H tout sous-R-module libre P de H, tel que $P \otimes_R K = H$. Soit P un réseau de H contenu dans $\mathrm{Lie}(\mathcal{E})$ tel que:

i) P est stable par v

ii) l'application exponentielle est définie sur P et réalise un isomorphisme du groupe additif rigide sus-jacent à P sur un sous-groupe ouvert rigide $\mathcal{U}$ de $\mathcal{E}_K$.

Par exemple, on peut prendre $P = p^r\mathrm{Lie}(\mathcal{E})$, avec $r = 1$ si $p \neq 2$ et $r = 2$ si $p = 2$. Le sous-groupe ouvert rigide $\mathcal{U}$ de $\mathcal{E}_K$ est nécessairement stable par ϕ et l'exponentielle transporte le Verschiebung V sur M en $\phi \mid \mathcal{U}$. Désormais on identifie (P, V) avec $(\mathcal{U}, \phi)$.

Pour $\lambda \in \mathbb{Q}$, posons $P_\lambda = H_\lambda \cap P$. Quitte alors à remplacer P par la somme directe de P_λ, on peut supposer que P vérifie de plus:

iii) $P = \bigoplus_{\lambda \in \mathbb{Q}} P_\lambda$, avec $P_\lambda = H_\lambda \cap P$.

Alors, en restriction à $\mathcal{U}$, la décomposition $T = \bigoplus T_\lambda$ du fibré tangent provient de la décomposition de P en produit des groupes additifs P_λ. Dans la suite de ce numéro, nous allons étudier les points rationnels de certains sous-espaces rigides de P. Rappelons en particulier que si Z est un espace analytique rigide lisse, partout de dimension n, $Z(K)$ a une structure naturelle de variété K-analytique, partout de dimension n au sens de ([3], §5).

5.2. Soit Y comme dans 4.2. Alors $P \cap Y$ est un sous-espace rigide ouvert de Y et un sous-espace rigide fermé de P. Les $P_\lambda \cap Y$ sont des sous-espaces rigides fermés de $P \cap Y$; $P \cap Y$ et les $P_\lambda \cap Y$ sont stables par V.

Proposition 5.2.1. *Supposons que l'origine 0 soit un point lisse de Y. Alors pour $\lambda \in \mathbb{Q}$, $P_\lambda \cap Y$ est lisse en 0, d'espace tangent $P_\lambda \cap T_Y(0)$.*

Soit $P'_\lambda = \bigoplus_{\mu \neq \lambda} P_\mu$ et notons $\pi_\lambda \colon \longrightarrow PP'_\lambda$ la projection de noyau P_λ. Alors $P_\lambda \cap Y$ est la fibre au-dessus de 0 de $\pi_\lambda \mid P \cap Y$. D'après 4.2.1 le rang de l'application linéaire tangente à $\pi_\lambda \mid P \cap Y$ est constant sur l'ouvert de lissité de $P \cap Y$. Comme K est de caractéristique 0, $\pi_\lambda \mid P \cap Y$ est une subimmersion au voisinage de 0 ([3], 5.10.6) et en particulier sa fibre au-dessus de 0, $P_\lambda \cap Y$ est lisse en 0. L'assertion sur l'espace tangent est par ailleurs immédiate et indépendante des questions de lissité.

5.3.

Proposition 5.3.1 *Soit P un R-module libre de rang fini, muni d'un endomorphisme $V\sigma_R^{-1}$-linéaire, ayant une seule pente $\lambda > 0$. Soit Z un sous-espace rigide analytique de P, ontenant 0, lisse en 0 de dimension n et stable par V. Alors il existe un entier $s \geq 0$ tel que $p^s P \cap Z(K)$ soit un sous-R-module de P, de rang n.*

Soit E l'espace tangent à Z en 0. Alors E est un sous-K-vectoriel de $P \otimes RK$, stable par V, de dimension n. La catégorie des K-vectoriels de dimension finie, munis d'un Verschiebung V étant semi-simple ([4], chap. IV), on peut trouver un supplémentaire E' de E dans $P \otimes_R K$, stable par V. La pente λ de V est égale à a/b, où a et b sont des entiers > 0, premiers entre eux. Comme le corps résiduel k de R est algébriquement clos, il résulte de [4], chap. IV, que l'on peut trouver une base $e_1, \ldots, e_n$ de E et une base $f_1, \ldots, f_m$ de E', telles que $V^b(e_i) = p^a e_i$, $i = 1, \ldots, n$ et $V^b(f_j) = p^a f_j$, $j = 1, \ldots, m$. Quitte à multiplier les e_i et les f_j par une puissance de p, on peut supposer que e_i et f_j sont dans P. Soient Q le R-module de base $e_1, \ldots, e_n$ et S le R-module de base $f_1, \ldots, f_m$. Quitte enfin à remplacer P par un sous-réseau, on peut supposer que $P = Q \oplus S$. Notons $\pi : P \to Q$ la projection sur le facteur Q. Vu le choix de $Q, \pi \mid Z$ est étale en 0. Quitte alors à remplacer P par $p^s P$, Q par $p^s Q$, pour $s \geq 0$ convenable, on peut supposer que $\pi \mid Z$ est un isomorphisme $Z \xrightarrow{\sim} Q$. Soient $x_1, \ldots, x_n$ les coordonnées de Q relatives à la base $e_1, \ldots, e_n$ et $y_1, \ldots, y_m$ les coordonnées de S relatives à la base $f_1, \ldots, f_m$. Alors Z admet une représentation paramétrique de la forme:

$$y_i = h_i(x_1, \ldots, x_n), \quad i = 1, \ldots, m$$

où $h_i = \sum_\alpha a_{i,\alpha} \underline{x}^\alpha$ est une série à coefficients dans R, tendant vers 0 quand $|\alpha| \to \infty$ et sans terme de degré ≤ 1.

Le fait que Z soit stable par V^b se traduit par les identités:

$$h_i(p^a \underline{x}^{\sigma^{-b}}) = p^a \big(h_i(\underline{x})\big)^{\sigma^{-b}}, \quad i = 1, \ldots, m.$$

Soit

$$\sum_\alpha p^{|\alpha| a} a_{i,\alpha} (\underline{x}^\alpha)^{\sigma^{-b}} = \sum_\alpha p^a a_{i,\alpha}^{\sigma^{-b}} (\underline{x}^\alpha)^{\sigma^{-b}},$$

soit encore:
$$\sum_{\alpha} p^{|\alpha|a} a_{i,\alpha}^{\sigma^b} \underline{x}^{\alpha} = \sum_{\alpha} p^a a_{i,\alpha} \underline{x}^{\alpha}.$$

D'où $p^{|\alpha|a} a_{i,\alpha}^{\sigma^b} = p^a a_{i,\alpha} \forall \alpha$. Comme $a > 0$ et $a_{i,\alpha} = 0$ pour $|\alpha| \leq 1$, on a $a_{i,\alpha} = 0$, donc $h_i = 0$. C'est dire, qu'avec les réductions déjà faites, $Z(K)$ coïncide avec Q, d'où la proposition.

Remarques 5.3.2.

i) Les calculs précédents restent valables lorsque $\lambda = 0$, mais ils permettent simplement de conclure que les séries $\sum_\alpha a_{i,\alpha} \underline{x}^\alpha$ sont à coefficients dans $\mathbb{Q}_p$ (autrement dit que, Z se descend sur $\mathbb{Q}_p$, au voisinage de zéro), ce qui n'implique pas une linéarisation de Z.

ii) La proposition 5.3.1 ne s'étend pas au cas où V agissant sur P admet plusieurs pentes > 0. Par exemple, supposons que P ait une base (e.f.g) telle que:

$$V(e) = pe \qquad \text{(pente 1)},$$

$$\left.\begin{array}{l} V(f) = g \\ V(g) = pf \end{array}\right\} \qquad \text{(pente 1/2)}$$

et soient x, y, z les coordonnées associées à la base e, f, g. Alors la sous-variété de P d'équation $x = yz$ est lisse, passe par l'origine, est stable par V et néanmoins n'est pas linéarisable près de 0.

5.4. Revenons au sous-espace rigide Y de $\mathcal{E}_K$. Lorsque 0 est un point lisse de Y, les espaces rigides $P_\lambda \cap Y$ sont lisses en 0 (5.2.1) et pour $\lambda > 0$, on peut leur appliquer 5.3.1. Si maintenant $y \in \Lambda$, est un point lisse quelconque de Y, on se ramène au cas précédent par translation par $-y$. En résumé des N° 4 et 5, on déduit le corollaire suivant:

Corollaire 5.4.1. *Soit Y l'adhérence rigide dans $\mathcal{E}_K$ d'un ensemble Λ de points $\mathcal{E}(R)$ indéfiniment p-divisibles. On suppose Y irréductible et on note U l'ouvert de lissité de Y. Soit λ une pente > 0 de V (17) et supposons que dans la décomposition (20) du fibré tangent T_U, $T_U \cap T_\lambda \mid U$ soit localement libre de rang n_λ. Alors, pour tout $y \in \Lambda \cap U(K)$, il existe un sous-R-module $H_{y,\lambda}$ de $H_\lambda \cap Lie(\mathcal{E})$, de rang n_λ, stable par V, contenu* *dans le domaine de définition de l'application exponentielle relative à* $\mathcal{E}_K$ *et tel que* $y + exp(H_{y,k})$ *soit contenu dans* Y *avec pour espace tangent en* $y(T_U \bigcap T_\lambda \mid U)(y)$.

6. Fin de la démonstration du théorème 3.5.1.

6.1. Commençons par un lemme "algébrique":

Lemme 6.1.1. *Soit* X *un sous-schéma fermé intègre d'une variété abélienne* A, X *et* A *étant définis sur un corps algébriquement clos* c. *On suppose que* X *contient une famille Zariski dense* X_λ, $\lambda \in \Lambda$, *de sous-schémas de la forme* $B_\lambda + a_\lambda$, *où* B_λ *est une sous-variété abélienne non nulle de* A *et* $a_\lambda \in A(c)$. *Alors* X *est stable par les translations d'une sous-variété abélienne non nulle* B *de* A.

(La démonstration qui suit a été modifiée sur épreuve). Pour chaque $\lambda \in \Lambda$, choisissons un point $a_\lambda \in X_\lambda(c)$. On a $X_\lambda = a_\lambda + B_\lambda$. Quitte à agrandir X_λ, ce qui ne change pas la propriété de densité, on peut supposer que B_λ est une sous-variété abélienne maximale de A telle que $a_\lambda + B_\lambda$ soit contenu dans X. Alors, d'après Bogomolov (Math. USSR Izvestija vol. 17 (1981) N° 1, Theorem 1 p. 58) les variétés abéliennes B_λ possibles sont en nombre fini. Comme X est irréductible, on se ramène, quitte à restreindre Λ, au cas où B_λ est une variété abélienne B indépendante de Λ. Mais alors B laisse stable les X_λ par translation, doc laisse stable X.

6.2. Nous en venons à la démonstration de 3.5.1.

Soient donc A un R-schéma abélien et $\mathcal{A}$ le schéma abélien formel complétion de A le long de sa fibre fermée A_o.

Soit $\sum$ un ensemble de points de $A(R) = \mathcal{A}(R)$. Comme A est propre sur R, l'adhérence formelle (4.1) $\mathcal{X}$ de $\sum$ dans A est algébrisable, c'est-à-dire est la complétion formelle d'un sous-schéma fermé X de A ([5] 5.1.8). Nécessairement X est l'adhérence de Zariski de $\sum$ dans A. Alors la fibre générique X_K de X est l'adhérence de Zariski de $\sum$ dans la fibre générique A_K de A, tandis que l'espace rigide $\mathcal{X}_K$, fibre générique de $\mathcal{X}$ est l'adhérence rigide de $\sum$ dans $\mathcal{A}_K$. Ainsi le foncteur "Gaga rigide analytique" transforme l'adhérence de Zariski X_K de $\sum$ dans A_K en l'adhérence rigide $\mathcal{X}_K$ de $\sum$

dans $\mathcal{A}_K$.

Puisque nous allons travailler avec les espaces rigides, nous changeons de notations et dèsignons par X_{alg} l'adhérence de Zariski de $\sum$ dans A_K et par X l'espace rigide associé, adhérence rigide de $\sum$ dans $\mathcal{A}_K$.

On suppose désormais que $\sum$ est formé de points indéfiniment p-divisibles, que X_{alg} est intègre et n'est stable par aucune tranlation non nulle de A_K.

Considérons la suite exacte de R-schémas formels en groupes introduite dans (1.5):

$$(9) \qquad 0 \to \mathcal{V} \to \mathcal{E} \xrightarrow{\pi} \mathcal{A} \to 0$$

et la suite exacte des K-groupes rigides associée:

$$(12) \qquad 0 \to \mathcal{V}_K \to \mathcal{E}_K \xrightarrow{\pi} \mathcal{A}_K \to 0.$$

D'après 3.3.1, on peut trouver un ensemble Λ de points indéfiniment p-divisibles de $\mathcal{E}(R)$ tel que $\pi(\Lambda) = \sum$. Soit Y l'adhérence rigide de Λ dans $\mathcal{E}_K$. Comme $\mathcal{E}_K$ est réunion d'un nombre fini d'ouverts affinoïdes, Y n'a qu'un nombre fini de composantes irréductibles $Y_i, i \in I$. Alors Y_i est l'adhérence rigide de $\Lambda_i = \Lambda \cap Y_i(R)$. Soit $\sum_i = \pi(\Lambda_i) \subset X(K) = X_{alg}(K)$. Comme X_{alg} est irréductible, il existe $i_o \in I$ tel que $\sum_{i_o}$ soit Zariski dense dans $X_{alg}(K)$. Pour démontrer 3.5.1, on peut, quitte à remplacer $\sum$ par $\sum_{i_o}$ et Y par Y_{i_o}, supposer Y irréductible.

Soit Λ' la partie de Λ contenue dans l'ouvert de lissité U de Y. Alors Y est aussi l'adhérence rigide de Λ' dans $\mathcal{E}_K$, donc X est aussi l'adhérence rigide de $\sum' = \pi(\Lambda')$ dans $\mathcal{A}_K$ et par suite X_{alg} est l'adhérence de Zariski de $\sum'$. Remplaçant $\sum$ par $\sum'$, on peut supposer $\Lambda \subset U(K)$.

6.3.

Lemme 6.3.1. *Soit*

$$(12') \qquad 0 \to L \otimes_R K \to H \to N \otimes_R K \to 0,$$

la suite exacte d'algèbres de Lie associée à (12) et soit H' un sous-K-vectoriel de $L \otimes_R K$, stable par le Verschiebung V de H. Alors H' est contenu dans la composante H_o de pente 0 de H.

Reprenons la suite exacte

$$(9') \qquad 0 \to L \to M \to N \to 0$$

associée à (9). Par tensorisation avec K on obtient (12′) et par réduction modulo p, on obtient

$$(10') \qquad 0 \to L/pL \to M/pM \to N/pN \to 0$$

qui d'après (1.1) s'identifie à la suite exacte:

$$0 \to H^o(A'_o, \Omega^1_{A'_o}) \to H^1_{DR}(A'_o) \to H^1(A'_o, O_{A'_o}) \to 0.$$

Les R-modules M et $H^1_{DR}(A'_o)$ sont munis des opérateurs F et V tels que $FV = VF = p$, et l'identification précédente est compatible avec ces opérateurs. Comme F annule $\Omega^1_{A'_o}$, F annule L/pL.

Notons M' le R-module $H' \cap M \subset H = M \otimes_R K$. Par hypothèse H' est stable par V, donc par F, et par suite M' est stable par F. D'autre part, M' est contenu dans L puisque H' est contenu dans $L \otimes_R K = \mathrm{Lie}\, \mathcal{V}_K$. Par construction M/M' est sans R-torsion, et il en est de même de L/M'. L'application $M'/pM' \to L/pL$ est alors injective, donc F est nul sur M'/pM'. C'est dire que l'opérateur F sur M' est divisible par p, donc V opérant sur M' est inversible et $H' = M' \otimes_R K$ est contenu dans la composante de pente 0.

Lemme 6.3.2. *Dans la décomposition (20) du fibré tangent à U:*

$$T_U = \oplus_{\lambda \in \mathbb{Q}} T_U \bigcap (T_\lambda \mid U),$$

ne figure que la pente $\lambda = 0$.

En effet, supposons qu'il existe $\lambda > 0$, tel que $T_U \cap T_\lambda \mid U$, soit non nul, donc localemment libre de rang $n_\lambda > 0$. D'après 5.4.1, pour tout $y \in \Lambda$, il existe un sous-R-module libre $H_{y,\lambda}$ de H_λ, de rang n_λ, stable par V, tel que $y + exp(H_{y,\lambda})$ soit contenu dans Y. Comme $\lambda > 0$, il résulte de 6.3.1 que $H_{y,\lambda}$ n'est pas contenu dans $\mathrm{Lie}\, \mathcal{V}_K$, donc $N_{y,\lambda} = (\mathrm{Lie}\, \pi)(H_{y,\lambda})$ est un sous-R-module non nul de $\mathrm{Lie}\, A_K$. Il en résulte que $\exp(N_{y,\lambda})$ engendre une sous-variété abélienne non nulle B_y dans A_K. Comme $\pi(y) + exp(N_{y,\lambda})$ est contenu dans $X(K) = X_{alg}(K), \pi(y) + B_y$ est contenu dans X_{alg}. L'ensemble $\sum$ des $\pi(y)$, $y \in \Lambda$ est Zariski dense dans X_{alg} et donc (6.1.1), X_{alg} est stable par les translations d'une sous-variété abélienne non nulle B de A_K, en contradiction avec les hypothèses faites sur X_{alg}.

6.4. Reprenons la suite exacte d'algèbres de Lie:

$$0 \to L \to M \to N \to 0$$

associée à (9) et soit $M = M_o \oplus M^o$ la décomposition de M en sa composante M_o où V est bijectif et sa composante M^o où V est topologiquement nilpotent. Alors $M_o \otimes_R K$ est la composante H_o de pente 0 de H, tandis qu'en réduction modulo p, M_o/pM_o est annulé par F, donc contenu dans L/pL. On peut donc trouver un R-module $\overline{L}$, facteur direct de M, tel que $\overline{L} \supset M_o$ et $\overline{L}/p\overline{L} = L/pL$ dans M/pM. D'aprés 1.6, $\overline{L}$ correspond à un certain relèvement formel $\overline{\mathcal{A}}$ de A_o, quotient de $\mathcal{E}$, de sorte que l'on a une suite exacte de R-schémas formels en groupes:

$$0 \to \overline{L} \to \mathcal{E} \xrightarrow{\overline{\pi}} \overline{\mathcal{A}} \to 0$$

(les relèvements $\overline{\mathcal{A}}$ que l'on obtient de cette façon sont ceux pour lesquels la suite exacte (16) de groupes p-divisibles est scindée).

Considérons alors le morphisme rigide $\mathcal{E}_K \to \overline{\mathcal{A}}_K$, fibre générique de $\overline{\pi}$. Par construction de $\overline{L}$, on a $\overline{L} \otimes_R K \supset H_o$ et il résulte alors de 6.3.2. que la restriction de $\overline{\pi}$ à l'ouvert de lissité U de Y a une application linéaire tangente nulle. Comme K est de caractéristique zéro, que Y est irréductible, réduit et contient des points rationnels, l'image de Y par $\overline{\pi}$ est un point $\overline{z}$ de $\overline{\mathcal{A}}(K) = \overline{\mathcal{A}}(R)$. Soit z_o la spécialisation de $\overline{z}$ dans $A_o(k) = A(k)$. Le diagramme commutatif:

$$\begin{array}{ccccc} & & & \overline{\mathcal{A}}(R) & \\ & & \overset{\overline{\pi}}{\nearrow} & & \searrow \\ Y(R) \to & \mathcal{E}(R) & & & A_o(k) \\ & & \underset{\pi}{\searrow} & & \nearrow \\ & & & \mathcal{A}(R) & \end{array}$$

montre $\pi Y(R)$ est une partie de $\mathcal{A}(R)$ qui se spécialise en $z_o \in A_o(k)$, en particulier $\sum = \pi(\Lambda)$ se spécialise en z_o, d'où le théorème 3.5.1.

7. Complément sur la démonstration de Bogomolov.

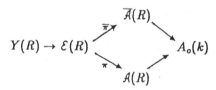

7.1. Soient c un corps algébriquement clos de caractéristique 0, A une variété abélienne définie sur c, X un sous-schéma fermé intègre de A qui n'est stable par aucune translation non nulle de A. Notons Z le fermé de X adhérence des sous-schémas de X de la forme $a + B$, où $a \in A(c)$ et B est une sous-variété abélienne non nulle de A. D'après 6.1.1, on a $Z \neq X$. Soit $U = X - Z$.

L'énoncé suivant, précise le résultat de Bogomolov et nous allons le démontrer par les mêmes techniques:

Proposition 7.1.1. *Soit p un nombre premier. Alors il existe un entier $N \geq 0$ tel que, $\forall_a \in A(c)$, la torsion p-primaire contenue dans $U + a$ est finie, de cardinal $\leq N$.*

Soit l un entier > 1 et notons l_A la multiplication par l dans A. Pour tout couple (a, b) de points de $A(c)$, nous allons construire par récurrence un certain sous-schéma fermé $U_{a,b}^{(n)}$ de U. Prenons $U_{a,b}^{(o)} = U$ et supposons avoir défini $U_{a,b}^{(m)}$, sous-schéma fermé de U. Soit $\overline{U_{a,b}^{(m)}}$ l'adhérence de $U_{a,b}^{(m)}$ dans X et définissons $U_{a,b}^{(m+1)}$ par la formule:

$$U_{a,b}^{(m+1)} + a = (U_{a,b}^{(m)} + a) \bigcap l_A^{-1}(\overline{U_{a,b}^{(m)}} + b).$$

Les $U_{a,b}^{(m)}$ forment une suite décroissante de sous-schémas fermés de U, donc stationnaire, de valeur $U_{a,b}$ pour $m \gg 0$. On a $l_A(U_{a,b}^{(m+1)} + a) \subset \overline{U_{a,b}^{(m)}} + b$ et donc $l_A(U_{a,b} + a) \subset \overline{U}_{a,b} + b$ et clairement, $U_{a,b}$ est le plus grand sous-schéma fermé de U possédant cette propriété.

Lemme 7.1.2. *Le schéma $U_{a,b}$ est fini et dépend de façon constructible de (a, b). En particulier, le cardinal de $U_{a,b}$ est borné par un entier N indépendant de (a, b).*

Le caractère constructible de $U_{a,b}$ est immédiat sur la définition, une fois noté que si (a, b) varie dans un fermé intègre Z de $A \times A$, alors pour m fixé, la formation de $\overline{U_{a,b}^{(m)}}$ commute au passage aux fibres au-dessus d'un ouvert non vide de Z. Il reste à voir que $U_{a,b}$ est fini. Soit $x \in A(c)$ tel que $(l - 1)x = a - b$ et soit $Y = U_{a,b} + a + x$. Alors $l_A(Y) \subset \overline{Y}$. Si Y' est la réunion des composantes irréductibles de Y de dimension maximum, on a aussi $l_A(Y') \subset \overline{Y'}$. Il existe alors une composante irréductible Y_i' de Y' et un entier $n > 0$, tels que $l_A^n(\overline{Y_i'}) = \overline{Y}_i'$. Donc $\overline{Y}_i'$ est le translaté d'une sous-variété abélienne de A (cf. [2] p. 703). Vu la définition de U, $\overline{Y}'_i$ est de dimension 0, donc Y et $U_{a,b}$ sont de dimension 0.

Prouvons maintenant 7.1.1. Soit L un sous-corps de c, de type fini sur $\mathbb{Q}$, tels que A, X et Z soient définis sur L. Notons $\overline{L}$ la clôture algébrique de L dans c et G le groupe de Galois de $\overline{L}$ sur L. D'après ([2], cor 2 p. 703), il existe un nombre entier $l > 1$, et $g \in G$, tel que si $x \in A(\overline{L})$ est d'ordre une puissance de p, alors ${}^g x = lx$. On étend g en un L-automorphisme noté encore g de c.

Si alors $a \in A(c)$ et si $u + a$ est un point d'ordre une puissance de p de $U + a$, on a ${}^g(u+a) = l(u+a)$ et d'autre part ${}^g(u+a) = {}^g u + {}^g a \in U + {}^g a$, donc $u \in U_{a,{}^g a}$ et on applique 7.1.2.

8. Points de torsion d'une sous-variété.

8.1. Dans ce numéro, nous démontrons le résultat énoncé dans l'introduction. Soient donc A une variété abélienne définie sur un corps algébriquement clos c de caractéristique 0, T le sous-groupe de torsion de $A(c)$ et X un sous-schéma fermé intègre de A tel que $T \cap X$ soit Zariski dense dans X. Il nous faut montrer que X est le translaté d'une sous-variété abélienne de A.

Quitte à passer au quotient par un sous-groupe de A, on peut supposer que X n'est stable par aucune translation non nulle de A, ce qui nous permet d'introduire le fermé strict Z de X (cf. 7.1) et l'ouvert complémentaire $U = X - Z$. Nous devons alors montrer que X est un point de A.

Par des réductions élémentaires détaillées dans ([9], 10.2 et 10.3), on se ramène au cas où A est un R-schéma abélien défini sur un anneau R du type décrit dans 1.2, de caractéristique résiduelle $p > 2$ et où X et U sont des sous-schémas de la fibre générique A_K de A. Soient $\overline{K}$ une clôture algébrique de K et $G = Gal(\overline{K}/K)$.

Procédant comme dans ([9], 5), on choisit une décomposition

$$T = T' \oplus T''$$

du G-module $T(\overline{K})$, où $T'(\overline{K}) = T'(K)$ est la torsion rationnelle, p-divisible (somme directe de la torsion d'ordre premier à p et de la torsion p-primaire p-divisible, rationnellle sur K) et où T'' est un supplémentaire (non canonique) de T' dans T. Rappelons ([9], 5.2.2) que l'action de G sur T'' est forte dans le sens suivant: $\forall M$ entier > 0, $\exists n$ entier > 0, tel que si

$x \in T'''(K)$ est d'ordre p^r avec $r > n$, alors l'orbite de x sous G a un cardinal $> M$.

Par ailleurs, d'après 7.1.1, la torsion p-primaire contenue dans $U + a$, $a \in A(\overline{K})$ est finie, bornée par un entier N indépendant de a.

8.2. Soit x un point de torsion de $U(\overline{K})$. On a $x = x' + x''$, avec $x' \in T'(K)$ et $x'' \in T'''(\overline{K})$. Alors x'' est un point de torsion p-primaire de $U - x'$. Comme U et x' sont définis sur K, toute l'orbite de x'' sous G est contenue dans $U - x'$. Puisque cette orbite a au plus N éléments, il existe n, indépendant de x'' tel que l'ordre de x'' divise p^n.

Soit X' l'image de x dans A par la multiplication par p^n. Alors, tout point de $T \cap U(\overline{K})$ a une image dans X' qui est contenue dans $T' \cap X'(K)$. Comme $T \cap U$ était Zariski dense dans U, $T' \cap X'$ est Zariski dense dans X'. Mais T' est constitué de points de $A(K) = A(R)$ indéfiniment p-divisibles. Il résulte alors de 3.5.2 que X' est le translaté d'une sous-variété abélienne de A; par suite il en est de même de X et donc X est réduit à un point.

9. Application aux courbes.

Soit A une variété abélienne définie sur un corps c algébriquement clos de caractéristique 0 et soit X une courbe fermée, intègre, non elliptique de A.

Proposition 9.1. *On suppose que la surface S de A engendrée par les différences $(x - x')$ entre points de X n'est pas une surface abélienne. Alors il existe un entier N tel que, $\forall_a \in A(c)$, la courbe $X + a$ contienne au plus N points de torsion.*

Soit $a \in A(c)$ tel que $X + a$ contienne un point de torsion t. Alors la courbe $X + a - t$ contient 0 et contient le même nombre de points de torsion que $X + a$. Pour établir 9.1, on peut donc se limiter à considérer les courbes translatées de X, de la forme $X - x$, avec $x \in X(c)$. Ces courbes forment une famille algébrique de courbes tracées sur S. Comme S n'est pas une variété abélienne, seulement deux cas sont possibles (cf. 6.1.1):

1er cas: il existe une courbe elliptique E de A qui laisse S stable par translation, auquel cas S est une surface elliptique de fibre E, de base une courbe non elliptique C de $B = A/E$. Comme C ne contient qu'un nombre fini de points de torsion, les points de torsion de S sont situés sur un nombre fini de courbes de genre 1, E_i, $i \in I$. Pour x variable dans $X(c)$, les intersections $E_i \cap (X - x)$ ont un cardinal borné, d'où le résultat.

2ème cas: Les translations qui laissent S stable sont en nombre fini. Alors S ne contient qu'un nombre fini de courbes de genre 1, E_i, $i \in I$ (6.1.1) et si Z est la réunion des E_i, l'ouvert $U = S - Z$ ne contient qu'un nombre fini de points de torsion et on conclut comme dans le ler cas.

Remarque 9.2.. La proposition 9.1 laisse ouvert le cas où S est une surface abélienne. Rappelons toutefois que pour presque tout nombre premier p, la torsion première à p, contenue dans $X + a$, est bornée indépendamment de a comme il résulte de ([9], 6.3.1).

Bibliographie

[1] R. Berger, R. Kiehl, E. Kunz, J.J. Nastold. Differentialrechnung in der analytischen Geometrie. Springer Lecture Notes in Math. 38, 1969.

[2] F. Bogomolov. Sur l'algébricité des représentations l-adiques. C. R. Acad. Sc. t. 290, p. 701–704, 1980.

[3] N. Bourbaki. Variétés différentielles et analytiques; fascicule de résultats. Hermann, 1967.

[4] M. Demazure. Lectures on p-divisible groups. Springer Lecture Notes in Math. 302, 1972.

[5] J. Dieudonné et A. Grothendieck. Eléments de Géométrie Algébrique II; Etude cohomologique des faisceaux cohérents. Pub. IHES N° 11, 1961.

[6] A. Grothendieck. Les schémas de Hilbert. Sém. Bourbaki N° 221, 1960/61.

[7] B. Mazur et W. Messing. Universal extensions and one dimensional crystalline cohomology. Springer Lecture Notes in Math. 370, 1974.

[8] M. Raynaud. Géométrie analytique rigide. Table ronde d'analyse non archimédienne. Bull. Soc. Math. France Mém. N° 39–40, 1974.

[9] M. Raynaud. Courbes sur une variété abélienne et points de torsion. A paraître dans Invent. Math.

Received June 8, 1982
Equipe associée au CNRS n° 653

Professor Michel Raynaud
Départment de Mathématiques
Université de Paris-Sud
Bat. 425, Centre d'Orsay
91405 Orsay, France

Euler and the Jacobians of Elliptic Curves

André Weil

To Igor Rostislavovich Shafarevich

Euler, most universal among mathematicians of his time and perhaps of all times, may be regarded as the grandfather of modern algebraic geometry, since he created the theory of elliptic integrals, whence proceeded Jacobi's theory of elliptic functions which in turn gave birth to Riemann's theory of algebraic functions and abelian integrals. As there is a direct line of succession from him to Chebyshev, he can also be counted as the grandfather of Russian mathematics, though none of his immediate pupils in Russia attained to any distinction in mathematics. Of his devotion to the welfare of science in Russia there can be no doubt, nor of the depth of his attachment to that country[1] and to its language, of which he acquired a good command during the early years of his stay in Saint Petersburg. Thus it seems not inappropriate that he should appear in a volume dedicated to Igor Shafarevich.

At the same time, since the present contribution is meant to be no more than a token of affectionate regard and admiration for a colleague whom I feel privileged to count as a friend, it may not be amiss for me to inject into it a personal note. Writing in 1947 a book on abelian varieties, I had been vexed at my inability to show that the jacobian of a curve defined over some field is itself defined over the same field. In my attempts to fill up this gap I discovered that a paper by Hermite ([2]) contained most of the formulas needed for treating one fairly typical special case, that of a curve given by an equation $u^2 = F(x)$, where F is of degree 4 (cf. [3]).

Hermite had obtained his formula in the course of an investigation of the invariants and covariants of binary biquadratic forms. Let

[1] Illusions about it, however, he had none. "Un pays où, quand on parle, on est pendu" is how he once called it (speaking to the Queen Dowager of Prussia after his arrival in Berlin in 1741), according to an anecdote preserved by Condorcet; cf. L. du Pasquier, *Léonard Euler et ses amis*, Paris, 1927, pp. 40–41.

$$(1) \qquad F(X,Y) = aX^4 + 4bX^3Y + 6cX^2Y^2 + 4b'XY^3 + a'Y^4$$

be such a form; its invariants, in the sense of classical invariant theory, are given by

$$(2) \qquad i = aa' - 4bb' + 3c^2, \quad j = aca' + 2bcb' - ab'^2 - a'b^2 - c^3.$$

Put $F(x) = F(x,1)$ and call C the curve of genus 1 defined by $u^2 = F(x)$. One can show that the jacobian $J(C)$ of C is the curve Γ given by

$$(3) \qquad \eta^2 = 4\xi^3 - i\xi - j$$

and that there is a rational mapping of C into Γ, given by

$$(4) \qquad (x,u) \to \left(\xi = \frac{g(x,1)}{u^2}, \quad \eta = \frac{h(x,1)}{u^3}\right)$$

where $g(X,Y)$, $h(X,Y)$ are the two covariants of the form $F(X,Y)$, as given by the classical theory. To this I added in 1954 (*loc. cit.*) that Γ and C may be regarded as the two components of a commutative algebraic group $G(C)$, with $\Gamma = J(C)$ as a subgroup of index 2, and with the point at infinity on Γ as the neutral element. The group-law being written additively, it induces the usual group-law on the "Weierstrass cubic" Γ, and also mappings $(M,P) \to M+P$ of $C \times \Gamma$ into C and $(M,N) \to M+N$ of $C \times C$ into Γ. The mapping (4) is then none else than $M \to 2M$. One defines a function φ on $G(C)$ by putting $\varphi(M) = x$ for $M = (x,u)$ on C, and $\varphi(P) = \xi$ for $P = (\xi,\eta)$ on Γ; we have then $-M = (x,-u)$, $-P = (\xi,-\eta)$, and in particular $\varphi(-M) = \varphi(M)$, $\varphi(-P) = \varphi(P)$.

Put $\omega = \frac{dx}{u}$; this is the differential of the first kind on C. For any given point P on Γ, the mapping $M \to M+P$ is an automorphism of C which leaves ω invariant; conversely all such automorphisms are of that form. Similarly, the automorphisms of C which change ω into $-\omega$ are the mappings $M \to P - M$. Thus, in an obvious sense, the points P of Γ may be said to parametrize these two families of automorphisms of C. As I found in 1954, explicit formulas for all these mappings can easily be derived from Hermite's results in [2].

Had I but known it, I could have found all those formulas in a paper of 1765 by Euler ([1(b)]), where I was surprised to discover them not long ago. This is to be explained here.

Notations being as above, let k be the ground field; for us it may be any field of characteristic other than 2 or 3; for Euler it was mostly the field of real numbers, occasionally the field of complex numbers, and just possibly (though Euler never says so) the field of rational numbers if he had the possible application to diophantine equations at the back of his mind. Put $K = k(P)$, where P is a point of Γ. For $M = (x, u)$ on C, put $x = \varphi(M)$ as before, and $u = \psi(M)$; also put $y = \varphi(P - M)$, $v = \psi(P - M)$. For a given P these are rational functions on C, the former being defined over k and the latter over K. As x, y are functions of order 2 on C, there is between them a relation $\Phi(x, y) = 0$, where Φ is a polynomial of degree 2 in each one of the variables x, y, with coefficients in K. Changing M into $P - M$ exchanges (x, u) with (y, v); therefore Φ is symmetric in x and y. Changing M, P into $-M, -P$ does not change x and y; therefore changing P into $-P$ can at most multiply Φ with a scalar; by choosing the scalar factor in Φ properly, we may thus assume that all its coefficients are polynomials in $\xi = \varphi(P)$.

Clearly the two roots of the equation $\Phi(x, Y) = 0$ are

$$y = \varphi(P - M) = \varphi(M - P), \quad y' = \varphi(-P - M) = \varphi(P + M).$$

As we have $u^2 = F(x)$, $v^2 = F(y)$, and as the mappings $M \to M \pm P$ leave $\omega = \frac{dx}{u}$ invariant, while the mappings $M \to \pm P - M$ change it into $-\omega$, one may thus, with Euler, interpret the above results by saying that $\Phi(x, y) = 0$ is an integral of either one of the differential equations

$$\frac{dx}{\sqrt{F(x)}} = \pm \frac{dy}{\sqrt{F(y)}}$$

or, more accurately perhaps, of the equation

$$\left(\frac{dy}{dx}\right)^2 = \frac{F(y)}{F(x)};$$

it is the "general integral", since it contains the arbitrary constant $\xi = \varphi(P)$. This is how Euler viewed the matter, from the time (at the end of 1751) when Fagnano's *Produzioni Matematiche* drew his attention upon the equation

$$\frac{dx}{\sqrt{1 - x^4}} = \frac{dy}{\sqrt{1 - y^4}} \tag{5}$$

and others of the same kind.

Having in fact discovered that the differential equation (5) has the general integral

$$x^2 + y^2 + c^2x^2y^2 = c^2 + 2xy\sqrt{1-c^4},$$

where c is an arbitrary constant ([1(a)], p. 63; cf. his letter to Goldbach of 30 May 1752, where this is introduced by the words "*Neulich bin ich auf curieuse Integrationen verfallen*"), and having derived from this the addition and multiplication formulas for the "lemniscatic" integral $\int \frac{dx}{\sqrt{1-x^4}}$, Euler proceeded to extend this to the most general elliptic case by seeking the general integral of the differential equation

$$(6) \qquad \pm\frac{dx}{\sqrt{F(x)}} = \pm\frac{dy}{\sqrt{F(y)}}$$

in the form of what he called a "canonical equation" $\Phi(x,y) = 0$, where Φ is of degree 2 in x and symmetric in x and y ([1(a)], pp. 71-72). In effect, as we have seen, this amounts to the determination of all the automorphisms of the curve $u^2 = F(x)$ which either leave invariant the differential $\frac{dx}{u}$ or change its sign. Having thus formulated the problem in [1(a)], he solved it completely, after a lapse of twelve years, with results which will now be described ([1(b)], pp. 321-326, 331, 341-344).

In Euler's notation, put

$$F(x) = A + 2Bx + Cx^2 + 2Dx^3 + Ex^4.$$

Write Φ with indeterminate coefficients as follows:

$$\Phi(x,y) = \alpha + 2\beta(x+y) + \gamma(x^2+y^2) + 2\delta xy + 2\epsilon xy(x+y) + \varsigma x^2y^2.$$

Differentiating the "canonical equation" $\Phi(x,y) = 0$, one must find a relation equivalent to (6). Now $\Phi(x,y)$ can be written as

$$\Phi(x,y) = P(x) + 2Q(x)y + R(x)y^2 = P(y) + 2Q(y)x + R(y)x^2,$$

where P, Q, R are the polynomials

$$P(X) = \alpha + 2\beta X + \gamma X^2, \quad Q(X) = \beta + \delta X + \epsilon X^2,$$
$$R(X) = \gamma + 2\epsilon X + \varsigma X^2.$$

Thus, for $\Phi(x,y)=0$, we have

$$(7)\qquad y=\frac{1}{R(x)}\Big[-Q(x)+\rho\sqrt{G(x)}\Big],\quad x=\frac{1}{R(y)}\Big[-Q(y)+\sigma\sqrt{G(y)}\Big]$$

with $G=Q^2-PR$, $\rho=\pm1$, $\sigma=\pm1$ ([1(b)], p. 321). At the same time, we have

$$\frac{1}{2}\frac{\partial\Phi}{\partial x}=Q(y)+R(y)x=\sigma\sqrt{G(y)},\quad \frac{1}{2}\frac{\partial\Phi}{\partial y}=Q(x)+R(x)y=\rho\sqrt{G(x)}$$

and therefore, differentiating the equation $\Phi(x,y)=0$,

$$\frac{dy}{dx}=-\frac{\dfrac{\partial\Phi}{\partial x}}{\dfrac{\partial\Phi}{\partial y}}=-\frac{\sigma}{\rho}\sqrt{\frac{G(y)}{G(x)}}.$$

This shows that $\Phi=0$ will be an integral of (6) if and only if G differs from F only by a scalar factor, i.e., if one can write $G=pF$, where p is such a factor. This gives a set of five equations in p and the coefficients α,β, etc. of Φ, beginning with $\beta^2-\alpha\gamma=Ap$ ([1(b)], p. 323; cf. [1(a)], p. 72).

Here appears Euler's extraordinary virtuosity as an algebraist; perhaps he had acquired some of that skill by handling diophantine equations. I need not go here into the details of his dazzling manipulation of the equations of that problem; suffice it to state the results ([1(b)],p. 326). Put

$$\theta=\tfrac{1}{p}[(\delta-\gamma)^2-\alpha\varsigma]$$

(I am writing θ where Euler writes M; otherwise all notations are Euler's). Then, up to an arbitrary scalar factor, the general solution of the system of equations in question is given by

$$(8)\begin{cases}\alpha=4(A\theta-B^2),\ \beta=2B(\theta-C)+4AD,\ \gamma=4AE-(\theta-C)^2,\\ \delta=\theta^2-C^2+4(AE+BD),\ \epsilon=2D(\theta-C)+4BE,\ \varsigma=4(E\theta-D^2),\\ p=4\theta(\theta-C)^2+16\theta(BD-AE)+16(AD^2-BCD+B^2E)\end{cases}$$

([1(b)], p. 326). In order to make the transition between Euler's notations and Hermite's, put

$$A=a',\quad B=2b',\quad C=6c,\quad D=2b,\quad E=a;$$

also, put $\theta = 4(\xi + c)$. Then the last equation in (8) gives

$$p = 64(4\xi^3 - i\xi - j),$$

where, as above, i, j are the invariants of the form $F(X,Y)$.

Put again

$$u = F(x)^{1/2}, \quad v = F(y)^{1/2}, \quad \eta = (4\xi^3 - i\xi - j)^{1/2}.$$

In order for (7) to determine a transformation $(x,u) \to (y,v)$ which changes $\frac{dx}{u}$ into $\frac{dy}{v}$, we must take $\sigma = -\rho$. Replacing then η by $\rho\eta$, we can rewrite (7) as follows:

$$(9) \qquad y = \frac{1}{R(x)}[-Q(x) + 8\eta u], \quad v = -\frac{1}{8\eta}[Q(y) + R(y)x].$$

This is therefore the general expression for the automorphisms of C which leave ω invariant; as expected, they are parametrized by the points $P = (\xi, \eta)$ of Γ.

Not only does Euler's paper thus contain the equation for Γ and explicit formulas for the mappings $(P, M) \to M + P$, but he also shows in effect that these mappings make up a simply transitive group on C, and he gives at the same time formulas for the mapping $(M, N) \to N - M$ of $C \times C$ into Γ. In order to explain this, let us make explicit the occurrence of θ in the coefficients of Φ by writing $\Phi(x, y, \theta)$ instead of $\Phi(x, y)$; this is a polynomial of degree 2 in each one of the three variables x, y, θ. Let $A = (a, m)$, $B = (b, n)$ be two points of C. If $P = B - A$, i.e., if (9) maps A onto B, one must have $\Phi(a, b, \theta) = 0$. Solving this for θ, Euler obtains, after a calculation (not quite an easy one) which he omits:

$$\theta = \frac{2}{(b-a)^2}\left[H(a,b) \pm \sqrt{F(a)F(b)}\right]$$

where H is the polynomial

$$H(a,b) = A + B(a+b) + \tfrac{1}{2}C(a^2+b^2) + Dab(a+b) + Ea^2b^2$$

([1(b)], p. 343). Clearly these values of θ must correspond to the points $P = B - A$, $P' = A + B$ on Γ. In order to determine the signs belonging to these two points, one need only observe that one sign must give for $B = A$

the point $P = \infty$ (the point at infinity on Γ) and therefore the value ∞ for θ, while the other sign will then give $P' = 2A$ and thus a finite value for θ. Therefore P, P' correspond respectively to the values of θ given by

$$\theta = \frac{2H(a,b) + 2mn}{(b-a)^2}, \quad \theta' = \frac{2H(a,b) - 2mn}{(b-a)^2}.$$

For $P = (\xi, \eta)$, $P' = (\xi', \eta')$, we have then $\xi = \frac{\theta}{4} - \frac{C}{6}$, $\xi' = \frac{\theta'}{4} - \frac{C}{6}$. As to η, η', they can now be obtained by using either one of the formulas (9) as it applies to the present case; η will be obtained by substituting a, m, b, for x, u, y in the first one of these formulas, using in the coefficients of Q and R the value of θ found above; for η' one has first to substitute θ' for θ in those coefficients, then $a, -m, b$ for x, u, y. This completes the solution found in 1765 by Euler for a problem first formulated in 1954.

References

[1] L. Euler: (a) *De integratione aequationis differentialis* $\frac{mdx}{\sqrt{(1-x^4)}} = \frac{ndy}{\sqrt{(1-y^4)}}$ (E 251/1753), *Opera Omnia, Series I*^{a}, vol. 20, pp. 58–79; (b) *Evolutio generalior formularum comparationi curvarum inservientium* (E 347/1765), *ibid.*, pp. 318–356.

[2] Ch. Hermite, *Sur la théorie des fonctions homogènes à deux indéterminées, Premier Mémoire*, J. Reine Angew. Math. 52 (1856), pp. 1–17 =Œuvres, t. I, pp. 350–371.

[3] A. Weil, *Remarques sur un mémoire d'Hermite*, Arch. Math. 5 (1954), pp. 197–202 = Œuvres (Coll. Papers), vol. 2, pp. 111–116.

April 23, 1982

Professor André Weil
School of Mathematics
Institute for Advanced Study
Princeton, New Jersey 08540

Progress in Mathematics

Edited by J. Coates and S. Helgason

Progress in Physics

Edited by A. Jaffe and D. Ruelle

- A collection of research-oriented monographs, reports, notes arising from lectures or seminars
- Quickly published concurrent with research
- Easily accessible through international distribution facilities
- Reasonably priced
- Reporting research developments combining original results with an expository treatment of the particular subject area
- A contribution to the international scientific community: for colleagues and for graduate students who are seeking current information and directions in their graduate and post-graduate work.

Manuscripts

Manuscripts should be no less than 100 and preferably no more than 500 pages in length.

They are reproduced by a photographic process and therefore must be typed with extreme care. Symbols not on the typewriter should be inserted by hand in indelible black ink. Corrections to the typescript should be made by pasting in the new text or painting out errors with white correction fluid.

The typescript is reduced slightly (75%) in size during reproduction; best results will not be obtained unless the text on any one page is kept within the overall limit of 6x9½ in (16x24 cm). On request, the publisher will supply special paper with the typing area outlined.

Manuscripts should be sent to the editors or directly to:
Birkhäuser Boston, Inc., P.O. Box 2007, Cambridge, Massachusetts 02139

PROGRESS IN MATHEMATICS

Already published

PM 1 Quadratic Forms in Infinite-Dimensional Vector Spaces
Herbert Gross
ISBN 3-7643-1111-8, 431 pages, paperback

PM 2 Singularités des systèmes différentiels de Gauss-Manin
Frédéric Pham
ISBN 3-7643-3002-3, 339 pages, paperback

PM 3 Vector Bundles on Complex Projective Spaces
C. Okonek, M. Schneider, H. Spindler
ISBN 3-7643-3000-7, 389 pages, paperback

PM 4 Complex Approximation, Proceedings, Quebec, Canada, July 3-8, 1978
Edited by Bernard Aupetit
ISBN 3-7643-3004-X, 128 pages, paperback

PM 5 The Radon Transform
Sigurdur Helgason
ISBN 3-7643-3006-6, 202 pages, paperback

PM 6 The Weil Representation, Maslov Index and Theta Series
Gérard Lion, Michèle Vergne
ISBN 3-7643-3007-4, 345 pages, paperback

PM 7 Vector Bundles and Differential Equations
Proceedings, Nice, France, June 12-17, 1979
Edited by André Hirschowitz
ISBN 3-7643-3022-8, 255 pages, paperback

PM 8 Dynamical Systems, C.I.M.E. Lectures, Bressanone, Italy, June 1978
John Guckenheimer, Jürgen Moser, Sheldon E. Newhouse
ISBN 3-7643-3024-4, 300 pages, paperback

PM 9 Linear Algebraic Groups
T. A. Springer
ISBN 3-7643-3029-5, 304 pages, hardcover

PM10 Ergodic Theory and Dynamical Systems I
A. Katok
ISBN 3-7643-3036-8, 352 pages, hardcover

PM11 18th Scandinavian Congress of Mathematicians, Aarhus, Denmark, 1980
Edited by Erik Balslev
ISBN 3-7643-3034-6, 528 pages, hardcover

PM12 Séminaire de Théorie des Nombres, Paris 1979-80
Edited by Marie-José Bertin
ISBN 3-7643-3035-X, 404 pages, hardcover

PM13 Topics in Harmonic Analysis on Homogeneous Spaces
Sigurdur Helgason
ISBN 3-7643-3051-1, 152 pages, hardcover

PM14 Manifolds and Lie Groups, Papers in Honor of Yozô Matsushima
Edited by J. Hano, A. Marimoto, S. Murakami, K. Okamoto, and H. Ozeki
ISBN 3-7643-3053-8, 476 pages, hardcover

PM15 Representations of Real Reductive Lie Groups
David A. Vogan, Jr.
ISBN 3-7643-3037-6, 776 pages, hardcover

PM16 Rational Homotopy Theory and Differential Forms
Phillip A. Griffiths, John W. Morgan
ISBN 3-7643-3041-4, 258 pages, hardcover

PM17 Triangular Products of Group Representations and their Applications
S.M. Vovsi
ISBN 3-7643-3062-7, 142 pages, hardcover

PM18 Géométrie Analytique Rigide et Applications
Jean Fresnel, Marius van der Put
ISBN 3-7643-3069-4, 232 pages, hardcover

PM19 Periods of Hilbert Modular Surfaces
Takayuki Oda
ISBN 3-7643-3084-8, 144 pages, hardcover

PM20 Arithmetic on Modular Curves
Glenn Stevens
ISBN 3-7643-3088-0, 236 pages, hardcover

PM21 Ergodic Theory and Dynamical Systems II
A. Katok, editor
ISBN 3-7643-3096-1, 226 pages, hardcover

PM22 Séminaire de Théorie des Nombres, Paris 1980-81
Marie-José Bertin, editor
ISBN 3-7643-3066-X, 374 pages, hardcover

PM23 Adeles and Algebraic Groups
A. Weil
ISBN 3-7643-3092-9, 138 pages, hardcover

PM24 Enumerative Geometry and Classical Algebraic Geometry
Patrick Le Barz, Yves Hervier, editors
ISBN 3-7643-3106-2, 260 pages, hardcover

PM25 Exterior Differential Systems and the Calculus of Variations
Phillip A. Griffiths
ISBN 3-7643-3103-8, 349 pages, hardcover

PM26 Number Theory Related to Fermat's Last Theorem
Neal Koblitz, editor
ISBN 3-7643-3104-6, 376 pages, hardcover

PM27 Differential Geometric Control Theory
Roger W. Brockett, Richard S. Millman, Hector J. Sussmann, editors
ISBN 3-7643-3091-0, 349 pages, hardcover

PM28 Tata Lectures on Theta I
David Mumford
ISBN 3-7643-3109-7, 254 pages, hardcover

PM29 Birational Geometry of Degenerations
Robert Friedman and David R. Morrison, editors
ISBN 3-7643-3111-9, 410 pages, hardcover

PM30 CR Submanifolds of Kaehlerian and Sasakian Manifolds
Kentaro Yano, Masahiro Kon
ISBN 3-7643-3119-4, 223 pages, hardcover

PM31 Approximations Diophantiennes et Nombres Transcendants
D. Bertrand and M. Waldschmidt, editors
ISBN 3-7643-3120-8, 349 pages, hardcover

PM32 Differential Geometry
Robert Brooks, Alfred Gray, Bruce L. Reinhart, editors
ISBN 3-7643-3134-8, 267 pages, hardcover

PM 33 Uniqueness and Non-uniqueness in the Cauchy Problem
Claude Zuily
ISBN 3-7643-3121-6, 185 pages, hardcover

PM 34 Systems of Microdifferential Equations
Masaki Kashiwara
ISBN 3-7643-3138-0, 182 pages, hardcover

PROGRESS IN PHYSICS

Already published

PPh1 Iterated Maps on the Interval as Dynamical Systems
Pierre Collet and Jean-Pierre Eckmann
ISBN 3-7643-3026-0, 256 pages, hardcover

PPh2 Vortices and Monopoles, Structure of Static Gauge Theories
Arthur Jaffe and Clifford Taubes
ISBN 3-7643-3025-2, 294 pages, hardcover

PPh3 Mathematics and Physics
Yu. I. Manin
ISBN 3-7643-3027-9, 112 pages, hardcover

PPh4 Lectures on Lepton Nucleon Scattering and Quantum Chromodynamics
W.B. Atwood, J.D. Bjorken, S.J. Brodsky, and R. Stroynowski
ISBN 3-7643-3079-1, 574 pages, hardcover

PPh5 Gauge Theories: Fundamental Interactions and Rigorous Results
P. Dita, V. Georgescu, R. Purice, editors
ISBN 3-7643-3095-3, 406 pages, hardcover

PPh6 Third Workshop on Grand Unification
University of North Carolina, Chapel Hill, April 15-17, 1982
P.H. Frampton, S.L. Glashow, and H. van Dam, editors
ISBN 3-7643-3105-4, 382 pages, hardcover

Printed in the United States
By Bookmasters